FOREST REGIONS AND PRINCIPAL TY

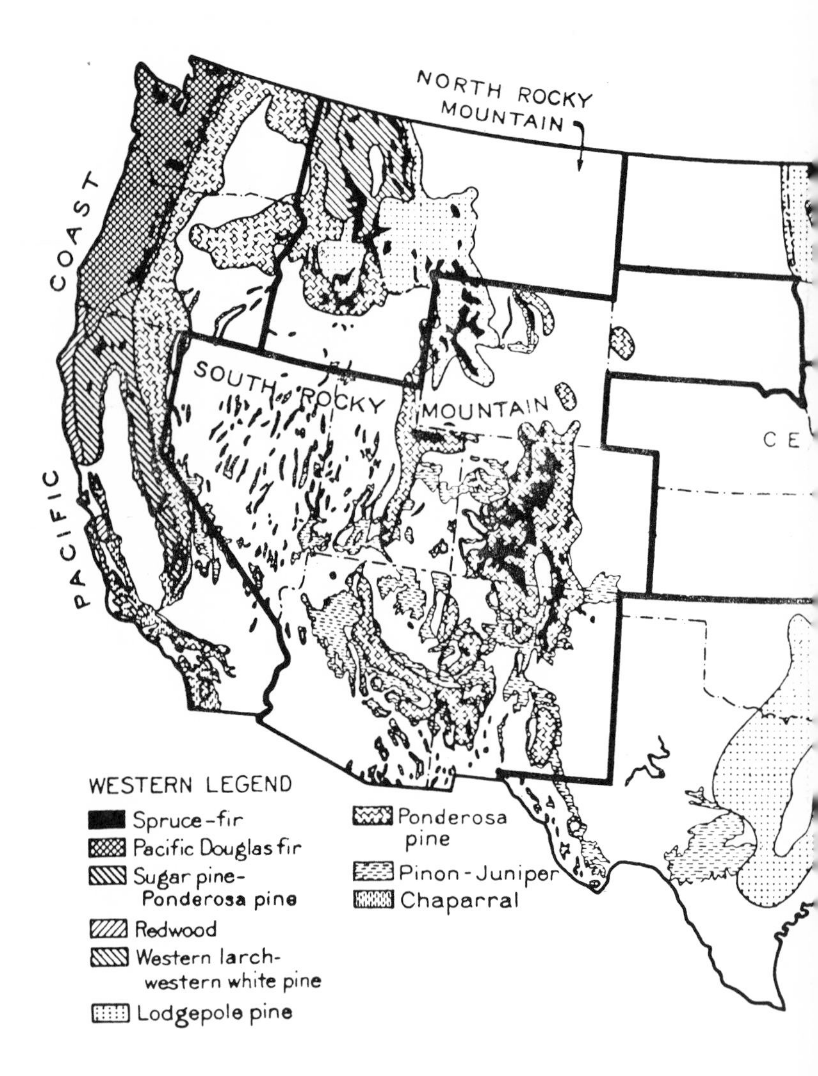

OF FORESTS IN THE UNITED STATES

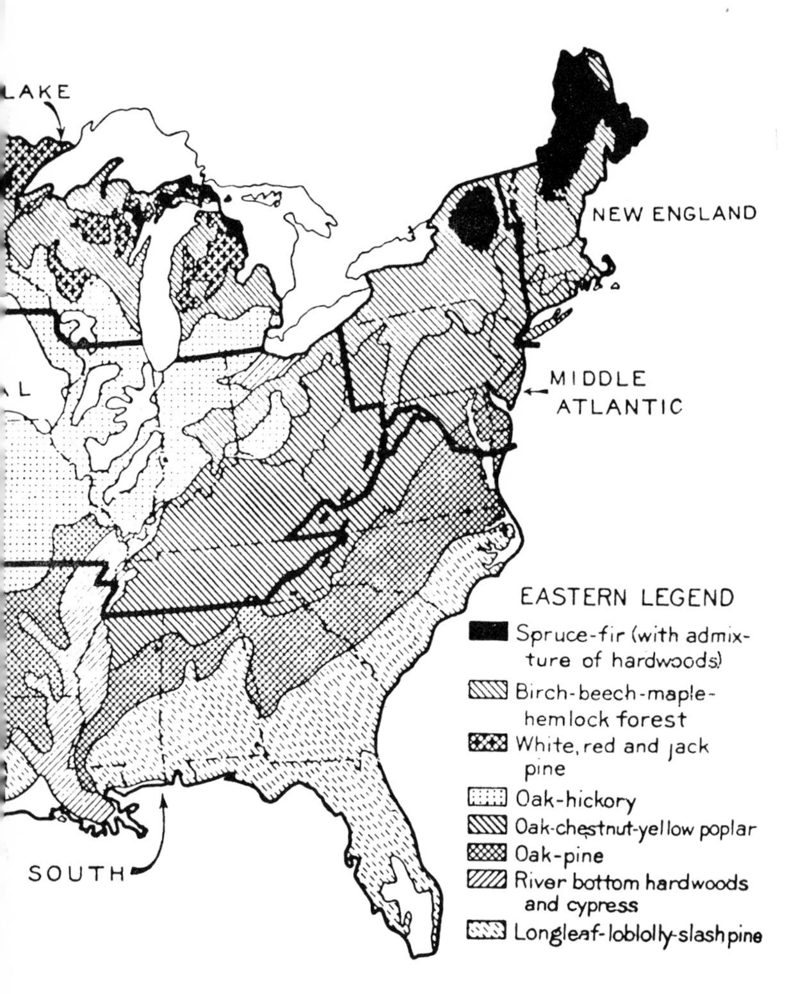

NORTH AMERICAN TREES

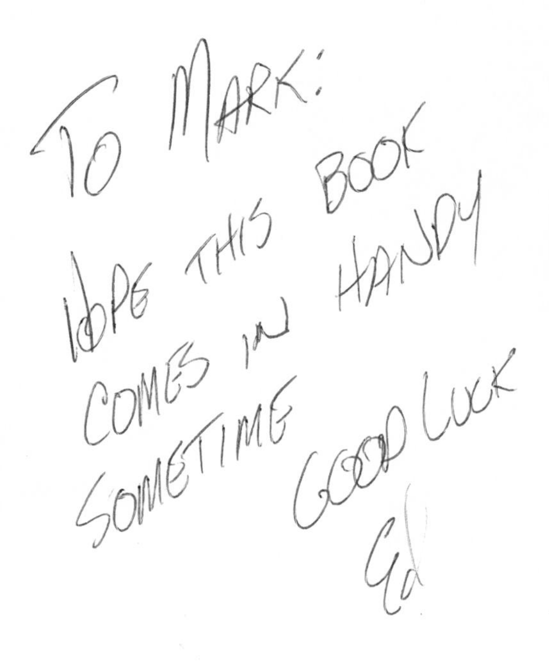
TO MARK:
HOPE THIS BOOK
COMES IN HANDY
SOMETIME
GOOD LUCK
Ed

A handbook designed for field use, with plates and distribution maps

THE MIT PRESS / Cambridge, Massachusett

Third Edition

NORTH AMERICAN TREES

(Exclusive of Mexico and Tropical United States)

Richard J. Preston, Jr. Professor of Forestry
North Carolina State University

ıd London, England

Library of Congress Cataloging in Publication Data

Preston, Richard Joseph, 1905-
North American trees (exclusive of Mexico and tropical Florida), a handbook designed for field use, with plates and distribution maps.

Includes index.
1. Trees--United States--Identification
2. Trees--Canada--Identification. I. Title
QK481.P68 1976 582'.1609'7

MIT Press paperback edition, 1976
Second printing, 1978

Library of Congress catalog card number: 76-40571
ISBN 0 262 66031 8 (paperback)

TO MY WIFE

BERNICE BOYNTON PRESTON

CONTENTS

PREFACE

Ever-increasing numbers of people are becoming interested in the trees and forests of North America. Regional manuals of our native trees are available for the different sections of the United States and Canada, but no simple yet complete up-to-date manual for the entire area has been available.

This book covers the trees of North America with the exception of Mexico and tropical Florida. Care has been taken to include all species native to this area (except for the 162 species of hawthorn and 20 usually shrubby willows that, for the most part, can be distinguished only by specialists) as well as naturalized or commonly planted exotic species.

North American Trees has been designed to meet the needs of an interested nontechnical public as well as those of students and scientists. Concise descriptions of botanical and silvical characters have been included for species of trees that are important or of general interest, while additional less important species are briefly described or included in the complete keys. The terminology has been kept as simple as possible without sacrificing scientific accuracy. Basic information on characters and techniques used in identifying trees appears in the introduction, and a comprehensive glossary defines necessary technical terms that might not be generally understood. It is believed that anyone interested in knowing how to identify the native trees will find this manual understandable and easily used.

This third edition follows the scientific and common names appearing in USDA Agriculture Handbook 41, *Check List of Native and Naturalized Trees of the United States*, 1953. This has resulted in numerous changes in nomenclature and accepted species. Distribution maps have been improved through use of the excellent maps prepared by Elbert L. Little, Jr., in USDA Misc. Publ. 1146, *Atlas of United States Trees*, Vol. 1, 1971. Keys have been revised and descriptions improved.

Included in this manual are 161 full pages of distribution maps and plates detailing descriptive characters for important species. Sixty-nine of these drawings are original, the work of Bruce Eastman or Robert Lorenz. Twelve were redrawn for the second edition by Tom Olive; three by Anne Kiser. The remainder of the drawings are from the following sources:

1. United States Forest Service. Publications of Sudworth and Timber Pines of the South. All the drawings of conifers except those on pages 8, 16, 20, 32, 54, 66, 80, and 108; also the drawings on pages 158, 170, 172, 184, 186, 210, 236, 300, 316, 332, 340, 350, 352, 356, and 370. The map showing Forest Regions also was secured from the United States Forest Service.

2. Otis, *Michigan Trees*. The drawings on pages 16, 20, 54, 66, 108, 122, 132, 136, 140, 142, 144, 148, 150, 152, 162, 164, 166, 168, 180, 190, 192, 196, 198, 216, 218, 222, 226, 228, 234, 238, 242, 246, 248, 250, 260, 262, 264, 272, 274, 276, 282, 286, 296, 298, 302, 306, 318, 326, 330, 346, 348, 350, 352, 360, and 366.

3. Illick, *Pennsylvania Trees*. The drawings on pages 240, 288, 290, 310, 322, 334, 336, 340, and 356.

4. Hanzlik, *Trees and Forests of Western United States*. The drawings on pages 8 and 80.

Both the author and publisher wish to thank the individuals and organizations involved for permission to make use of these drawings.

Richard J. Preston, Jr.

INTRODUCTION

NATURAL RELATIONSHIP OF TREES

Trees, with the exception of tropical tree ferns, belong to one of two major plant groups: the forms with naked seeds commonly subtended by a scale being known as Gymnosperms, and the forms containing an ovary that encloses the ovules being known as Angiosperms.

Two of the four orders of Gymnosperms contain trees that grow in temperate climates, and by far the most important of these are the cone-bearing trees such as the pines, spruces, and cedars known as conifers (*Coniferales*). The members of this group have distinctive needlelike, linear, or scalelike leaves and are often known either as evergreens (in spite of the fact that some shed their needles annually) or as softwoods (although several have hard dense wood). Fifteen genera of conifers containing 95 species are native to North America. These and 13 exotic species are treated in the first section of this book.

The other order of Gymnosperms present in tree form is represented by a single surviving species, the ginkgo or maidenhair tree (*Ginkgo biloba* L.) of China. This interesting tree is extensively planted through all but the coldest parts of the United States and is remarkably free from disease or insects and very tolerant of city smoke. Ginkgo forms a slender, spirelike tree (Fig. 1) with fan-shaped, fernlike, deciduous leaves borne either spirally on the twigs or at the ends of spurlike shoots. The flowers are dioecious and of interest botanically because of the free-swimming antherozoids. The fruit is orange-yellow to green, about one inch in diameter, and resembles a plum with its fleshy, ill-smelling, outer pulp enclosing a large, edible, silvery pit.

The Angiosperms likewise divide into two groups—the Monocotyledons and the Dicotyledons. The first of these, while containing many tree forms in the tropics, is represented in this text by one palm, three palmettos, and seven yuccas. These monocotyledonous plants are characterized by parallel major veins in the leaf, flower parts in threes or sixes, and fibrovascular bundles scattered through the stem. While trees in this group are important in the tropics, they are of little use in the temperate zones.

GINKGO. MAIDENHAIR TREE

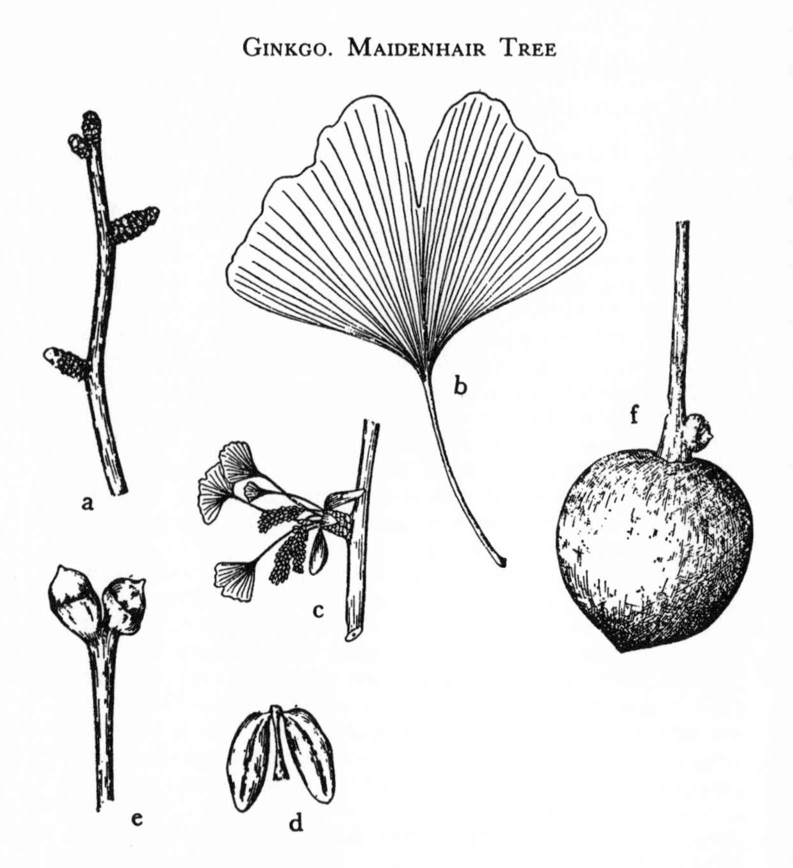

Fig. 1—(*a*) Winter twig, x ½; (*b*) leaf, x ½; (*c*) staminate flowering branchlet, x ½; (*d*) staminate flower, enlarged; (*e*) pistillate flower, enlarged; (*f*) fruit, x 1.

The Dicotyledons contain our most common and widespread trees such as the oaks, maples, and elms, and they are frequently referred to as broadleaf trees or hardwoods. They differ from the Monocotyledons in having prominent, branched (pinnate or palmate), major veins in the leaves; floral parts usually in fours or fives; and fibrovascular bundles disposed in a ring around a pith. While most of the dicotyledonous trees are deciduous (drop their leaves each fall), many are evergreen and have leaves that persist for two or more years. In this text 123 genera of Dicotyledons are covered, and 316 species are either briefly described or included in the complete keys.

FOREST REGIONS OF NORTH AMERICA

As a result of differing conditions of moisture, temperature, and soil, the forests may vary from one section of the country to another, both in the species of trees present and in the type or character of the forest. In North America six broad, natural forest regions can be easily recognized. Two are restricted to the eastern part, two to the western part, and two are transcontinental. These regions are described in the sections immediately following. Foresters generally subdivide these six major regions into a larger number.

Within each region are several smaller natural associations or "forest types." These are usually defined and named by one or more dominating tree species, such as the oak-hickory type in the Central Hardwoods region or the lodgepole pine type in the Rocky Mountain region. More than 150 such forest types are recognized in the United States alone. The regions will be briefly described as well as the important species found in each. The principal forest types for the United States are shown on the Forest Service map inside the front and back covers.

Tropical Forest Region

In southern Florida and extreme southern Texas, Arizona, and California are forests made up largely of tropical species. While unimportant, this small area of tropical forests contains many interesting botanical forms. The tropical forest in Florida and Texas is humid in character and is the northernmost range for many species typical of the deep tropics. Arizona and California, on the other hand, have a desert climate with largely endemic species.

Southern Forest Region

This region covers the coastal area from Maryland to Texas, extending north along the river bottoms into Missouri, Arkansas, and Oklahoma. The area is characterized by abundant rainfall and mild temperatures and is one of the most important timber-producing areas on the continent. The seven species of pine are found largely on the uplands or flatlands, while baldcypress, Atlantic white-cedar, and numerous hardwoods are typical of the extensive swamps and bottomlands. Among the important hardwoods are the sweetgum, tupelo gums, numerous oaks and hickories, beech, river birch, and several ashes and elms.

Central Hardwood Forest Region

There has been a greater decrease in the amount of forested land in this region than in any other due to the large percentage of fertile agricultural land. However, large quantities of high-grade hardwood lumber still remain. This region embraces the central portion of the eastern United States (except for the Appalachian Mountains) and extends into southern Canada. It is characterized by a very large number of species and many types. The climate is humid and generally moderate. Oaks are the dominant species, with hickories, ashes, elms, maples, gum, beech, yellow-poplar, walnut, cottonwood, sycamore, and dogwood also important.

Northern Forest Region

This is by all odds the largest of the forest regions, extending from Georgia north along the Appalachian Mountains to New York, New England, and the northern portions of the Lake States in the United States and north in Canada to the limits of tree growth. In Canada this region is transcontinental, extending in a continuous belt across the northern limits of the Great Plains and through the interior of Alaska. The southeastern part of this region is characterized by highly valuable timber species such as white, red, and jack pine; hemlock; red and white spruce; birches; beech; basswood; maples; and red oak. The northern transcontinental portion of this region is enormous but is largely composed of small-sized trees of white and black spruce, balsam fir, larch, paper birch, aspen, balsam poplar, and willows.

Rocky Mountain Forest Region

This vast region is spread over the mountains and high plateaus of the western interior from Mexico into northern Canada, where it merges with the Northern Forest region. It is bordered on the east by the Great Plains and on the west by the Pacific Coast forest. The forests are not continuous in this region, but are limited to the more humid higher areas interspersed between large treeless stretches occupying the arid lowlands.

The forests characteristically are grouped in conspicuous life zones or belts determined by the moisture, temperature, and topography. These zones tend to increase in altitude as they become more remote from polar regions, so that trees typical of the Northern Forest in northern Canada at elevations of a few hundred feet above sea level might be found at elevations of 8,000 feet in Montana, 10,500 feet in Colorado, and 12,500 feet in Arizona. In the higher, colder, and more humid belt, often called the Canadian zone, are dense forests of Engelmann spruce, alpine fir, western white and lodgepole pine, aspen and balsam poplar, and many other species closely resembling those found in the Northern Forest. Below this belt, in the transition zone, the characteristic tree is ponderosa pine, with less important species such as western larch, Douglas-fir, various oaks, narrowleaf cottonwood, and water birch. This zone is warmer and drier, being intermediate between the humid Canadian zone and the arid Sonoran zone. The Sonoran zone is semitropical and ranges from high plains and foothills in the south to low plains and valleys further north; it is typified by pinyon-juniper forests, broad-leaved cottonwoods, mesquite, and numerous other shrubby tree species.

Pacific Coast Forest Region

Many highly important, very large trees growing in dense stands characterize this region, which contains the major portion of our remaining saw timber. This region, bounded on the east by the Rocky Mountain Forest, includes the Pacific Coast states plus the coastal portions of British Columbia and Alaska. Numerous forest types are recognized, including one or more of such species as the coast redwood, Douglas-fir, ponderosa pine, sugar pine, Jeffrey pine, western hemlock, true firs, cedars, spruces, cypresses, red alder, maples, myrtle, and oaks. While some of the southern portions are dry, the great part of the coastal forests lies in a zone of heavy rainfall. The largest trees in the world (giant sequoia), the tallest trees in the world (redwood with a maximum height of 372 feet), and the most important timber species in the world (Douglas-fir) are found in this region.

TREE CHARACTERS

A brief discussion of the structural and silvical characters, an understanding of which is necessary in identifying trees, is presented here to aid students without previous botanical training.

Definition of a tree. There is no clear-cut line of demarcation between a tree and a shrub, and it is often impossible to place a plant definitely in one group or the other. Frequently a species, treelike under favorable environmental conditions, will be shrub-like over most of its range. In general, height, form, and diameter must be taken into account in determining the classification of a doubtful form. In this manual a tree is defined as a woody plant having one well-defined stem, a more or less definitely formed crown, a minimum height of fifteen to twenty feet, and a diameter of not less than two inches.

Terminology. While appearing cumbersome to the beginner, the use of technical terms in describing characters is often necessary for a concise, accurate description. These technical terms have been avoided wherever their omission did not impair meaning, and a glossary explaining such terms has been included. The student should familiarize himself with the more commonly used terms early in his study of the trees.

Name. Most species of trees have been given one or many common names, which usually describe some character of the tree and are easily learned by students. Unfortunately, these common names have many limitations. Some are merely local, others apply to two or more entirely different species, and some apply to trees belonging to different genera. Because of this, while common names may be useful and convenient, it is essential that each species have a definite, individual name that can be accepted throughout the world and cannot be applied to any other species.

Botanists and scientists as a whole have agreed that these scientific names should be in Latin, as this is a dead language and not subject to change. Botanists have further agreed that the name of a tree should consist of three parts: (1) a generic name, which is capitalized, refers the species to the group to which it belongs; (2) a specific epithet, beginning with a small letter, refers to the single species; followed by (3) the full or abbreviated name of the authority or person first describing the plant. Thus,

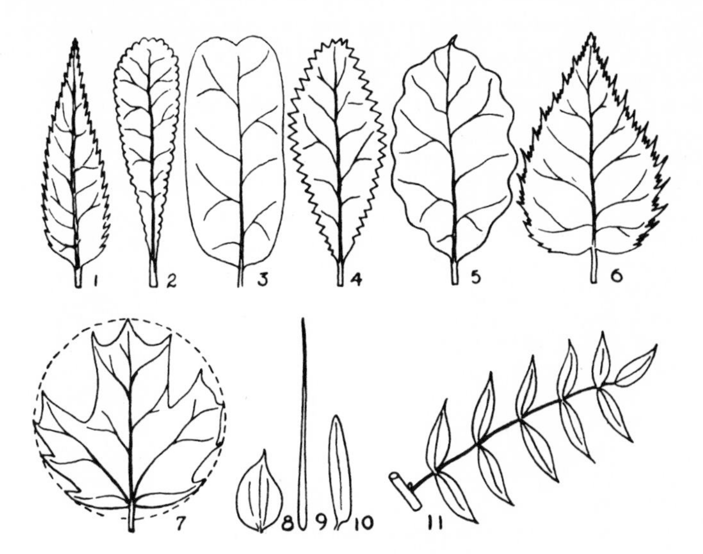

Fig. 2—LEAF PATTERNS

Leaf Shapes

1-Lanceolate. 2-Oblanceolate. 3-Oblong. 4-Elliptic. 5-Oval. 6-Ovate. 7-Orbicular. 8-Scalelike. 9-Acicular. 10-Linear.

Leaf Margins

1-Serrate. 2-Crenate. 3-Entire. 4-Dentate. 5-Sinuate. 6-Doubly serrate. 7-Lobed.

Leaf Apices

1-Acuminate. 2-Rounded. 3-Emarginate. 4-Obtuse. 5-Mucronate. 6-Acute.

Leaf Bases

1-Obtuse. 2-Cuneate. 3-Rounded. 4-Acute. 5-Rounded. 6-Cordate. 7-Truncate.

Leaf Types

1 to 10-Simple leaves. 11-Pinnately compound leaf.

the scientific name of ponderosa pine is *Pinus ponderosa* Laws. When a variety of a species is recognized, the varietal name follows the specific name; in the case of Arizona pine (a southern variety of ponderosa pine) the scientific name is *Pinus ponderosa* var. *arizonica* (Engelm.) Shaw. When the names of two authorities are given, one appearing in parentheses, this indicates that the species was first described by the authority in

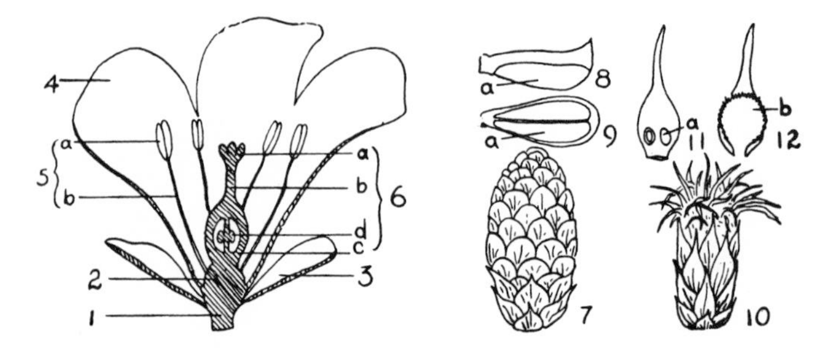

Fig. 3—FLOWER STRUCTURE

Perfect Flower — Flowers of Pine

1-Peduncle. 2-Receptacle. 3-Sepal (Calyx). 4-Petal (Corolla). 5-Stamen; a-Anther, b-Filament. 6-Pistil; a-Stigma, b-Style, c-Ovary, d-Ovule. 7-Staminate conelet. 8 and 9-Stamen, or pollen-bearing scale, showing side and lower surfaces. 10-Pistillate conelet. 11 and 12-Pistillate scale showing inner and outer surfaces respectively; a-Ovule, b-Bract.

the parentheses, but in a different group of the same rank or in a different rank, and that the second author was the one effecting the change. This is the case with the western hemlock, *Tsuga heterophylla* (Raf.) Sarg.

Habit. This refers to the general appearance of a tree, usually as seen from a distance. The size; appearance and form of the trunk; shape, density, and size of crown; and the number, size, and direction of growth of the branches are all factors helpful in distinguishing trees. In the conifers there is typically a trunk that extends to the top of the tree without dividing (*excurrent*), while most of the hardwoods have the trunk breaking up into several large branches (*deliquescent*). The crown and branches may vary greatly in the same species, depending upon whether the tree is growing in the open or in a dense stand where it is shaded on the sides by its neighbors.

Leaves. Since leaves display characteristic patterns, they are probably the most useful organs in identifying trees (Fig. 2).

Leaves consist of an expanded portion or *blade*, a supporting stalk or *petiole*, and small leaflike or scaly structures (*stipules*) attached in pairs at the base of the petiole. Leaves having no stipules are termed *estipulate*, and those without petioles are called *sessile*. Evergreen species can be identified by their leaves throughout the year, while species that shed their leaves annually (*deciduous*) must be identified by other characters during the winter months. Leaves may be arranged *alternately* (with

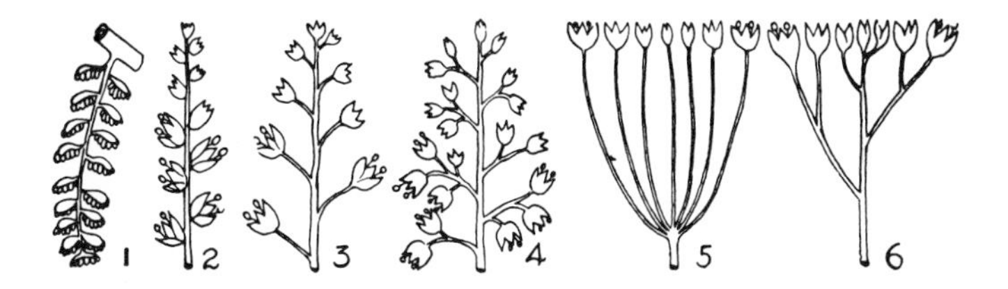

Fig. 4—TYPES OF INFLORESCENCES

1-Ament. 2-Spike. 3-Raceme. 4-Panicle. 5-Umbel. 6-Corymb.

one leaf attached to the twig at a certain point), *oppositely* (where two leaves emerge at opposite sides from the same place on a twig), *whorled* (where more than two leaves emerge from one point of the twig), or *fascicled* (where a number of leaves emerge in a cluster or bundle). They may be *simple* and consist of a single blade or expanded portion, or *compound* and made up of several individual leaflets. If the leaflets in a compound leaf are arranged along each side of a common axis (*rachis*), the leaf is said to be *pinnately compound;* while if the leaflets all arise from the apex of the petiole, the leaf is termed *palmately compound*. Other characters used in identification are the shapes and types of margin, apex, and base. Texture, color, and surface, whether smooth or hairy, are also useful characters.

Flowers. All trees have flowers, although frequently they are small and inconspicuous. These are the reproductive organs by means of which the species is perpetuated. Floral characters are the most accurate means of identifying many trees (Fig. 3), although they are little used in the field because the period of bloom is so short. Flowers vary greatly in form, structure, and size. A *complete* flower (Fig. 3) is usually made up of leaflike *sepals* (*calyx*), often brightly colored *petals* (*corolla*), *stamens* (the male organs that bear the *pollen* in saclike *anthers*), and a *pistil* (the female organ consisting of a terminal *stigma* that catches the pollen, a *style*, and an *ovary*). The ovary may consist of one or more compartments (*cells*) and contain one to many *ovules* that later mature into seeds.

If the ovary is inserted on top of the other flower parts, it is *superior*, while if it appears below the origin of sepals, petals, and stamens, it is *inferior*. Flowers may be *perfect* (contain both stamens and pistil) or *imperfect* (contain one sex but not the other). Plants having imperfect flowers are termed either *monoecious* (both sexes present in different flowers on the same plant) or *dioecious* (each sex borne on a different plant). When the only functioning sex organs in an imperfect flower are stamens, the flower is termed *staminate;* while one in which the pistil is the active organ is *pistillate*. Plants that bear some perfect and some imperfect flowers are termed *polygamous*. Flowers may appear singly or in clusters (*inflorescences*, Fig. 4). In Gymnosperms the pistil is replaced by a seed-bearing scale that does not enclose the seed.

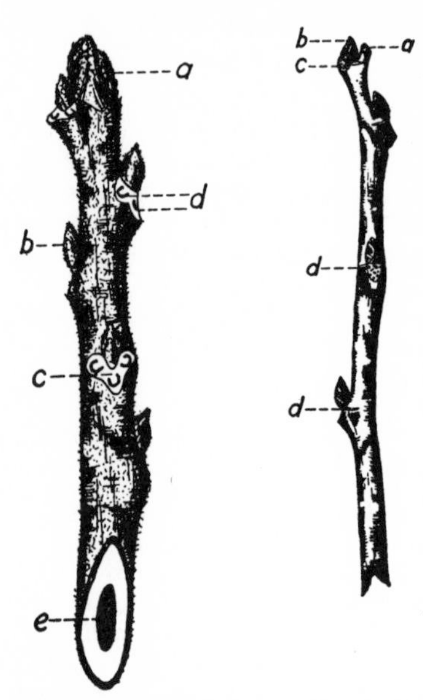

Fig. 5—(Left) Winter Twig of Black Walnut. (*a*) terminal bud; (*b*) lateral bud; (*c*) leaf scar; (*d*) bundle scars; (*e*) pith.

(Right) Winter Twig of Red Mulberry. (*a*) tip scar; (*b*) lateral bud; (*c*) leaf scar; (*d*) stipule scars.

Fruit. While varying greatly in type and appearance, fruits are very useful in identification. The different kinds of fruits are pictured and discussed in the text and defined in the glossary. The seeds included in the fruit contain the embryonic plant. To the layman these are generally of secondary utility in identification.

Twigs. The color, stoutness, central pith, or surface coverings and markings may be very usetul in identifying trees, especially during the winter months (Fig. 5). As buds, stipules, and leaves fall off, they frequently leave characteristic scars on the twig. Buds are conspicuous on most twigs and helpful in identification. The shape, size, color, number of scales, arrangement, and the presence or absence of a terminal bud are important diagnostic characters.

Bark. The appearance of bark, while varying greatly with age and environment, is often a helpful character in identification. Color and thickness of the bark and whether it is furrowed, scaly, or smooth as well as its taste are commonly helpful features.

Wood. The character of the wood forms a separate means of identification that is more technical and difficult than the use of external characters. In this manual only the outstanding wood characters are given, such as the weight, color, and arrangement of large pores (whether *ring-porous* in a definite ringlike zone or *diffuse-porous* and scattered throughout the wood). A statement as to importance and use of the wood is also included.

Silvical characters. The tolerance, sites, associates, reproduction, enemies, roots, life zones, and altitudinal distributions of species are often helpful in identification and have been included wherever possible.

In identifying trees it must be remembered that characters are variable and often overlap with those of closely related trees. Wherever possible, identification should not be based on a single character but on as many as are available.

KEY TO THE GENERA

BASED LARGELY ON LEAF CHARACTERS

1. Seeds not in ovary; flowers unisexual; leaves fan-shaped, linear, needlelike, or scalelike . . . . GYMNOSPERMS (see A, below)
1. Seeds borne in a closed ovary that becomes the fruit at maturity . . . . ANGIOSPERMS
   2. Leaves narrow with parallel veins; flower parts in 3's or 6's; stems without central pith or annual wood layers; embryo with 1 cotyledon . . . . MONOCOTYLEDONS (see B, p. xxiv)
   2. Leaves generally broad or rarely wanting, with netted veins; flower parts usually in 4's or 5's; stems with pith and annual wood layers; embryo with 2 cotyledons . . . . DICOTYLEDONS (see C, p. xxiv)

A. GYMNOSPERMS

1. Leaves fan-shaped with numerous fine veins radiating from petiole, deciduous; twigs with spur shoots; fruit a drupelike seed . . . . GINKGO, p. xii
1. Leaves linear, needle-shaped, or scalelike, with 1 vein, usually persistent . . . . CONIFERS, 2
   2. Leaves needle-shaped, in fascicles of 1-5 each enclosed at base by a sheath . . . . PINUS, p. 2
   2. Leaves single, linear or scalelike 3
3. Leaves mostly linear; leaves and fruit scales spirally arranged 4
3. Leaves mostly scalelike; leaves and fruit scales opposite 14
   4. Fruit a woody or papery cone of seed-bearing scales 5
   4. Fruit a single seed partially or wholly surrounded by a fleshy aril 13
5. Cone scales thin, terminally attached; seed terminally winged, 2 to each scale . 6
5. Cone scales peltate; seed laterally winged, 2 to many per scale 11
   6. Twigs with spur shoots; cones upright 7
   6. Twigs without spur shoots; cones upright or pendent, maturing in 1 year . . 8
7. Needles deciduous; cones maturing in 1 year . . . . LARIX, p. 42
7. Needles persistent; cones maturing in 2 or 3 years . . . . CEDRUS, p. 43
   8. Base of leaves persistent on twigs as peglike projections (sterigmata). . . . 9
   8. Leaves not leaving peglike bases on twig upon falling 10
9. Leaves sessile, 4-angled or flattened . . . . PICEA, p. 48
9. Leaves stalked, flat or rounded in cross section . . . . TSUGA, p. 64
   10. Leaves sessile; cones erect, scales deciduous from axis at maturity; buds rounded, resinous (except *A. bracteata*) . . . . ABIES, p. 72
   10. Leaves narrowed at base into stalk; cones pendent with exserted bracts; buds pointed, nonresinous . . . . PSEUDOTSUGA, p. 63
11. Leaves deciduous (persistent in tropical *T. mucronatum*); 2 seeds under each scale; southeastern . . . . TAXODIUM, p. 93
11. Leaves persistent; 2-9 seeds per scale; Pacific 12
    12. Leaves on lateral shoots linear, 2-ranked. . . . . SEQUOIA, p. 89
    12. Leaves ovate, appressed . . . . SEQUOIADENDRON, p. 91
13. Leaves 1/2-1 inch long; fruit or seed partially enclosed in scarlet aril, maturing in 1 year . . . . TAXUS, p. 119
13. Leaves 1-3 1/2 inches long; fruit a drupelike seed enclosed in green or purple fleshy covering; maturing in 2 years . . . . TORREYA, p. 117
    14. Cone berrylike; leaves scalelike or awl-shaped, both types usually present; seed wingless . . . . JUNIPERUS, p. 106
    14. Cone woody or leathery; leaves scalelike; seed winged 15

15. Cones globose with peltate scales; 2-20 seeds per scale 16
15. Cones oblong, thin-scaled; 2-3 seeds per scale 17
   16. Cone woody, over 1/2 inch across, of 6-12 scales; 6-20 seeds per scale; branchlets angled or rounded . . . . CUPRESSUS, p. 98
   16. Cone leathery, 1/4-1/2 inch across, of 4-8 scales; 1-5 seeds per scale; branchlets flattened . . . . CHAMAECYPARIS, p. 98
17. Cones 3/4-1 1/2 inches long, 6-scaled . . . . LIBOCEDRUS, p. 95
17. Cones 1/3-1/2 inch long, 6-12 scaled. . . . . THUJA, p. 97

B. MONOCOTYLEDONS

Leaves with Parallel Veins, Alternate, Simple, and Persistent

1. Leaves lanceolate and long; fruit a large capsule . . . . YUCCA, p. 369
1. Leaves fan-shaped, large; fruit a drupelike berry 2
   2. Leaf stalks armed with spines; southwestern . . . . WASHINGTONIA, p. 371
   2. Leaf stalks unarmed; southeastern . . . . SABAL, p. 371

C. DICOTYLEDONS

Leaves with Netted Veins or Rarely Absent

a. Leaves and buds opposite
   b. Leaves compound . . . . I, p. xxiv
   b. Leaves simple . . . . II, p. xxiv
a. Leaves and buds alternate
   c. Leaves compound . . . . III, p. xxv
   c. Leaves simple
      d. Leaves persistent. . . . . IV, p. xxxi
      d. Leaves deciduous . . . . V, p. xxvii

I. LEAVES OPPOSITE, COMPOUND, AND DECIDUOUS

1. Leaves palmately compound; fruit a capsule . . . . AESCULUS, p. 308
1. Leaves pinnately compound or 3-foliate 2
   2. Leaflets entire or finely toothed; fruit a samara . . . . . FRAXINUS, p. 342
   2. Leaflets deeply and sharply toothed 3
3. Flowers yellow, showy; fruit a linear capsule 4-8 inches long; southwestern . . . . TECOMA, p. 363
3. Flowers small; fruit a double samara or drupe 4
   4. Fruit a double samara in racemes; terminal bud present . . . . ACER NEGUNDO, p. 307
   4. Fruit a small drupe in dense clusters; terminal bud absent, lateral scaly . . . . SAMBUCUS, p. 365

II. LEAVES OPPOSITE AND SIMPLE

1. Leaves persistent, entire, thick; fruit a drupe; southeastern coastal plain . . . . OSMANTHUS, p. 357
1. Leaves deciduous, not leathery 2
   2. Leaves palmately lobed; fruit a double samara . . . . ACER, p. 294
   2. Leaves not lobed, fruit not a double samara. 3
3. Leaves entire 4
3. Leaves toothed 11

4. Leaves heart-shaped, 5-12 inches long 5
4. Leaves linear to ovate, less than 5 inches long 6
5. Leaves 6-10 inches long, densely hairy below; flowers purple; capsule leathery, ovoid, 1-2 inches long . . . . PAULOWNIA, p. 361
5. Leaves 8-12 inches long, slightly hairy below; flowers white; capsule linear, 8-20 inches long . . . . CATALPA, p. 361
6. Leaves linear to lanceolate; fruit a linear capsule 7-12 inches long; southwestern . . . . CHILOPSIS, p. 359
6. Leaves suborbicular to elliptic 7
7. Exfoliating bark; showy fall flowers . . . . LAGERSTROEMIA, p. 325
7. Bark not exfoliating; spring flowers 8
8. Stipules present and persistent; terminal bud absent, laterals minute; fruit head of nutlike capsules . . . . CEPHALANTHUS, p. 363
8. Stipules absent; terminal bud present; fruit not a capsule 9
9. Fruit a samara, terminal wing; southwestern . FRAXINUS ANOMALA, p. 345
9. Fruit a drupe 10
10. Leaves with arcuate venation; flowers small in heads; leaf scars, 3 bundle traces . . . . CORNUS, p. 331
10. Leaves with pinnate venation; flowers white in panicles; leaf scar, 1 bundle trace . . . . CHIONANTHUS, p. 357
11. Buds enclosed by 2 large, valvate scales; fruit a drupe; petioles grooved . . . . VIBURNUM, p. 367
11. Buds many-scaled; petioles round in cross section 12
12. Leaves closely and finely serrate; lateral buds solitary; fruit a capsule in scarlet aril . . . . EUONYMUS, p. 293
12. Leaves remotely and finely serrate; lateral buds several, superposed; fruit a drupe . . . . FORESTIERA, p. 357

III. LEAVES ALTERNATE AND COMPOUND

1. Fruit a legume; leaves 1 to many pinnate . . . . LEGUMINOSAE (key to), p. 267
1. Fruit not a legume 2
2. Leaves persistent; fruit a small drupe 3
2. Leaves deciduous 4
3. Leaflets spatulate, 1/4 inch long; Texas PISTACIA, p. 309
3. Leaflets lanceolate, 2-3 inches long; southeastern coastal plain . . . . SAPINDUS MARGINATUS, p. 313
4. Twigs armed with spines 5
4. Twigs unarmed 6
5. Leaves doubly pinnate; fruit a berry . . . . ARALIA, p. 323
5. Leaves once pinnate; fruit a capsule . . . . ZANTHOXYLUM, p. 285
6. Leaves doubly pinnate; fruit a drupe 7
6. Leaves once pinnate 8
7. Leaflets 1-2 inches long; drupe yellow, 1/2-3/4 inches . . . . . . MELIA, p. 287
7. Leaflets 1/4 inch long; drupe red, 1/4 inch . . . . BURSERA, p. 287
8. Leaves 3-foliate; fruit a winged samara . . . . PTELEA, p. 285
8. Leaves 5 to many foliate 9
9. Pith conspicuously chambered; fruit a nutlike drupe with indehiscent husk . . . . JUGLANS, p. 134
9. Pith solid; fruit not as above 10
10. Leaves 1 1/2-4 feet long; leaflets with glands on coarse teeth; fruit a samara with twisted wing . . . . AILANTHUS, p. 287
10. Leaves smaller or leaflets without glands 11
11. Stipules or stipular scars present; leaflets sharply serrate; buds resinous; fruit a pome . . . . SORBUS, p. 249
11. Stipules absent; buds not resinous; fruit not a pome 12
12. Leaflets entire; fruit a yellow berry . . . . SAPINDUS, p. 313

12. Leaflets toothed (rarely entire in *Rhus*) 13
13. Buds large; fruit a nutlike drupe . . . . CARYA, p. 138
13. Buds small; fruit berrylike or capsular 14
14. Juice milky; fruit a berrylike drupe . . . . RHUS, p. 289
14. Juice watery; fruit a capsule 15
15. Capsule leathery; southwestern . . . . UNGNADIA, p. 313
15. Capsule papery; ornamental . . . . KOELREUTERIA, p. 313

IV. LEAVES ALTERNATE, SIMPLE, AND PERSISTENT

1. Leaves lobed 2
1. Leaves entire or toothed 4
2. Leaves 1/3-1/2 inch long; fruit a long-tipped achene; southwestern . . . . COWANIA, p. 258
2. Leaves 1 inch or longer 3
3. Leaves leathery; fruit an acorn. . . . . QUERCUS (in part), p. 174
3. Leaves thick; fruit a hairy capsule; southwestern. . . . . FREMONTODENDRON, p. 321
4. Lower surface of leaves conspicuously dotted with scales or glands 5
4. Lower surface of leaf without scales or glands 8
5. Leaves covered with yellow scales below; fruit a bur; West Coast. . . . . CASTANOPSIS, p. 171
5. Leaves glandular-dotted below; fruit a capsule or drupe 6
6. Leaves oblanceolate to oblong, 2-4 inches long, aromatic; fruit a wax-covered drupe . . . . MYRICA, p. 135
6. Leaves lanceolate, 1 1/2-2 inches long, nonaromatic; southeastern coastal plain 7
7. Fruit a 2-4 winged, indehiscent capsule . . . . CLIFTONIA, p. 293
7. Fruit a dehiscent, oblong capsule . . . . LYONIA, p. 333
8. Fruit a linear achene tipped with a long, hairy style; western . . . . CERCOCARPUS, p. 255
8. Fruit not as above 9
9. Fruit an acorn in a cup; leaves leathery 10
9. Fruit not an acorn 11
10. Fruit cup scaly; leaves not woolly below . . . . . QUERCUS (in part), p. 174
10. Fruit cup spiny; leaves woolly below; Pacific Coast . . . . LITHOCARPUS, p. 173
11. Leaves entire (rarely toothed on vigorous shoots) 12
11. Leaves toothed 25
12. Leaves aromatic 13
12. Leaves not aromatic 15
13. Fruit a woody capsule; exotic ornamental . . . . EUCALYPTUS, p. 329
13. Fruit a drupelike berry 14
14. Calyx lobes persistent; southeastern . . . . PERSEA, p. 239
14. Calyx lobes deciduous; California and Oregon . . UMBELLULARIA, p. 237
15. Sap milky; fruit a multiple . . . . FICUS, p. 229
15. Sap watery; fruit simple or aggregate 16
16. Fruit an aggregate of follicles. 17
16. Fruit simple 18
17. Leaves 5-8 inches long, red-woolly below. . . . . MAGNOLIA, p. 233
17. Leaves 3-4 inches long, glabrous . . . . ILLICIUM, p. 231
18. Flowers showy; fruit 5-celled capsule; southeastern 19
18. Flowers small; fruit fleshy or small 2-celled capsule 20
19. Leaves 3-4 inches long, flat, elliptic-lanceolate; twigs slender; fruit less than 1/4 inch long . . . . KALMIA, p. 335
19. Leaves 4-12 inches long, revolute, oblong; twigs stout; fruit 1/2 inch long . . . . RHODODENDRON, p. 335

20. Twigs spiny or spinescent; fruit a drupe 21
20. Twigs not spiny or spinescent 22
21. Leaves 1-3 inches; drupe black, 1/2 inch . . . . BUMELIA, p. 339
21. Leaves 1/2-3/4 inch; drupe red, 1/5 inch . . . . CONDALIA, p. 317
22. Fruit a 2-celled capsule 1/8 inch long; leaves oblong, 2-3 inches long; southeastern coast . . . . CYRILLA, p. 293
22. Fruit a fleshy drupe or berry 23
23. Stipules or their scars present; fruit red to brown . . . . ILEX, p. 291
23. Stipules or their scars absent 24
24. Fruit black, lustrous; southeastern . . . . VACCINIUM, p. 337
24. Fruit orange, glandular-coated; western . . . . ARBUTUS, p. 333
25. Leaves 3-nerved from base, 1-1 1/2 inches long; flowers in corymbs; fruit a dry 3-lobed drupe; California . . . . CEANOTHUS, p. 317
25. Leaves with pinnate veins; flowers and fruit not as above 26
26. Stipules absent; southeastern coastal plain 27
26. Stipules or their scars present 28
27. Leaves 1-2 1/2 inches long, entire or obscurely toothed; flowers small, white; fruit a berry 1/4 inch long . . . . VACCINIUM, p. 337
27. Leaves 3-6 inches long, finely or remotely toothed; flowers showy, 2 1/2 inches across; fruit a capsule . . . . GORDONIA, p. 323
28. Leaves narrowly lanceolate; fruit a woody capsule 1/4 inch long; southwestern . . . . VAUQUELINIA, p. 245
28. Leaves broader; fruit a drupe or pome 29
29. Fruit, obovoid pome 1/3 inch; California . . . . PHOTINIA, p. 258
29. Fruit a drupe; flowers and fruit not in panicles 30
30. Twigs spinescent; terminal bud absent; leaves 1-1 1/2 inches long, mostly suborbicular with glandular spiny teeth; southwestern . . . . RHAMNUS CROCEA, p. 317
30. Twigs not spinescent; leaves 2-4 inches long, lanceolate to ovate 31
31. Terminal bud absent; leaves 2-ranked; fruit 1/6-1/5 inch, yellow brown . . . . TREMA, p. 221
31. Terminal bud present; leaves not 2-ranked; fruit 1/4 inch or larger, red to black 32
32. Fruit 1/4 inch long; flowers in cymes or solitary; leaf scar with single bundle trace . . . . ILEX, p. 291
32. Fruit 1/2-2/3 inch long; flowers in racemes; leaf scars with 3 or more bundle traces . . . . PRUNUS (in part), p. 257

V. LEAVES ALTERNATE, SIMPLE, AND DECIDUOUS

A. Leaves Scalelike, Inconspicuous, or Wanting (p. xxix)
B. Leaves Lobed (p. xxx)
C. Leaves Entire (p. xxx)
D. Leaves Toothed (p. xxxi)

A. LEAVES SCALELIKE, INCONSPICUOUS, OR WANTING

1. Plants succulent (Cacti), with clustered spines 2
1. Plants treelike, woody 3
2. Branches columnar; seed dark . . . . CEREUS, p. 324
2. Branches jointed; seed pale . . . . OPUNTIA, p. 325
3. Leaves scalelike, feathery; fruit a capsule . . . . TAMARIX, p. 308
3. Trees appearing leafless; southwestern 4
4. Twigs with black, cushionlike processes . . . . CANOTIA, p. 293
4. Twigs without cushionlike processes 5

5. Twigs with lateral spines; fruit a legume . . . . DALEA, p. 266
5. Twigs ending in spine; fruit a berry . . . . KOEBERLINIA, p. 321

B. LEAVES LOBED

1. Leaves 4-lobed; bud large with valvate scales; fruit an aggregate of samaras . . . . LIRIODENDRON, p. 235
1. Leaves not 4-lobed; bud scales not valvate 2
   2. Leaves entire or 1-2 lobed, all types usually present; aromatic; fruit a blue drupe on red stalk . . . . SASSAFRAS, p. 239
   2. Leaves more than 2-lobed, not strongly aromatic 3
3. Leaves palmately 3-7 lobed; fruit in multiple heads 4
3. Leaves not palmately 3-7 lobed; fruit not in heads 5
   4. Leaves finely toothed; terminal bud scaly; twigs with corky wings; fruit of capsules . . . . LIQUIDAMBAR, p. 241
   4. Leaves coarsely toothed; terminal bud absent, lateral enclosed in enlarged leaf petiole; fruit of achenes; bark characteristically mottled . . . . PLATANUS, p. 243
5. Terminal bud absent; sap milky; fruit a berrylike multiple of drupes 6
5. Terminal bud present; sap watery; fruit simple 7
   6. Leaves velvety-hairy below; fruit globose, orange . BROUSSONETIA, p. 227
   6. Leaves smooth to hairy below; fruit ovoid, purple, dark red, or white . . . . MORUS, p. 227
7. Twigs with sharp thorns; fruit a small, red pome . . . . . . CRATAEGUS, p. 253
7. Twigs unarmed (sometimes spinescent in *Malus*) 8
   8. Leaf scars with many bundle traces; fruit an acorn . . . . QUERCUS, p. 174
   8. Leaf scars with 3 bundle traces 9
9. Leaves toothed; fruit a pome; buds small . . . . MALUS, p. 244
9. Leaf lobes entire; fruit a capsule; buds large . . . . POPULUS, p. 120

C. LEAVES ENTIRE

1. Leaves, twigs, and drupes silvery-scurfy 2
1. Plants not silvery-scurfy 3
   2. Leaves elliptic to oblong . . . . SHEPHERDIA, p. 329
   2. Leaves lanceolate . . . . ELAEAGNUS, p. 329
3. Leaves heart-shaped; fruit a legume . . . . CERCIS, p. 273
3. Leaves not heart-shaped; fruit not a legume 4
   4. Sap milky; twigs with sharp axillary spines; fruit a globose multiple of drupes 4-5 inches across . . . . MACLURA, p. 229
   4. Sap watery; twigs and fruit not as above 5
5. Leaves obovate-lanceolate, 10-12 inches long; stipules absent; buds minute; fruit a berry 3-5 inches long . . . . ASIMINA, p. 237
5. Leaves and fruit not as above 6
   6. Terminal bud present, conspicuous 7
   6. Terminal bud absent; lateral buds often minute 14
7. Terminal bud large, with single caplike scale; leaves 4-30 inches long; fruit conelike aggregate . . . . MAGNOLIA, p. 230
7. Terminal bud scaly; leaves 1-7 inches long; fruit simple 8
   8. Stipules present; fruit an acorn . . . . QUERCUS, p. 174
   8. Stipules absent; fruit not an acorn 9
9. Leaves with arcuate veins; petioles long and slender; flowers small in cymes; fruit a drupe . . . . CORNUS, p. 331
9. Leaves with pinnate veins; petioles short 10

10. Twigs with white lenticels, aromatic if crushed; leaves obtuse, oval to obovate, 4-6 inches long; fruit, dry drupe 1/8 inch in loose panicles . . . . COTINUS, p. 309
10. Twigs, leaves, and fruit different 11
11. Leaves lanceolate, fruit a dry drupe, twigs with conspicuous circular lenticels, southeastern and rare . . . . LEITNERIA, p. 147
11. Leaves oblong to obovate 12
12. Leaves oblong to obovate, 2-7 inches long; pith diaphragmed; fruit an oblong, fleshy drupe NYSSA, p. 327
12. Leaves oblong, 2-4 inches long; pith solid; fruit a capsule; southeastern coastal plain 13
13. Leaves 3-4 inches long; hairy below, early deciduous; twigs without lenticels; fruit a globose, 3-5 celled capsule 3/8 inch long . . . . ELLIOTTIA, p. 333
13. Leaves 2-3 inches long, smooth below, tardily deciduous; twigs with lenticels; fruit 2-celled, capsulelike, indehiscent, 1/8 inch . . . . CYRILLA, p. 293
14. Leaves unequal at base; pith chambered at nodes; fruit a dry drupe . . . . CELTIS, p. 220
14. Leaves equal at base; pith solid; fruit not a dry drupe 15
15. Stipules or their scars present 16
15. Stipules absent 19
16. Buds, single caplike scale; fruit a capsule . . . . SALIX, p. 130
16. Buds scaly or naked; fruit a drupe 17
17. Twigs stout with thick pith; sap milky . . . . RHUS, p. 289
17. Twigs slender, often spinescent; sap watery 18
18. Leaves 1/2-3/4 inch long; drupe 1/8 inch; southwestern . . . . CONDALIA, p. 317
18. Leaves 1-7 inches long; drupe 1/4-1/2 inch . . . . RHAMNUS, p. 315
19. Leaves stellate-hairy below; buds naked, hairy; fruit a drupe 1/3 inch long in persistent calyx; southeastern coastal plain . . . . STYRAX, p. 343
19. Leaves not stellate-hairy; buds scaly 20
20. Twigs usually spinescent; leaves often fascicled on spur shoots; drupe black . . . . BUMELIA, p. 339
20. Twigs not spinescent; spur shoots absent 21
21. Leaves glabrous; fruit a large berry subtended by woody persistent calyx; bark conspicuously blocky . . . . DIOSPYROS, p. 341
21. Leaves woolly below; fruit a red drupe subtended by hairy calyx; bark thin, scaly; southwestern. . . . . CORDIA, p. 359

D. LEAVES TOOTHED

1. Buds naked, terminal present; leaves unequal at base, oval, crenately toothed; flowers yellow, appearing in fall; fruit a capsule . . . . . . HAMAMELIS, p. 241
1. Buds scaly (sometimes naked in *Rhamnus*); flowers in spring 2
2. Fruit a nut in bur or cup; leaves mostly penniveined. 3
2. Fruit not in bur or cup 5
3. Buds lance-shaped, 3/4-1 inch long, terminal present; nut triangular in spiny bur; bark thin, smooth, blue-gray . . . . FAGUS, p. 167
3. Buds ovoid to globose 1/8-1/4 inch long; fruit rounded 4
4. Buds with 2-3 visible scales, terminal absent; leaves with sharp teeth; fruit in spiny bur . . . . CASTANEA, p. 169
4. Buds many-scaled, terminal present; leaves with rounded or short teeth; fruit in scaly cup . . . . QUERCUS, p. 174
5. Fruit a nut or nutlet in strobile or leafy involucre; terminal bud absent 6
5. Fruit a samara, capsule, pome, or drupaceous 10
6. Nutlet in persistent woody strobile; buds with 2 visible scales . . . . ALNUS, p. 156
6. Nut or nutlet with leafy involucre or strobile; buds many-scaled. 7

7. Nutlets winged in deciduous strobile; bark with horizontal lenticels, often papery . . . . BETULA, p. 146
7. Nut or nutlets unwinged in involucres; bark not papery 8
   8. Nut large, edible, single; shrubby . . . . CORYLUS, p. 165
   8. Nutlet small in loose leafy strobile; small trees 9
9. Nutlet enclosed in bladderlike sac; bark thin, gray, shreddy . . OSTRYA, p. 163
9. Nutlet subtended by 3-lobed leafy involucre; bark smooth, blue-gray; stem fluted . . . . CARPINUS, p. 165
   10. Leaves unequal at base; fruit a samara or woody drupe; terminal bud absent 11
   10. Leaves equal at base; fruit a capsule, pome, or drupaceous 14
11. Fruit a winged samara; leaves mostly doubly serrate . . . . ULMUS, p. 214
11. Fruit a drupe; leaves singly serrate 12
   12. Winter buds naked; fruit juicy; twigs often spinescent . . RHAMNUS, p. 315
   12. Winter buds scaly; fruit woody; twigs not spinescent 13
13. Leaves 3-veined; buds ovoid; fruit globose . . . . CELTIS, p. 220
13. Leaves many-veined; buds minute, subglobose; fruit with conspicuous knoblike processes . . . . PLANERA, p. 221
   14. Fruit a capsule 15
   14. Fruit a pome or drupaceous 17
15. Buds small, partially embedded in bark, terminal absent; flowers and fruit in panicles . . . . OXYDENDRUM, p. 337
15. Buds over 1/8 inch long, not embedded; flowers and fruit in aments 16
   16. Buds scaly, terminal present . . . . POPULUS, p. 120
   16. Buds, single caplike scale, terminal absent . . . . SALIX, p. 131
17. Leaf base heart-shaped; fruit nutlike drupe attached to persistent, leafy bract . . . . TILIA, p. 319
17. Leaf base not heart-shaped; no persistent, leafy bract 18
   18. Fruit fleshy; pith solid 19
   18. Fruit dry drupe; pith chambered 26
19. Fruit a berrylike multiple; sap milky; terminal bud absent . . MORUS, p. 227
19. Fruit single; sap watery 20
   20. Terminal bud absent; leaves, twigs, and drupes stellate-pubescent . . . . STYRAX, p. 345
   20. Terminal bud present; not stellate-pubescent 21
21. Leaf scars, 1 bundle trace; leaves simply toothed . . . . ILEX, p. 291
21. Leaf scars with 3 traces; leaves mostly doubly toothed 22
   22. Fruit a drupe; leaves mostly lanceolate . . . . PRUNUS, p. 256
   22. Fruit a pome; leaves mostly oblong to ovate 23
23. Twigs with sharp thorns . . . . CRATAEGUS, p. 253
23. Twigs without thorns (often spinescent) 24
   24. Fruit pyriform . . . . PYRUS, p. 346
   24. Fruit globose or nearly so 25
25. Pome over 1/2 inch long; buds ovoid to globose . . . . MALUS, p. 244
25. Pome smaller; buds acuminate . . . . AMELANCHIER, p. 251
   26. Leaves leathery; fruit brown, 1/2 inch long; crowned with persistent calyx . . . . SYMPLOCOS, p. 341
   26. Leaves thin; fruit nutlike, 2-4-winged, about 2 inches long . . . . HALESIA, p. 341

HANDBOOK OF NORTH AMERICAN TREES

PINACEAE

The Pines

Characteristics of the Genus *Pinus* L.

HABIT. Evergreen trees with straight, unbranched, cylindrical trunks, and whorled, spreading branches.

LEAVES. Needlelike; fascicles or bundles of 2-5 (1 in singleleaf pinyon); enclosed in bud by 6-12 scales forming a persistent or soon-deciduous basal sheath; usually with several lines of stomata on each surface; juvenile leaves on young shoots differ, being spirally arranged, single, and scalelike.

FLOWERS. Monoecious; male, or pollen-producing, consisting of spirally arranged, sessile anthers, yellow, orange, or scarlet; female, or cone- and seed-producing, small, conelike bodies consisting of numerous spirally arranged, 2-ovuled scales, each subtended by a small bract.

FRUIT. A cone, usually pendent, composed of the hardened, woody scales of the flower; scales more or less thickened on the exposed terminal surface (the apophysis) with the ends of the growth of the previous year appearing as a terminal or dorsal, brown protuberance or scar (the umbo), which is often armed with a prickle; maturing in 2 (rarely 3) seasons. Seed: 2 borne at the base on inner face of each fertile scale; thin, terminal, papery wing or wingless.

BUDS. Variable in size, shape, and color; covered by fringed or papery-margined overlapping scales; these component scales each protecting a tiny bud that after the main bud unfolds, develops into a fascicle of leaves or occasionally into a female flower.

WOOD. Among our most important trees; properties extremely variable; numerous, large, easily visible resin ducts; resinous scent. Some species produce naval stores and edible nuts in addition to lumber.

GENERAL. The largest and most important genus of conifers, including about 80 species widely scattered over the Northern Hemisphere; 35 species are native to the United States and Canada; the North American species can be conveniently grouped into the soft or white pines and hard, pitch, or yellow pines.

KEY TO THE SPECIES OF PINES

1. Needles in clusters of 5 (except Pinyons), cluster sheath deciduous, absent on mature needles; 1 fibrovascular bundle in needle cross section; cone scales without prickles (except Foxtails) . . . . Soft Pines
   2. Needles in clusters of 1-4; cones globose, few-scaled; seed large and edible, without wing; western . . . . Pinyon Pines
      3. Needles single; Idaho, Nevada, Utah, Arizona, California . . . . *P. monophylla*, Singleleaf Pinyon, p. 15
      3. Needles predominantly in 2's; Colorado, Utah . . . *P. edulis*, Pinyon, p. 15
      3. Needles in 3's; southern Arizona, New Mexico, Texas . . . . *P. cembroides*, Mexican Pinyon, p. 15
      3. Needles predominantly in 4's; southern California . . . . *P. quadrifolia*, Parry Pinyon, p. 15
   2. Needles in clusters of 5; cones many-scaled.
      4. Cone scales thin, without prickles; seeds long-winged; needles slender, 2-4 inches long, persistent 2-4 years . . . . White Pines
         5. Cones 4-8 inches, scales flexible; eastern Canada to Iowa, Illinois, Georgia . . . . *P. strobus*, Eastern White Pine, p. 9
         5. Cones 5-11 inches (most about 8 inches) long; scales flexible; western Canada to Montana and California . . . . *P. monticola*, Western White Pine, p. 9
         5. Cones 12-18 inches long; scales rigid; Oregon, California, Nevada . . . . *P. lambertiana*, Sugar Pine, p. 7
      4. Cone scales thick; needles stout, 1-3 inches long, persistent 5-17 years; western.
         6. Cone scales without prickles; seed wing absent or short; needles 1 1/2-3 inches, persistent 5-8 years . . . . Stone Pines
            7. Cones 1 1/2-3 inches long, subglobose, remaining closed; alpine, southern Canada to California and Wyoming . . . . *P. albicaulis*, Whitebark Pine, p. 11
            7. Cones 3-10 inches long, subcylindrical, opening at maturity; through west . . . . *P. flexilis*, Limber Pine, p. 11
         6. Cone scales with prickles; seed with long wing; needles 1-1 1/2 inches long, resin-dotted, persistent 10-17 years . . . . Foxtail Pines
            8. Cones with long, slender prickles; alpine, Colorado to California, south . . . . *P. aristata*, Bristlecone Pine, p. 13
            8. Cones with minute, incurved prickles; California alpine . . . . *P. balfouriana*, Foxtail Pine, p. 13
1. Needles in clusters of 2 or 3 (5 in Torrey and Arizona Pine), cluster sheath persistent (deciduous in Chihuahua Pine); 2 fibrovascular bundles in cross section of needle; cone scales usually armed with prickles . . . . Hard Pines
   9. Eastern Hard Pines (northeastern and central states).
      10. Needles in 3's or 2's and 3's, 3-5 inches long, twisted; cones 1 1/2-3 1/2 inches long, persistent on branch.
         11. Needles in 3's, rigid; cones ovoid with stout prickles; New Brunswick and Ontario to Georgia . . . . *P. rigida*, Pitch Pine, p. 21
         11. Needles in 2's and 3's, soft and flexible; cones oblong with weak prickles; New York to Missouri and south . . . . *P. echinata*, Shortleaf Pine, p. 23
      10. Needles in 2's.
         12. Needles 4-6 inches long; cones symmetrical.
            13. Cones 2-3 1/2 inches, persistent several years, falling entire; scales yellow-brown, short spine; widely planted exotic . . . . *P. nigra*, Austrian Pine, p. 17
            13. Cones 1 1/2-2 1/2 inches long, falling in 1 year, leaving basal scales on limb; scales red-brown, unarmed; southern Canada to Minnesota and West Virginia . . . . . *P. resinosa*, Red Pine, p. 17

12. Needles 1-3 inches long.
    14. Bark distinctly orange-colored; cones 1 1/2-2 1/2 inches long, unsymmetrical, opening at maturity, falling when ripe; scales often pyramidally thickened; widely planted and naturalized in Northeast . . . . *P. sylvestris*, Scotch Pine, p. 17
    14. Bark not distinctly orange-colored; cones long persistent; scales not pyramidally thickened.
        15. Cones 2-3 1/2 inches long, heavy; scales, much thickened ends, armed with conspicuous sharp spurs; Appalachian Mountains, New Jersey to Georgia . . . . *P. pungens*, Table-Mountain Pine, p. 29
        15. Cones 1 1/2-2 inches long, thin-scaled, slender or minute prickles.
            16. Cones strongly incurved, commonly remaining closed; scales irregularly developed; Canada to New York, Indiana, Wisconsin, Minnesota . . . . *P. banksiana*, Jack Pine, p. 19
            16. Cones symmetrical, opening at maturity; New York to Indiana, south to Georgia, Mississippi . . . . *P. virginiana*, Virginia Pine, p. 19

9. Southern Yellow Pines (Gulf and South Atlantic states).
  17. Needles 6-18 inches long, in 3's (2's and 3's in Slash Pine).
    18. Cones 2-2 1/2 inches long and broad, often remaining closed, long persistent; needles 6-8 inches long; New Jersey to Florida . . . . *P. serotina*, Pond Pine, p. 21
    18. Cones 2-10 inches long, much longer than broad, opening at maturity and not persistent.
        19. Needles 8-18 inches; cone 6-10 inches; conspicuous thick twigs, large silvery-white buds; Virginia to Florida, Texas . . . . *P. palustris*, Longleaf Pine, p. 27
        19. Needles 6-12 inches long; cone 2-6 inches long; twigs not conspicuously stout; buds red-brown.
            20. Needles in 3's, 6-9 inches; cone light brown, stout spine; New Jersey to Oklahoma, Texas . . . . *P. taeda*, Loblolly Pine, p. 25
            20. Needles in 2's and 3's, 8-12 inches; cone chocolate-brown, small spine or prickle; South Carolina to Louisiana . . . . *P. elliottii*, Slash Pine, p. 29
  17. Needles 1 1/2-5 inches, in 2's (2's and 3's, Shortleaf Pine).
    21. Needles 3-5 inches long, in 2's and 3's, widely distributed on dry soils; New York to Missouri and south . . . . *P. echinata*, Shortleaf Pine, p. 23
    21. Needles 1 1/2-3 1/2 inches long, in 2's.
        22. Cone 2-3 1/2 inches long, often remaining closed and persistent for years; dry sandy soils in Florida . *P. clausa*, Sand Pine, p. 25
        22. Cone 1/2-2 inches long, opening at maturity and falling in few years; coastal bottomlands, South Carolina to Louisiana . . . . *P. glabra*, Spruce Pine, p. 23

9. Western Yellow Pines (region west of Great Plains).
  23. Needles in 5's.
    24. Needles 5-7 inches long; cones thin-scaled, 2-3 1/2 inches long; Arizona and New Mexico . . . . *P. ponderosa arizonica*, Arizona Pine, p. 31
    24. Needles 8-13 inches long; cones thick-scaled, 4-6 inches; southern California . . . . *P. torreyana*, Torrey Pine, p. 41
  23. Needles in 2's and 3's, or 3's.
    25. Cones massive, 6-14 inches long, clawlike spines; needles in 3's.

26. Cones yellow-brown, 10-14 inches long; seeds shorter than wings; southern and lower California . . . . *P. coulteri,* Coulter Pine, p. 37

26. Cones red-brown, 6-10 inches long; seeds longer than wings; dry California foothills . . . . *P. sabiniana,* Digger Pine, p. 37

25. Cones not massive; without clawlike spines.

27. Cones unsymmetrical with thickened scales, often remaining closed and persistent for years.

28. Needles in 2's, 1-3 inches; cone under 2 inches.

29. Cones with knoblike scales and long prickles.

30. Needles 1-1 1/2 inches long; small contorted tree of coastal area . . . . *P. contorta,* Shore Pine, p. 35

30. Needles 1-3 inches; straight, slender, mountain tree throughout West . . . . *P. contorta latifolia,* Lodgepole Pine, p. 35

29. Cones with thin scales and minute prickle; Canada and northeastern United States . . . . *P. banksiana,* Jack Pine, p. 19

28. Needles 3-7 inches long; cones 2-6 inches long; Oregon, California.

31. Needles in 2's; cones 2-3 1/2 inches long, with stout, spurlike prickles . . . . *P. muricata,* Bishop Pine, p. 41

31. Needles in 3's; cones 3-7 inches long, without spurlike prickles.

32. Thickened cone scales flattened, pyramidal; dry foothills, California, Oregon . . . . *P. attenuata,* Knobcone Pine, p. 39

32. Thickened cone scales rounded, domelike; California coast . . . . . *P. radiata,* Monterey Pine, p. 39

27. Cones symmetrical, thin-scaled.

33. Needles in 3's, 2-4 inches long, cluster sheath deciduous; cone 1 1/2-2 inches long, maturing in 3 years, long-stalked, often remaining closed, long persistent; southern Arizona, southern New Mexico . . *P. leiophylla,* Chihuahua Pine, p. 35

33. Needles over 4 inches, sheath persistent; cone 2-15 inches, maturing in 2 years, short-stalked, opening, deciduous, basal scales remaining on twig.

34. Cones 2-3 1/2 inches long; needles 4-6 inches; rare and local in Washoe County, Nevada . . . . *P. washoensis* M. & S., Washoe Pine, p. 33

34. Cones 3-15 inches long; needles 4-15 inches.

35. Cones 5-15 inches long; needles 4-9 inches, persistent 6-9 years; twigs purplish, lemonlike odor; buds not covered with resin droplets; California, Oregon . . . . *P. jeffreyi,* Jeffrey Pine, p. 33

35. Cones 3-6 inches long; needles persistent 2-3 years; twigs orange to red, turpentine odor; buds often covered with droplets.

36. Needles 4-11 inches (mostly 5-7 inches) long, in 2's and 3's, yellow-green; throughout West. . . . . *P. ponderosa,* Ponderosa Pine, p. 31

36. Needles 8-15 inches, in 3's (rarely 2-5), dark green; Arizona, New Mexico . . . . *P. engelmannii,* Apache Pine, p. 33

Sugar Pine

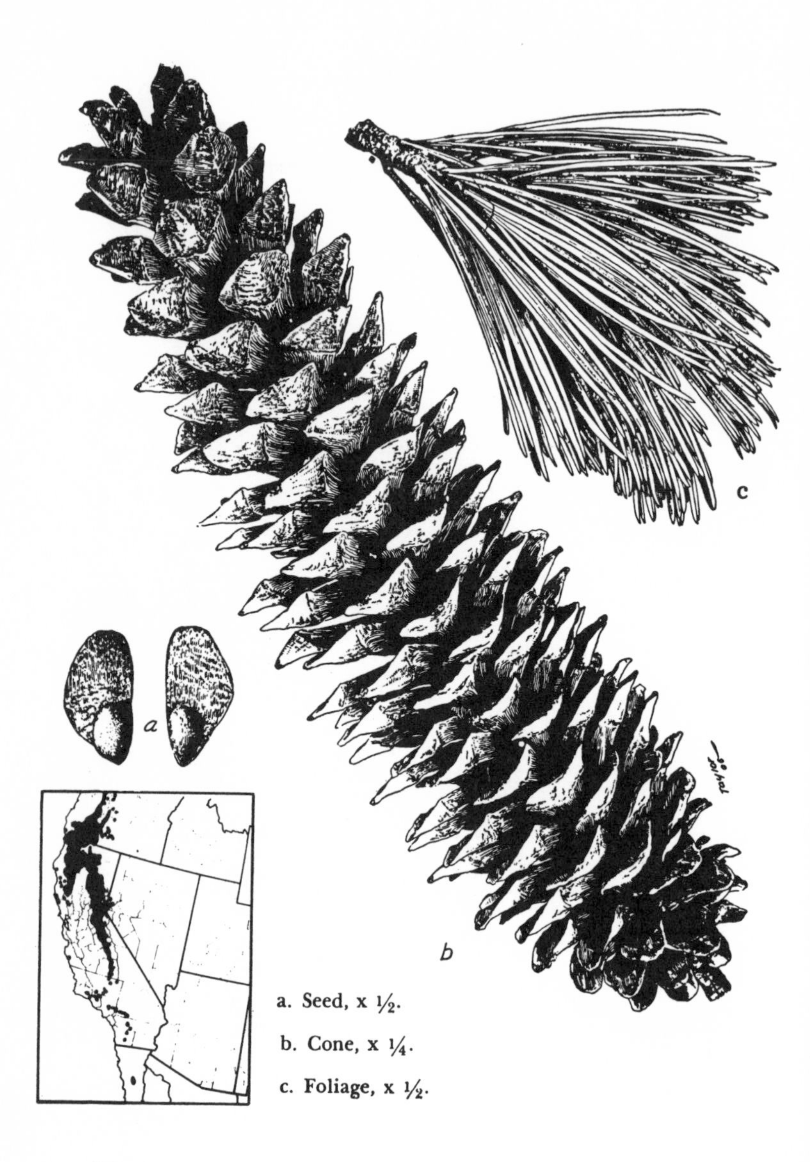

a. Seed, x ½.

b. Cone, x ¼.

c. Foliage, x ½.

PINACEAE

Sugar Pine

Pinus lambertiana Dougl.

HABIT. The largest American pine, commonly 175-200 feet high and 3-5 feet in diameter (max. 246 by 10 feet); on good sites with a long, clear bole and a short crown of large, often contorted, horizontal branches.

LEAVES. In fascicles of 5; 2-4 inches long; stouter than in white pine; twisted; blue- to gray-green; persistent 2-3 years; sheath deciduous; margin with minute teeth; often silvery with conspicuous white lines of stomata.

FLOWERS. Male yellow; female bright pink with purple scale margins.

FRUIT. Cones long-stalked; 10-26 inches long and 4-5 inches in diameter; scales slightly thicker and more rigid than in white pine, unarmed and with terminal umbo. Seed: 1/2 inch long; dark brown to black; wing 1-1 1/2 inches long.

TWIGS. Slender to stout; at first rusty-pubescent, later smooth and orange-brown. Winter buds: 1/3 inch long; sharp-pointed; chestnut-brown.

BARK. On young stems thin, smooth, and gray-green; on mature boles 1 1/2-4 inches thick, in thick platelike ridges covered with purplish to reddish scales.

WOOD. Very important; similar to white pine but coarser in texture; sash, door, pattern work, etc.; a sweet, sugarlike substance called pinite, exudes from wounds and gives the tree its name.

SILVICAL CHARACTERS. Intolerant; maturity reached in 200-350 years (extreme age 623 years); reproduction generally sparse; tree windfirm with well-developed tap and lateral root system; blister rust and bark beetles cause damage.

HABITAT. Transition zone; altitudinal range 3,000-10,000 feet; best development on west slopes of Sierra Nevada between 4,500-5,500 feet; in mixed stands with ponderosa and Jeffrey pines, Douglas-fir, and other conifers; on cooler and moister sites than associated pines.

White Pine

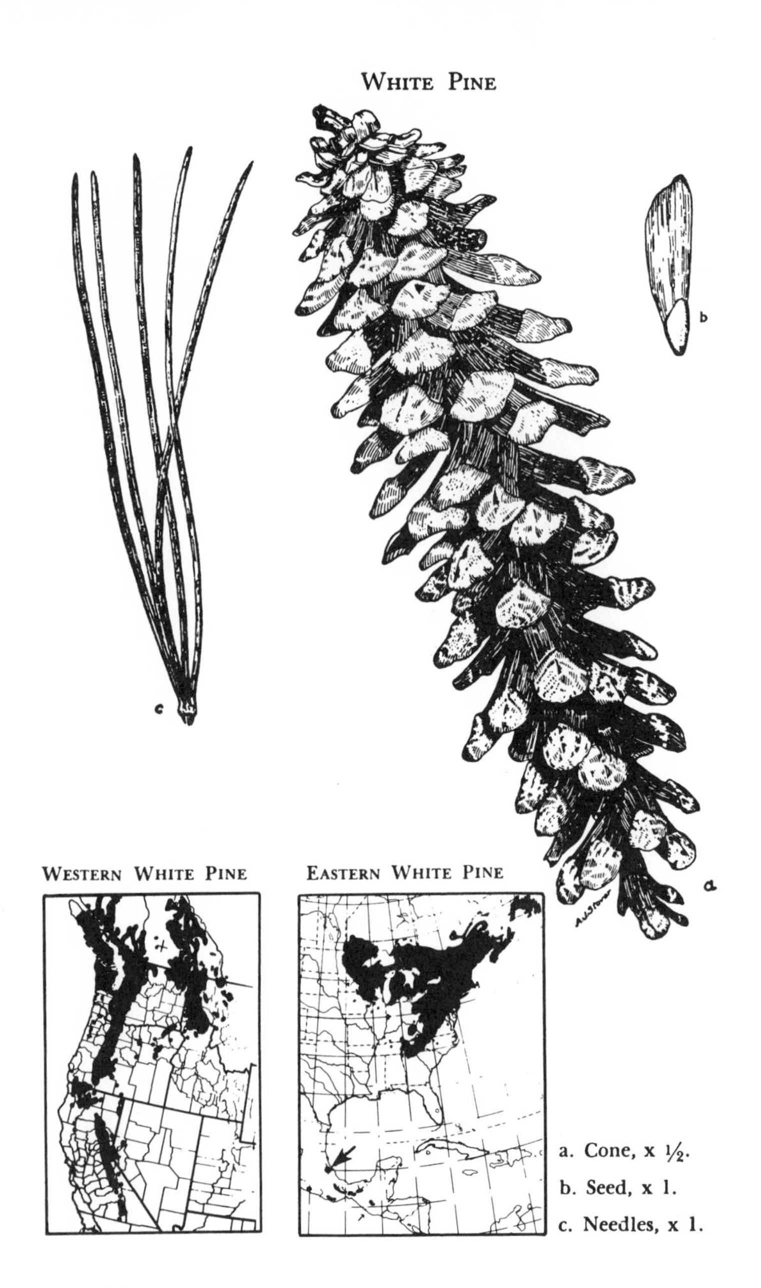

a. Cone, x ½.
b. Seed, x 1.
c. Needles, x 1.

Western White Pine

Pinus monticola Dougl.

HABIT. A tree 90-180 feet high and 2 1/2 feet in diameter (max. 200 by 8 feet); on good sites a tall bole with narrow, symmetrical crown and slender, drooping branches.

LEAVES. In fascicles of 5; 2-4 inches long; slender; twisted; blue-green; persistent 3-4 years; sheath deciduous; margin with minute teeth; all sides marked by stomata.

FLOWERS. Male yellow; female red-purple in clusters.

FRUIT. Cones long-stalked; 5-15 (mostly 8-11) inches long; narrow; scales thin, unarmed, with terminal umbo. Seed: 1/4 inch long; wing about 1 inch long; red-brown.

TWIGS. Slender; at first rusty-pubescent, later smooth and red-brown to purple-brown. Winter buds: 1/2 inch long; oblong-ovoid.

BARK. On young stems thin, smooth, and light gray; on mature trees rarely over 1 1/2 inches thick in square or rectangular, dark gray plates.

WOOD. Very important; soft, light in weight, and not strong; light brown heartwood; similar to eastern white pine; planing mill products, building, construction, patterns, etc.

SILVICAL CHARACTERS. Tolerant when young, becoming intolerant with age; maturity reached in 200-350 years (extreme age, 400-500 years); reproduction generally sparse; tree windfirm with well-developed tap and lateral root system; fire, blister rust, and bark beetles cause damage.

HABITAT. Canadian zone; altitudinal range 2,000-7,000 feet; best on rich, porous, moist soils; often forming extensive pure forests, in mixtures elsewhere with other conifers.

* * *

Eastern White Pine

Pinus strobus L.

Until about 1890 this species was the most important forest tree in North America. It is still an important species and very similar in appearance to western white pine, from which it can be distinguished by its distribution, cone length, which varies from 4-8 (mostly about 5) inches long, and needle persistence of 2 (rarely 3) years.

Limber Pine

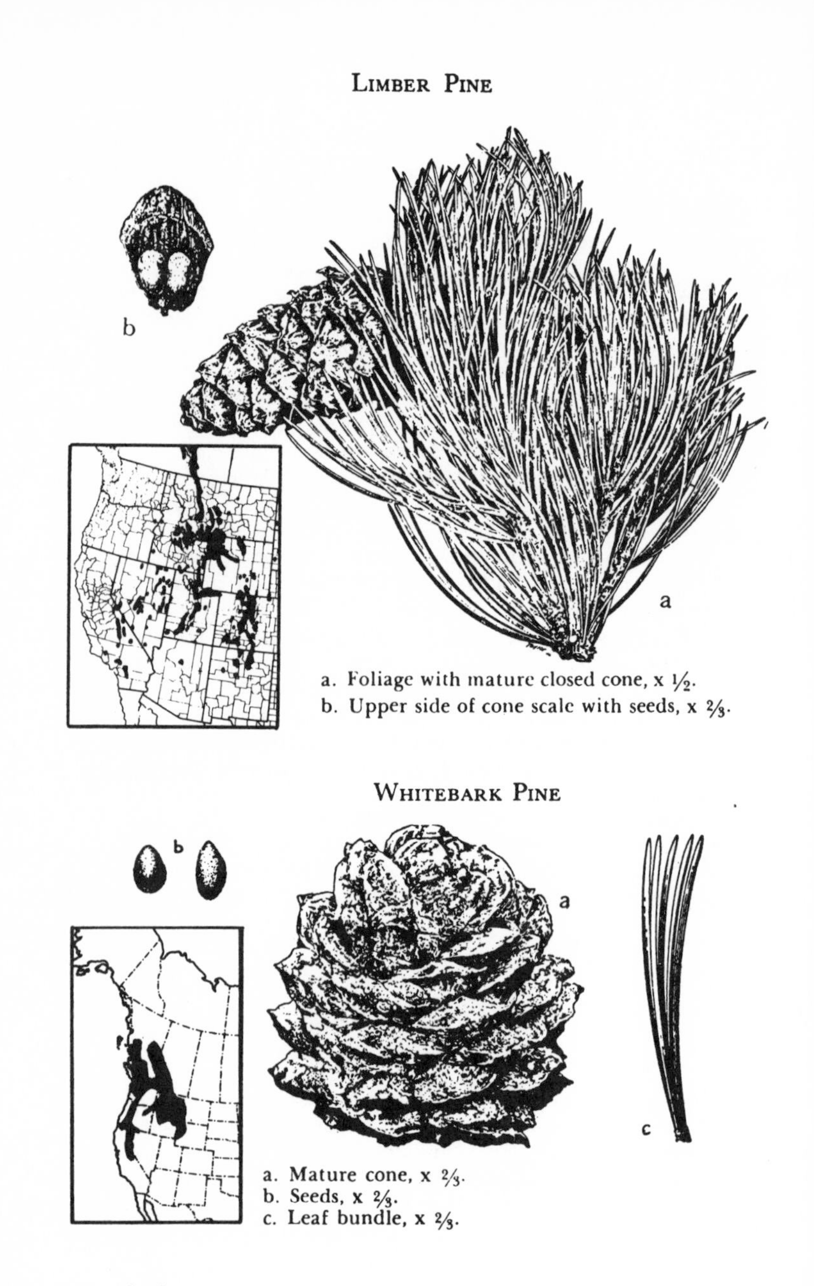

a. Foliage with mature closed cone, x ½.
b. Upper side of cone scale with seeds, x ⅔.

Whitebark Pine

a. Mature cone, x ⅔.
b. Seeds, x ⅔.
c. Leaf bundle, x ⅔.

PINACEAE

Limber Pine

Pinus flexilis James

HABIT. A tree 25-50 feet high (max. 85 by 6 1/3 feet); crown broad, open, with large, plumelike, often drooping branches.

LEAVES. In fascicles of 5; 1 1/2-3 inches long; stout; rigid; dark green; persistent 5-6 years; sheath deciduous; margins with minute teeth; marked on all sides by rows of stomata.

FLOWERS. Male red; female clustered, red-purple.

FRUIT. Cones short-stalked; 3-10 inches long; subcylindrical; open at maturity; scales greatly thickened and often slightly reflexed, with terminal unarmed umbo. Seed: 1/3-1/2 inch long; wingless; thick, light brown shell. The variety *reflexa* Engelm. (*P. strobiformis* Engelm.) of the Mexican border has strongly reflexed cone scales.

TWIGS. Stout and tough; smooth and silver-white or gray. Winter buds: 1/3-1/2 inch long; broad-ovoid and pointed.

BARK. Characteristically thin, smooth, white-gray.

WOOD. Unimportant; moderately light and soft; close-grained; used locally for mine props, railroad ties, etc.

SILVICAL CHARACTERS. Very intolerant; growth slow; maturity reached in 200-300 years; tree very windfirm with taproot; fire, blister rust, and bark beetles cause damage.

HABITAT. Upper Sonoran to Hudsonian zones; altitudinal range 4,000-11,500 feet; adapted to wide variety of sites, but typical of summits and rocky foothills.

* * *

Whitebark Pine

Pinus albicaulis Engelm.

A subalpine tree almost indistinguishable from limber pine except for its very characteristic cones.

FRUIT. Cones short-stalked; 1 1/2-3 inches long; subglobose; closed at maturity and opening by disintegrating at the axis; purple-brown; scales thick with stout, pointed umbos.

Bristlecone Pine

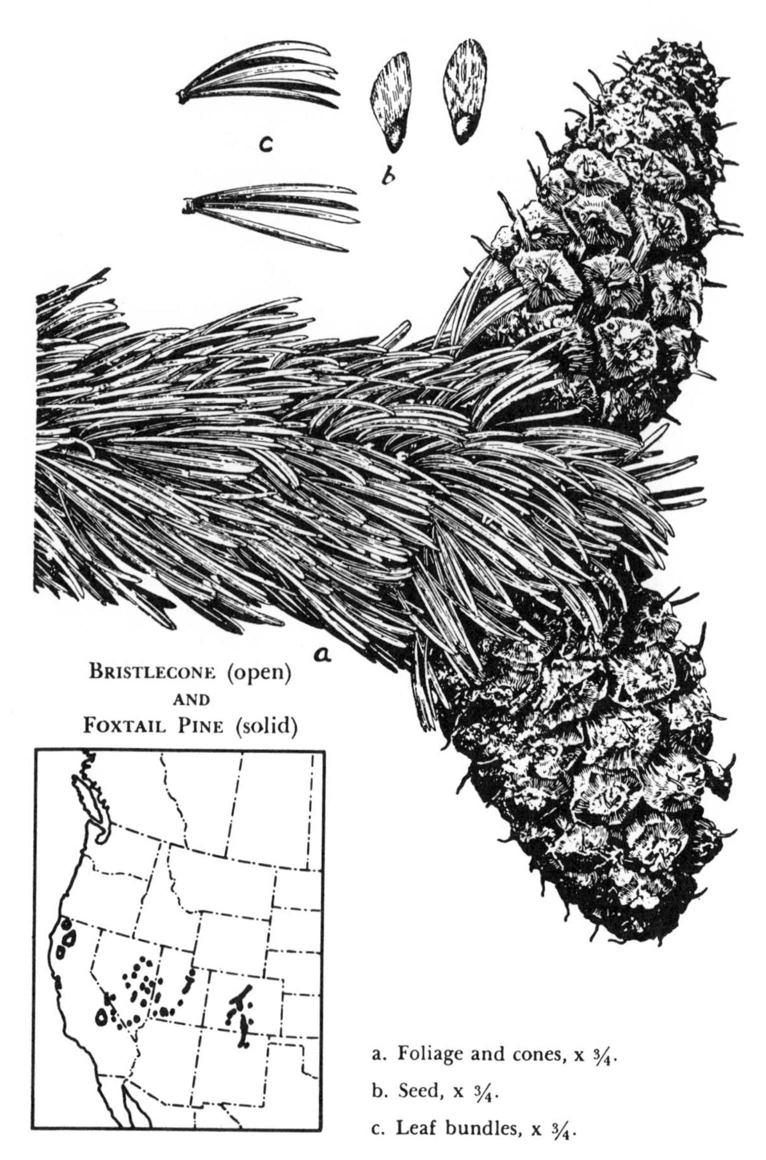

a. Foliage and cones, x ¾.

b. Seed, x ¾.

c. Leaf bundles, x ¾.

Bristlecone Pine

Pinus aristata Engelm.

HABIT. A subalpine tree 30-40 feet high and 1-2 feet in diameter (max. 60 by 3 feet); bole short, stocky, and commonly malformed; crown dense, irregular, bushy in appearance, and frequently clothing the stem nearly to the ground.

LEAVES. In fascicles of 5; 1-1 1/2 inches long; stout and curved; deep green; persistent for 10-17 years; in dense, often appressed, clusters; sheath deciduous; lustrous on back, marked on lower or ventral surfaces by numerous rows of stomata; usually showing conspicuous whitish exudations of resin.

FLOWERS. Male dark orange-red; female purple.

FRUIT. Cones short-stalked, 3-3 1/2 inches long; ovoid-oblong; open at maturity; scales thick, with dark chocolate-brown apophysis; umbo dorsal, with long, bristlelike, fragile, incurved prickle. Seed: 1/4 inch long; long terminal wing.

TWIGS. Stout; orange-colored, becoming nearly black; long tufts of foliage at ends. Winter buds: 1/3 inch long; brown.

BARK. Thin, smooth, and gray-white on young stems; 1/2-3/4 inch thick on mature trunks; red-brown and furrowed.

WOOD. Unimportant; moderately soft; heartwood pale red-brown; used locally for fuel and mine props.

SILVICAL CHARACTERS. Very intolerant; growth slow; maturity reached in 200-250 years, attains great age, some trees over 4,000 years old and possibly the oldest living organisms; reproduction sparse and scattered; tree windfirm.

HABITAT. From typical Hudsonian to Transition zones; altitudinal range 7,500-10,800 feet; typical of exposed sites; usually in mixture with limber pine, fir, and spruce.

* * *

Foxtail Pine

Pinus balfouriana Grev. & Balf.

This California subalpine species closely resembles *Pinus aristata* except for its cones which are dark purple, 3-5 inches long, and armed with a minute, incurved, deciduous prickle.

Mexican Pinyon

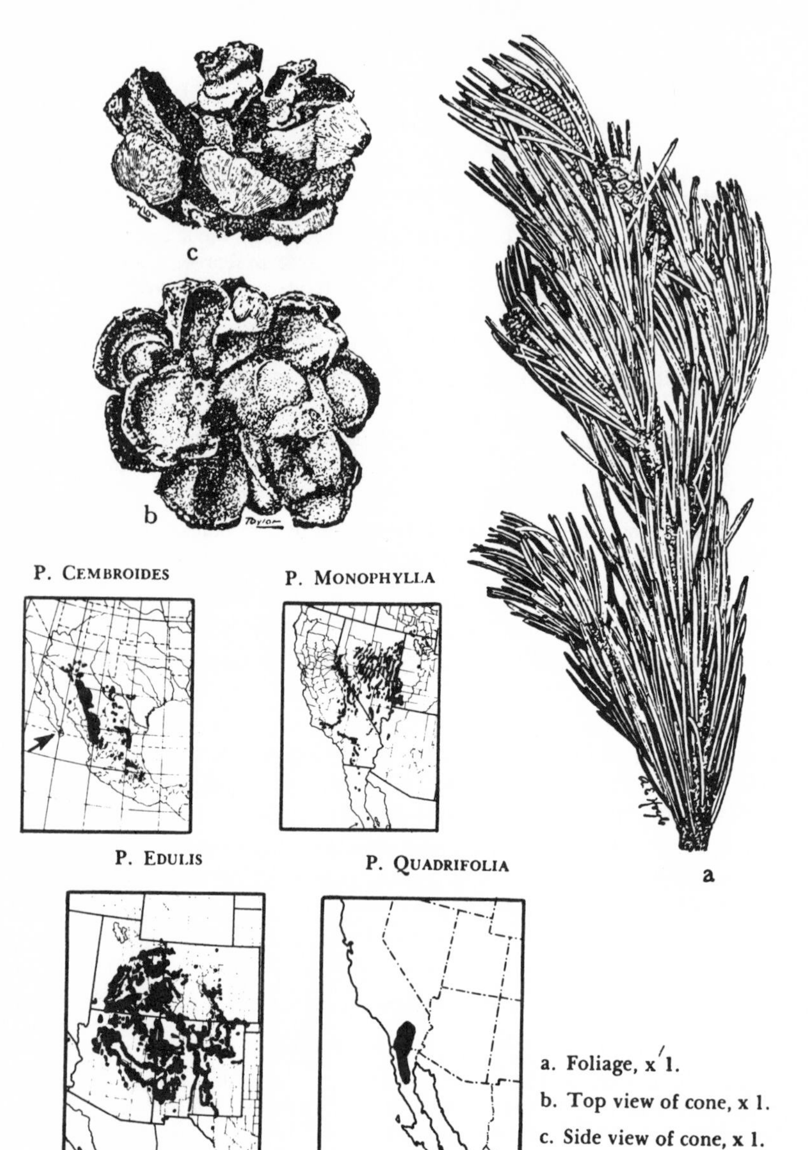

a. Foliage, x 1.
b. Top view of cone, x 1.
c. Side view of cone, x 1.

PINACEAE

Mexican Pinyon. Nut Pine
Pinus cembroides Zucc.

HABIT. A tree 10-40 feet high and 1-2 feet in diameter (max. 50 by 3 feet); bole often divided with a spreading, rounded crown, giving the tree a bushy appearance.

LEAVES. In fascicles of 3, or 2 and 3 (rarely 1-5); 1-2 inches long; moderately slender (less than 1 mm); blue-green; incurved; sharp-pointed; persistent 3-4 years; margins with minute teeth; fascicle sheath only partially deciduous.

FLOWERS. Male yellow in crowded clusters; female red.

FRUIT. Cone 1-2 1/2 inches long; ovoid to globose; scales few and unarmed. Seed: 1/2-3/4 inch long; wingless; edible; hard-shelled.

TWIGS. Moderately stout; orange-colored to brown.

BARK. Rarely over 1/2 inch thick on mature trees; divided into ridges separated by shallow fissures; light red-brown.

WOOD. Unimportant; rather hard; fine-textured; light yellow heartwood; used locally for fuel, ties, and posts.

SILVICAL CHARACTERS. Very intolerant; growth slow; maturity reached in 250-350 years; reproduction generally sparse and scattered; tree very windfirm.

HABITAT. Upper Sonoran zone; arid, gravelly slopes; usually in mixture with junipers and scrub oaks.

* * *

The following species are very similar to *Pinus cembroides* but differ from it in distribution and the following characters:

Pinyon. Colorado Pinyon
Pinus edulis Engelm. (*Pinus cembroides* var. *edulis* Voss)

LEAVES. In fascicles of 2 (rarely 1 or 3); over 1 mm thick; yellow-green; persistent 3-9 years; margins entire.

FRUIT. Seed thin-shelled, easily cracked with teeth.

Singleleaf Pinyon
Pinus monophylla T. & F. (*Pinus cembroides* var. *monophylla* Voss)

LEAVES. Single (rarely in 2's); over 1 mm thick; pale green; persistent 4-12 years; margins entire.

Parry Pinyon
Pinus quadrifolia Parl. (*Pinus cembroides* var. *parryana* Voss)

LEAVES. In fascicles of 4 (rarely 3-5); over 1 mm thick; blue-green; persistent 3 years; margins entire or toothed.

Red Pine

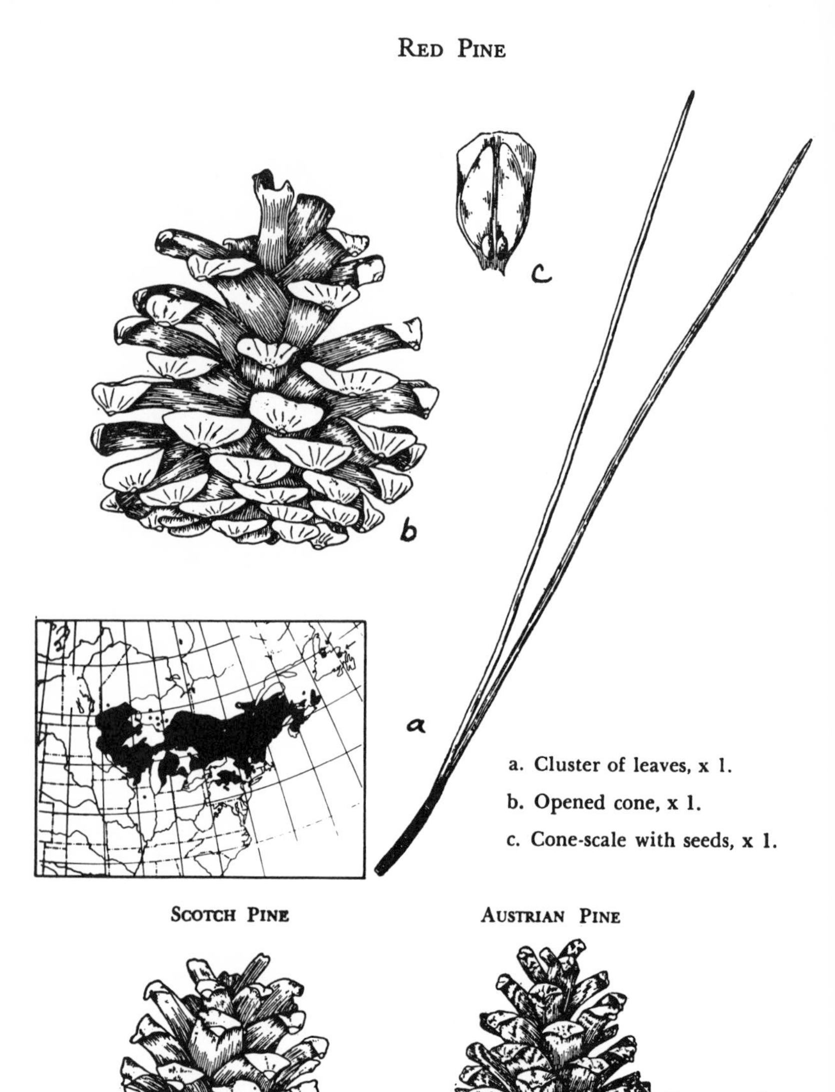

a. Cluster of leaves, x 1.

b. Opened cone, x 1.

c. Cone-scale with seeds, x 1.

Scotch Pine

Cone, x 1.

Austrian Pine

Cone, x ½.

Red Pine

Pinus resinosa Ait.

HABIT. A tree 60-80 feet high and 2-3 feet in diameter (max. 120 by 5 feet); bole symmetrical, clear; crown broad, rounded, open.

LEAVES. In fascicles of 2; 4-6 inches long; slender; straight; brittle; dark yellow-green; persistent 4-5 years; margins with minute teeth; basal sheath persistent.

FLOWERS. Male purple; female scarlet.

FRUIT. Subsessile; 1 1/2-2 1/4 inches long; ovoid-conic; chestnut-brown; umbo dorsal and unarmed; in falling leaving basal scales on twig. Seed: 3/16 inch long; wing 2/3 inch long.

TWIGS. Stout; rough; orange to red-brown. Winter buds: 1/2-3/4 inch long; scales red-brown with white fringed margins.

BARK. 1-1 1/2 inches thick; light red-brown; broken into scaly plates.

WOOD. Important; rather light and soft; heartwood light red-brown; used for construction, ties, etc.

SILVICAL CHARACTERS. Intermediate in tolerance; moderate rate of growth; extreme age about 350 years; roots wide-spreading with taproot; reproduction vigorous; hardy and resistant to enemies.

HABITAT. Light, sandy soils; in pure stands, or mixed with white pine on better sites and jack pine on poorer.

* * *

Two related Eurasian pines have been widely planted:

Scotch pine, *Pinus sylvestris* L., has become naturalized in the Northeast. This species is easily identified by its twisted needles, 1 1/2-3 inches long, in clusters of 2; by yellow-brown cones 1-2 1/2 inches long with flat or pyramidally thickened scales; and by bark that is distinctly orange in color.

Austrian pine, *Pinus nigra* Arnold, has escaped from cultivation. It resembles red pine in its needles, but can be distinguished by its dark brown bark; silvery-white buds; and yellow-brown cones 2-3 inches long, armed with a short spine, and falling intact from the twigs.

Jack Pine

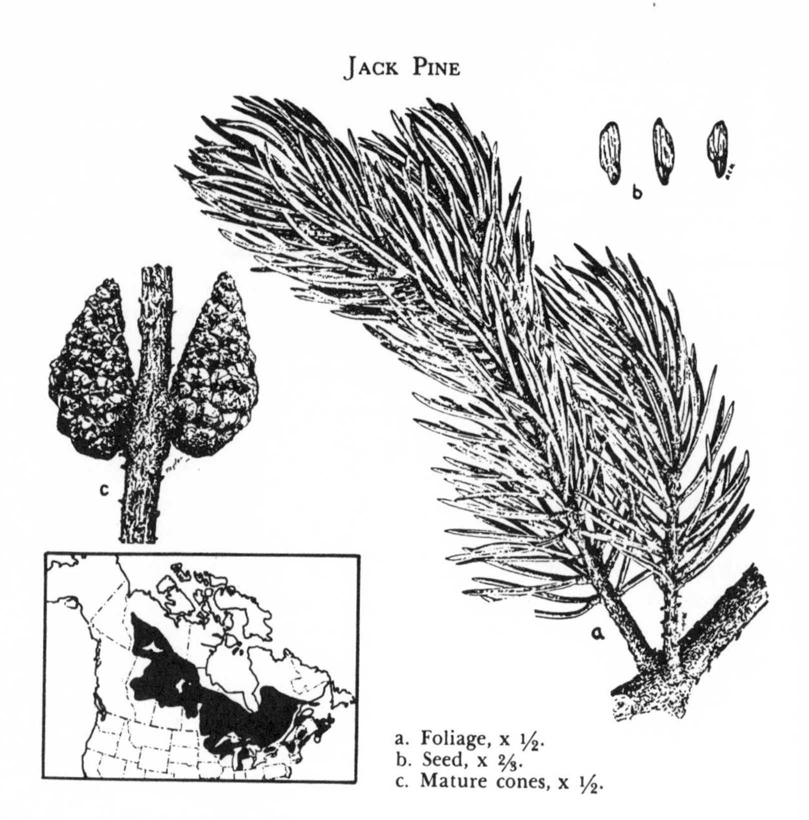

a. Foliage, x 1/2.
b. Seed, x 2/3.
c. Mature cones, x 1/2.

Virginia Pine

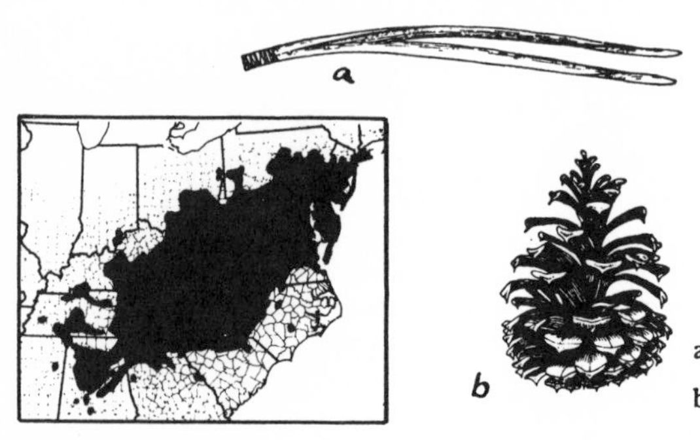

a. Cluster of leaves, x 2/3.
b. Cone, x 1/3.

PINACEAE

Jack Pine

Pinus banksiana Lamb.

HABIT. A tree rarely 60-80 feet high (max. 90 by 2 feet); bole often crooked; crown open, irregular, scraggly.
LEAVES. In fascicles of 2; 3/4-1 1/2 inches long; divergent; stout; twisted; yellow-green; persistent 2-3 years; margins with minute teeth; basal sheath persistent.
FLOWERS. Male yellow; female dark purple; clustered.
FRUIT. Sessile; 1 1/2-2 inches long; oblong-conic; unsymmetrical; often remaining closed and on tree for years; scales irregularly developed, unarmed or with minute, deciduous prickle. Seeds: 1/12 inch long; black; wing about 1/3 inch long.
TWIGS. Thin, tough, flexible; becoming rough and red-brown. Winter buds: 1/4 inch long; ovoid; pale cinnamon-brown; very resinous.
BARK. Thin; dark red-brown; shallowly ridged.
WOOD. Moderately important; light and soft; heartwood light brown; used for pulp, fuel, posts, etc.
SILVICAL CHARACTERS. Very intolerant; short-lived; widespreading roots with a taproot; reproduction vigorous.
HABITAT. Essentially Canadian, extending farther north than other pines; dry, sterile soils; mostly in pure stands.

* * *

Virginia Pine

Pinus virginiana Mill.

This species is similar to jack pine in habit, twigs, bark, wood, silvical characters, and site; it differs from jack pine in distribution and the following characters:

LEAVES. In fascicles of 2; 1 1/2-3 inches long; divergent; stout; twisted; gray-green; persistent 3-4 years; margins with minute teeth; short-needled trees hardly distinguishable from jack pine.
FLOWERS. Male orange-brown; female pale green tinged with rose.
FRUIT. Subsessile; 1 1/2-3 inches long; ovoid-conic; symmetrical; opening at maturity but persisting on twigs 3-4 years; scales thin, flat, with persistent prickle and deep purple inner lip. Seed: 1/4 inch long, pale brown, with wing about 1 inch long.

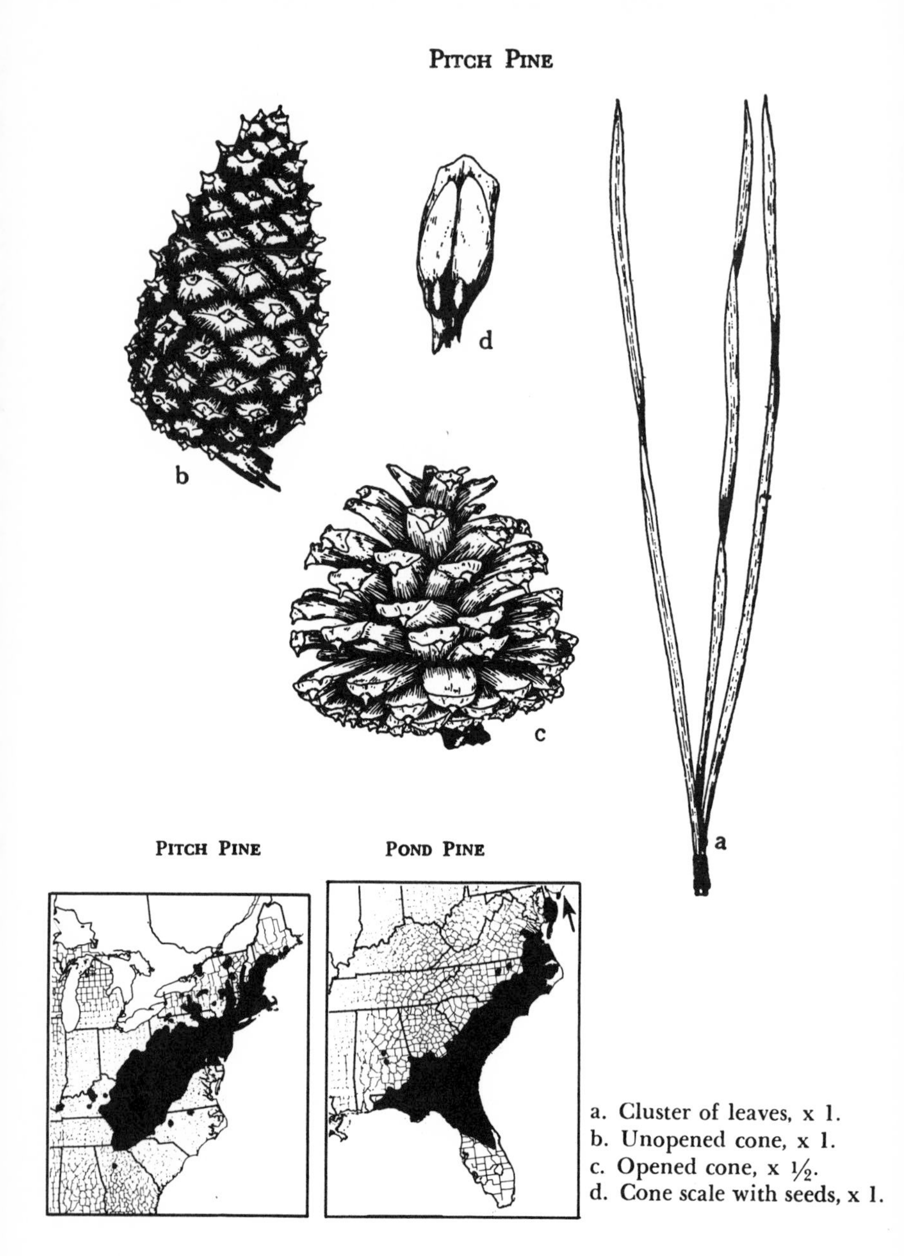

a. Cluster of leaves, x 1.
b. Unopened cone, x 1.
c. Opened cone, x ½.
d. Cone scale with seeds, x 1.

Pitch Pine

Pinus rigida Mill.

HABIT. A tree 40-60 feet high and 1-2 feet in diameter (max. 100 by 3 feet); on good sites; straight bole; open, irregular crown of large, nearly horizontal branches.

LEAVES. In fascicles of 3; 3-5 inches long; stout; stiff; usually twisted; standing out at nearly right angles to twig; yellow-green; often produced on trunk; persistent 2-3 years; margins with minute teeth; basal sheath persistent.

FLOWERS. Male yellow; female red-green.

FRUIT. Short-stalked; 1 1/2-3 1/2 inches long; ovoid-conic; opening at maturity but persistent for many years; scales armed with short, rigid prickle. Seeds: 1/4 inch long; dull black; winged.

TWIGS. Stout; rough; gray-brown. Winter buds: 1/2-3/4 inch long; scales chestnut-brown and fringed; resinous.

BARK. Thick; dark red-brown; flat plates separated by furrows.

WOOD. Of little importance; light and soft; heartwood light brown; used for lumber, ties, props, fuel.

SILVICAL CHARACTERS. Very intolerant; short-lived; taproot in youth; reproduction vigorous; young trees produce sprouts.

HABITAT. Typically dry, sterile sites; pure or mixed stands, with scrub oak, red maple, gray birch, etc.

* * *

Pond Pine

Pinus serotina Michx.
[*Pinus rigida* var. *serotina* (Michx.) Loud.]

This closely related southern species found in swamps and low, wet flats differs from pitch pine in having flexible needles 6-8 inches long and nearly globose cones armed with weak, deciduous prickles that remain closed for several years. Pond pine closely resembles *P. taeda* when persistent cones are lacking but may be distinguished by its very resinous buds, less upright needle orientation, and frequent presence of trunk sprouts. It has remarkable ability to sprout following fire.

Shortleaf Pine

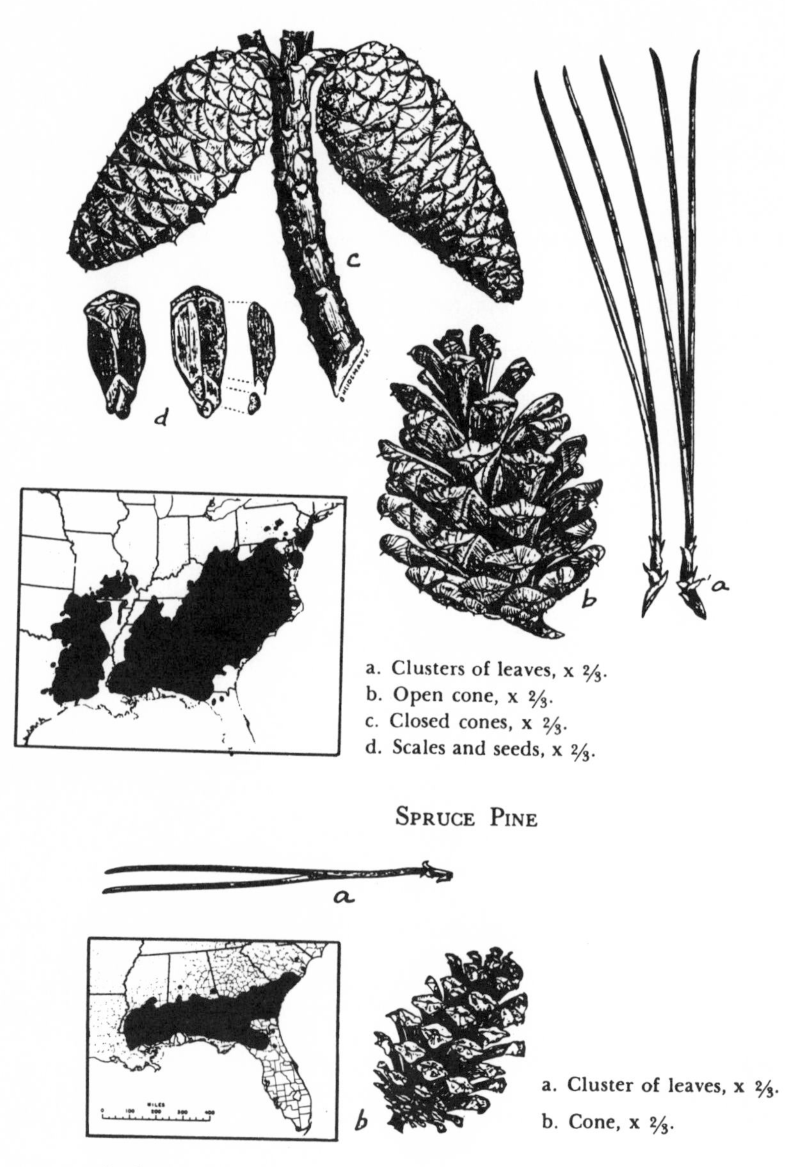

a. Clusters of leaves, x 2/3.
b. Open cone, x 2/3.
c. Closed cones, x 2/3.
d. Scales and seeds, x 2/3.

Spruce Pine

a. Cluster of leaves, x 2/3.
b. Cone, x 2/3.

PINACEAE

Shortleaf Pine

Pinus echinata Mill.

HABIT. A tree 80-100 feet high and 2-3 feet in diameter (max. 130 by 4 feet); long, clear bole; narrow, pyramidal crown.

LEAVES. In fascicles of 2 and 3; 3-5 inches long; slender; flexible; yellow-green; persistent 2-4 years; often produced on trunk; basal sheath persistent.

FLOWERS. Male pale purple; female pale rose color.

FRUIT. Nearly sessile; 1 1/2-2 1/2 inches long; oblong to conical; opening at maturity and persistent several years; red-brown; scales thin, armed with small sharp prickle. Seeds: 3/16 inch long; brown with black markings; wing 1/2 inch long.

TWIGS. Slender; flexible; roughened; red-brown. Winter buds: 1/4 inch long; red- to gray-brown; not highly resinous.

BARK. Thin; nearly black and scaly on young trees; later becoming red-brown and scaly-plated; characteristic resin holes or pockets.

WOOD. Very important; intermediate between longleaf and western pines in strength and hardness; many uses.

SILVICAL CHARACTERS. Intolerant; growth very rapid on good sites; deep taproot system; reproduction aggressive; young trees capable of producing sprouts if stems killed.

HABITAT. Typical of dry, light soils in pure or mixed stands.

* * *

Spruce Pine

Pinus glabra Walt.

Nowhere common, this unimportant medium-sized tree is found scattered among hardwoods on moist sandy loams. It is very tolerant and resembles shortleaf pine.

LEAVES. In fascicles of 2; 1 1/2-3 inches long; slender; dark green.

FRUIT. Similar to shortleaf, but more globose and smaller, with weak, small, often deciduous prickle.

BARK. Thin; gray; smooth on twigs and young stems.

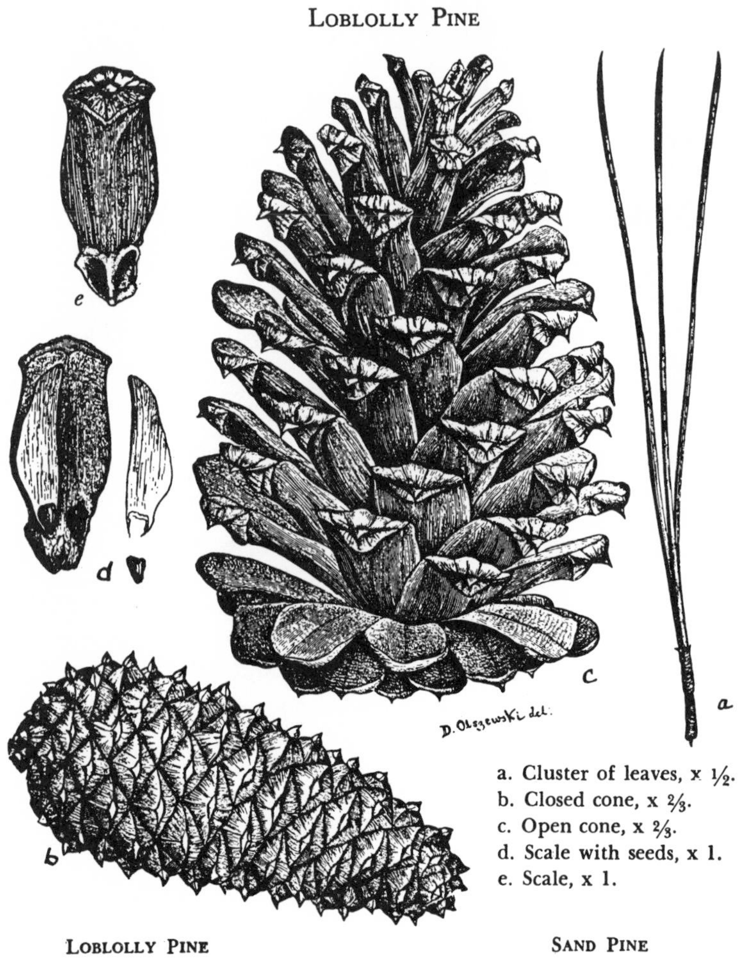

a. Cluster of leaves, x ½.
b. Closed cone, x ⅔.
c. Open cone, x ⅔.
d. Scale with seeds, x 1.
e. Scale, x 1.

LOBLOLLY PINE

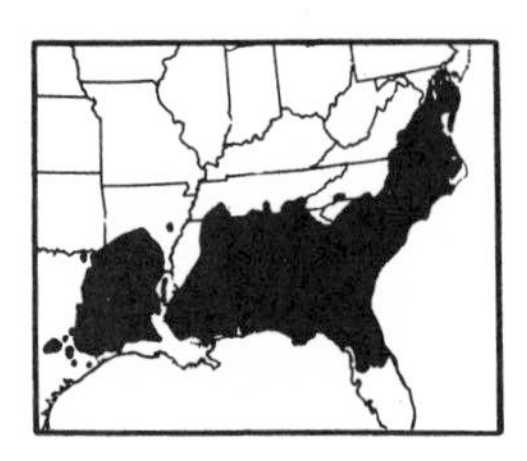

SAND PINE

PINACEAE

Loblolly Pine
Pinus taeda L.

HABIT. A tree 90-110 feet high and 2-3 feet in diameter (max. 150 by 5 1/2 feet); long, clear bole; crown rounded, dense.

LEAVES. In fascicles of 3 (rarely 2); 6-9 inches long; slender; yellow-green; persistent 3 years; basal sheath persistent.

FLOWERS. Male yellow; female short-stalked, yellow.

FRUIT. Sessile; 2 1/2-6 inches long; typically oblong-conic; opening at maturity and falling next year; dull, pale, red-brown; scales thin, armed with a stout, sharp spine. Seeds: 1/4 inch long; brown with black markings; wing 3/4 inch long.

TWIGS. Moderately slender; roughened; yellow- to red-brown. Winter buds: 1/2 inch long; red-brown scales with free tips.

BARK. Thin, becoming 3/4-2 inches thick; scaly and nearly black on young trees, becoming bright red-brown and scaly-plated.

WOOD. Important; intermediate between longleaf and western pines in strength and hardness; marketed as shortleaf pine.

SILVICAL CHARACTERS. Most tolerant of important southern pines; growth very rapid; windfirm; reproduction very aggressive; not resistant to fire; in pure or mixed stands.

HABITAT. Typical of wet bottomlands; aggressive on drier cutover or abandoned areas when protected from fire; in pure stands or mixed with pines and hardwoods.

* * *

Sand Pine
Pinus clausa (Chapm.) Vasey

This unimportant small tree is common in Florida on very dry, sterile, sandy soils; it comes in aggressively after fires. It is easily identified by its needles and characteristic cones.

LEAVES. In fascicles of 2; 2-3 1/2 inches long; slender; dark green.

FRUIT. Short-stalked; 2-3 1/2 inches long; some cones remaining closed and persistent for years; yellow-brown; scales armed with short prickle.

Longleaf Pine

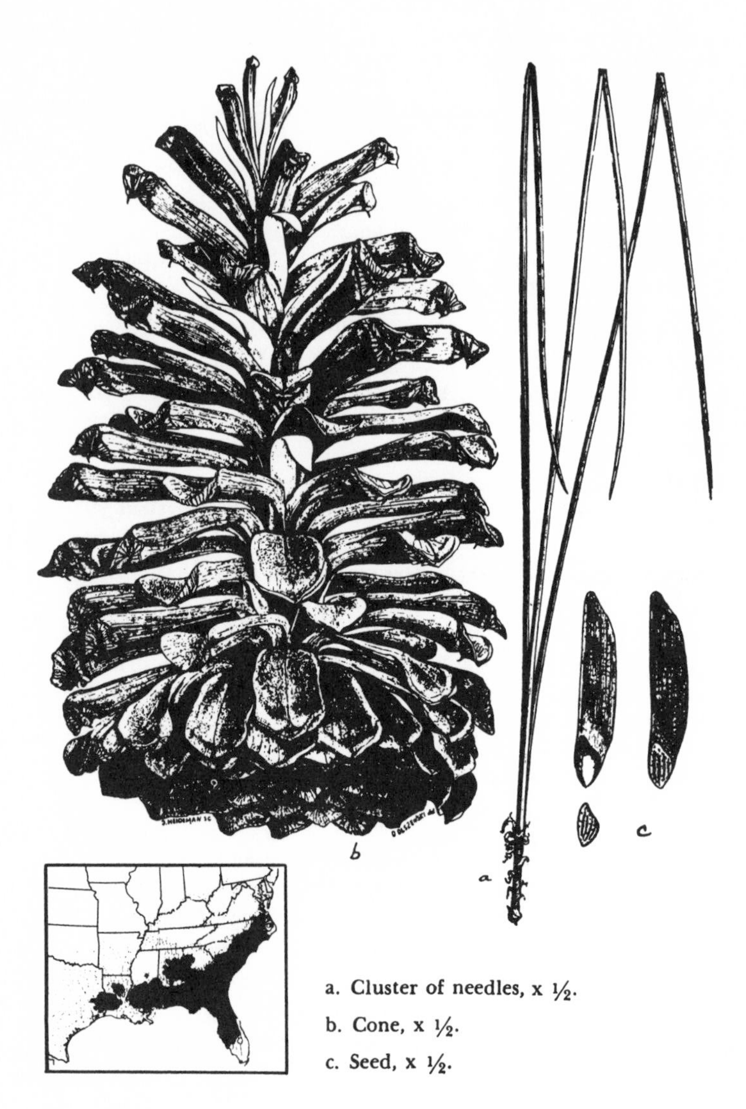

a. Cluster of needles, x ½.

b. Cone, x ½.

c. Seed, x ½.

Longleaf Pine

Pinus palustris Mill. (*Pinus australis* Michx.)

HABIT. A medium-sized tree 80-120 feet high and 2-3 feet in diameter (max. 150 by 4 feet); long, clear bole; small, open crown with needles in dense tufts at ends of branches.

LEAVES. In fascicles of 3; 8-18 inches long; slender; flexible; bright green; persistent 2 years; basal sheath persistent, 1/3-1/2 inch long.

FLOWERS. Male dark rose-purple; female dark purple.

FRUIT. Nearly sessile; 6-10 inches long; narrowly ovoid-cylindric; opening at maturity; in falling, leaves basal scales attached to twig; red-brown; scales thin, with an incurved prickle. Seeds: 1/2 inch long; pale with dark blotches; wings 1 1/2 inches long.

TWIGS. Very stout, the terminal leader up to 1 inch in diameter; roughened by leaf bases; orange-brown. Winter buds: 1/2-1 inch long; conspicuously white, covered by silvery, lustrous, fringed scales that make the bud remarkably resistant to fire.

BARK. Thin, becoming 3/4 to 1 1/2 inches thick on old stems; orange-brown; in rough, scaly plates.

WOOD. Very important; heavy, hard, and strong; used for construction timbers, lumber, flooring, boxes, ties, posts, poles, etc.; the chief source of naval stores.

SILVICAL CHARACTERS. Very intolerant; growth generally slow; maturity reached in about 150 years with extreme age of 300 years; practically no above-ground growth for 3-6 years while taproot becoming established; windfirm with deep taproot system; reproduction vigorous and plentiful; very resistant to fire and disease.

HABITAT. On variety of sites but typical of dry, sterile, sandy sites; usually in pure stands or with scrub oak. The only tree that will produce timber over much of its range.

GENERAL. A hybrid between longleaf and loblolly pines has been named *Pinus* X *sondereggeri* H. H. Chap. This has been reported as not uncommon in Louisiana, North Carolina, and Texas.

Slash Pine

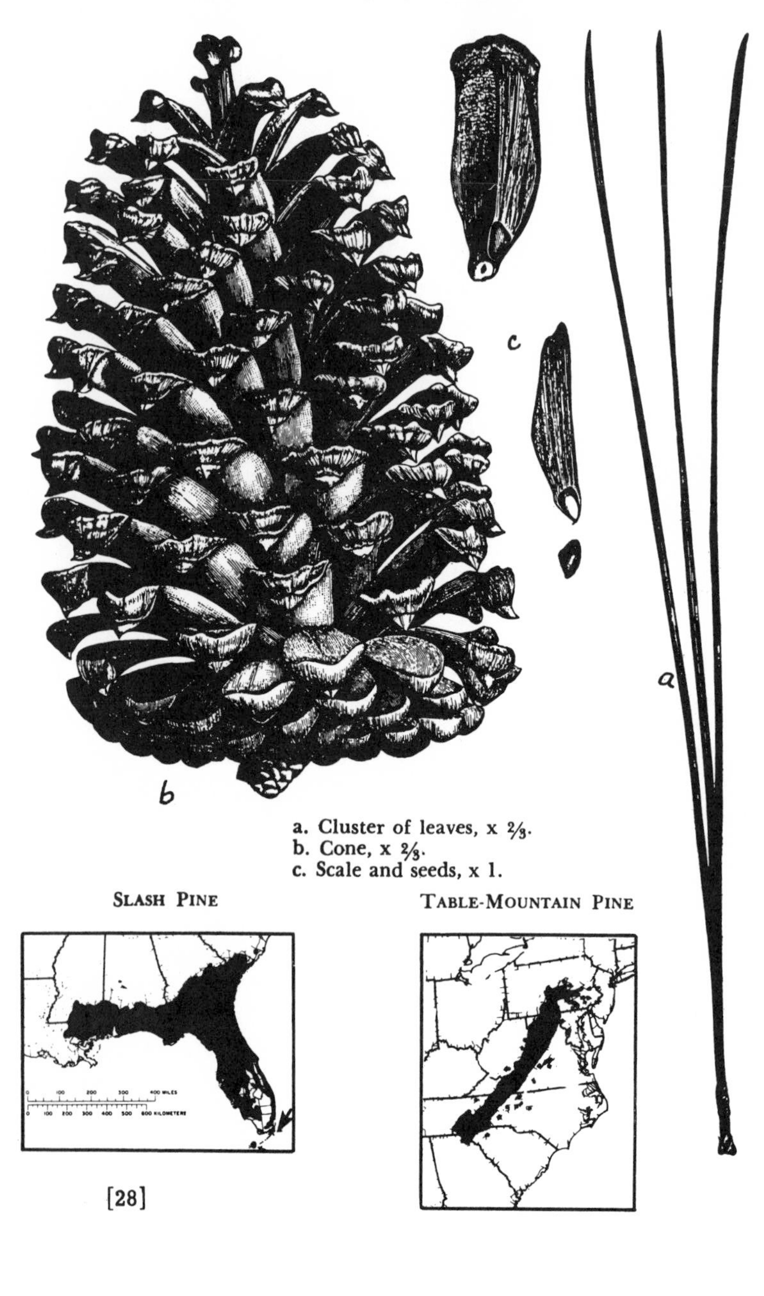

a. Cluster of leaves, x ⅔.
b. Cone, x ⅔.
c. Scale and seeds, x 1.

PINACEAE

Slash Pine

Pinus elliottii Engelm. (*Pinus caribaea* Morelet)

HABIT. A tree 80-100 feet high and 2-3 feet in diameter; long, clear bole; crown rounded, dense.

LEAVES. In fascicles of 2 and 3; 8-12 inches long; stout; glossy, dark green; persistent 2 years; basal sheath persistent.

FLOWERS. Male dark purple; female pink.

FRUIT. Stalked; 2 1/2-6 inches long; ovoid-conic; opening at maturity and falling following year; chocolate-brown, shining as if varnished; scales thin, armed with small sharp spine. Seeds: 1/4 inch long; black; wing about 1 inch long.

TWIGS. Stout; roughened by leaf bases; orange-brown. Winter buds: 1/2-3/4 inch long; scales silvery-brown.

BARK. Thin, becoming 3/4-1 1/2 inches thick; orange to purple-brown plated; peeling off in thin, papery, characteristic layers.

WOOD. Important; heavy, hard, and strong; marketed as longleaf pine and used for same purposes; important source of naval stores.

SILVICAL CHARACTERS. Intolerant, but less so than longleaf; growth very rapid; windfirm on deep soils; reproduction aggressive; not resistant to fire.

HABITAT. Old stands in low, wet sites because of fire damage; aggressive on cutover areas if protected; pure or mixed stands. The southern Florida variety with very dense wood and grass stage seedlings is designated *P. elliottii* var. *densa* L. & D. This species was incorrectly listed as *Pinus caribaea* Morelet, a species of the Bahama Islands.

* * *

Table-Mountain Pine

Pinus pungens Lamb.

This unimportant species, characteristic of the tablelands of the Appalachians, is a small, intolerant tree. It is easily identified by its needles and characteristic cones.

LEAVES. In fascicles of 2; 1 1/2-3 1/2 inches long; yellow-green; rigid; twisted.

FRUIT. Sessile; 2 1/2-3 1/2 inches long; light brown; ovoid; serotinous and persistent for years; heavy; thick scales armed with conspicuous, sharp, hooked spines or claws.

Ponderosa Pine

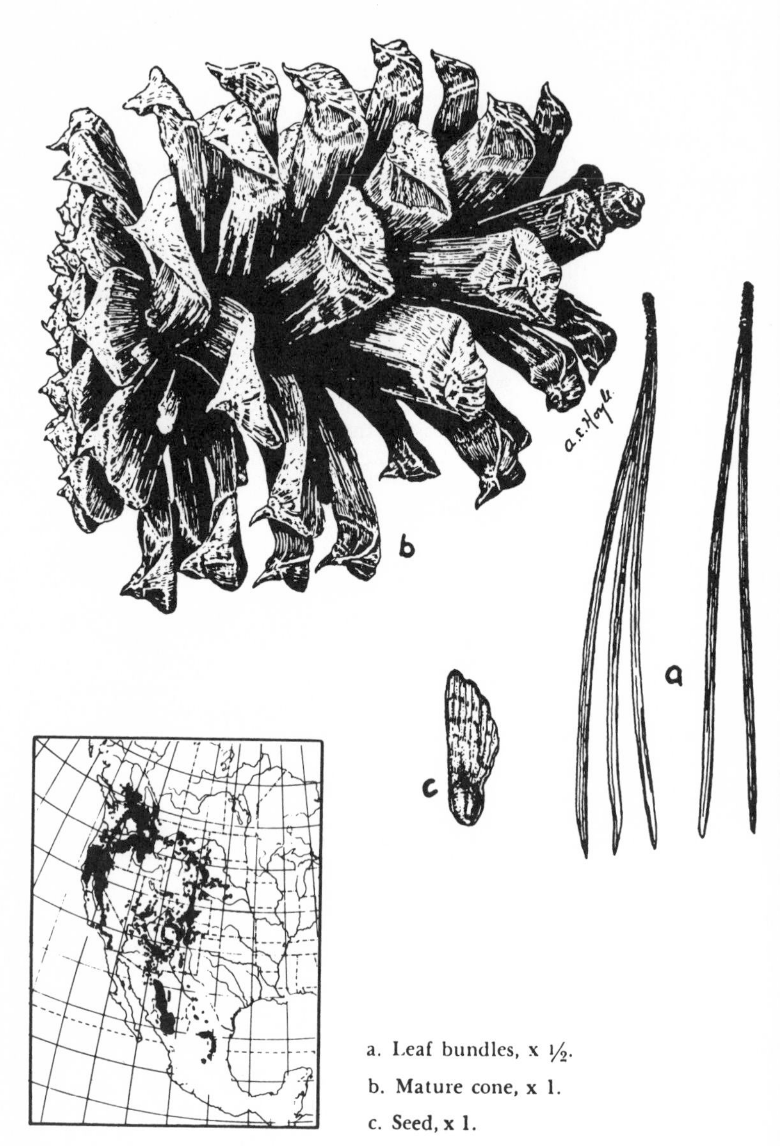

a. Leaf bundles, x ½.

b. Mature cone, x 1.

c. Seed, x 1.

PINACEAE

Ponderosa Pine. Western Yellow Pine

Pinus ponderosa Laws.

HABIT. A tree 150-180 feet high and 3-4 feet in diameter (max. 232 by 8 feet on Pacific Coast); bole symmetrical, clear; crown short, conical or flat-topped.

LEAVES. In fascicles of 3, or 2 and 3 (rarely from 1 to 5); 3-11 inches long (mostly 4-7 inches); stout; dark to yellow-green; persistent 2-7 years (usually 4-6 years); cross section shows 2-5 resin ducts; basal sheath 1/4-3/4 inch long, persistent.

FLOWERS. Male yellow; female red, clustered or paired.

FRUIT. Subsessile; 2 1/2-6 inches long; ovoid; open at maturity; basal scales remaining attached to twig when cones shed; scales thin, armed with short prickles. Seed: 1/4 inch long; brown-purple, often mottled; wing 1 inch long.

TWIGS. Stout; orange-colored; turpentine odor when bruised. Winter buds: about 1/2 inch long; often resinous.

BARK. Black and furrowed on young trees; on old trunks 2-4 inches thick; yellow-brown to cinnamon-red; in large, flat plates.

WOOD. Very important; rather light and soft; sapwood very thick with properties of white pine; heartwood light brown; uses include construction, planing mill products, ties, etc.

SILVICAL CHARACTERS. Intolerant; growth slow; maturity in 350-500 years (extreme age 660 years); reproduction vigorous; long taproot; fire and bark beetles cause damage.

HABITAT. Transition zone; large altitudinal range, 2,000-8,000 feet; exceedingly drouth resistant; in open pure stands or more commonly the most abundant tree in mixed coniferous stands.

GENERAL. The Rocky Mountain variety is designated *P. ponderosa* var. *scopulorum* Engelm. Arizona pine, *P. ponderosa* var. *arizonica* (Engelm.) Shaw, of southern New Mexico, Arizona, and northern Mexico, differs in having needles in fascicles of 5 and cones less than 3 1/2 inches long.

Jeffrey Pine

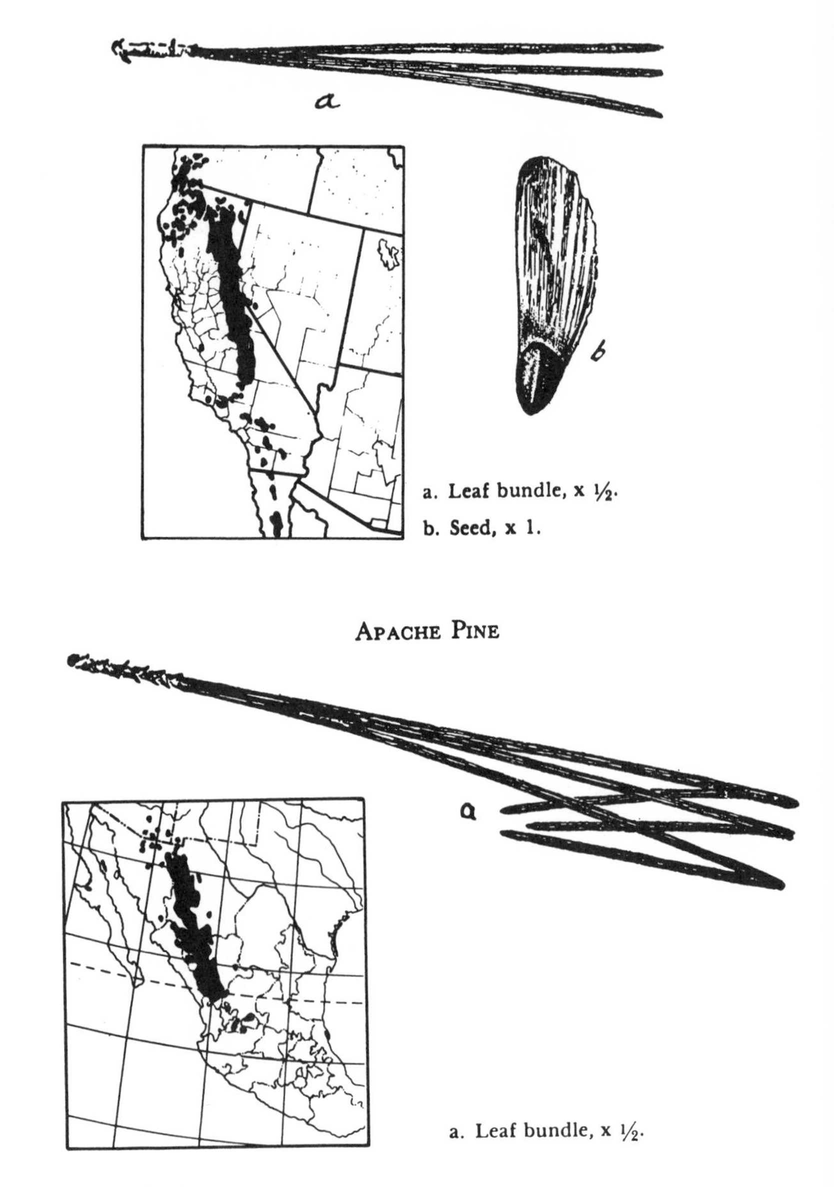

a. Leaf bundle, x ½.

b. Seed, x 1.

Apache Pine

a. Leaf bundle, x ½.

Jeffrey Pine

Pinus jeffreyi Grev. & Balf. (*Pinus ponderosa* var. *jeffreyi* Vasey)

A tree similar in most of its characteristics to *Pinus ponderosa* and differing from it in the following ways:

HABIT. A tree 100-180 feet high and 4-6 feet in diameter, not quite reaching the proportions of ponderosa pine; lumber is sold as ponderosa.

LEAVES. In fascicles of 3, or 2 and 3; 4-9 inches long; blue-green; persistent 6-9 years.

FRUIT. 5-15 inches long; scales thin, armed with a stout, long, incurved prickle.

TWIGS. Stout; purple; pineapplelike odor when bruised. Winter buds: not resinous.

HABITAT. Endures greater extremes of climate than ponderosa; mixed with ponderosa pine and other conifers.

* * *

Apache Pine

Pinus engelmannii Carr. (*Pinus apacheca* Lemm.)

A tree similar in most of its characteristics to *Pinus ponderosa*, and differing from it in the following ways:

HABIT. A tree 50-60 feet high and 1-2 feet in diameter (max. 75 by 3 feet).

LEAVES. In fascicles of 3 (occasionally 2-5); 8-15 inches long (mostly about 10 inches); dark green; conspicuously fringed margins; persistent 2 years; cross section shows 11-14 resin ducts; basal sheath 3/4-1 inch long.

HABITAT. Transition zone; altitudinal range 5,500-8,200 feet; otherwise similar to ponderosa pine.

GENERAL. Some authors consider this tree a synonym of *Pinus ponderosa;* it is characterized during its first few years by development of a very deep taproot and little height growth.

* * *

Washoe Pine

Pinus washoensis Mason & Stock.

This rare and local species was described from Washoe County, Nevada, in 1945. It is related to *P. jeffreyi*.

Lodgepole Pine

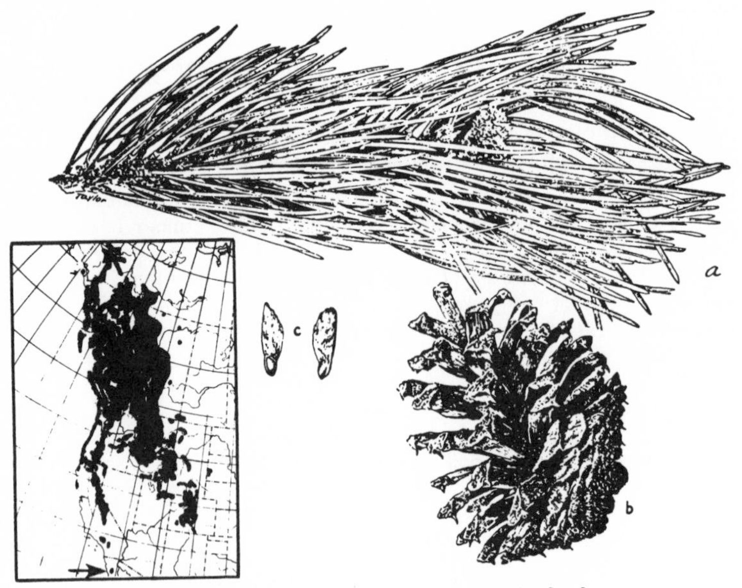

(a) Foliage, x ⅔; (b) Mature cone, x ⅔; (c) Seed, x ⅔.

Chihuahua Pine

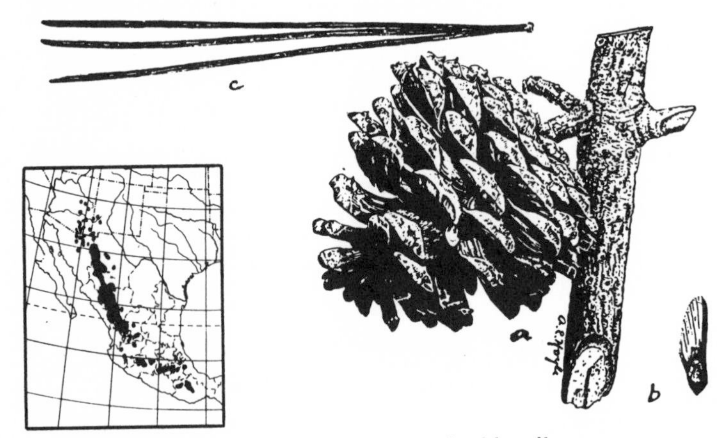

(a) Mature cone, x ⅔; (b) Seed, x ⅔; (c) Leaf bundle, x ⅔.

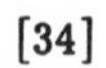

PINACEAE

Lodgepole Pine

Pinus contorta Dougl.

HABIT. A tree 70-80 feet high and 1 1/4-2 1/2 feet in diameter (max. 150 by 3 feet); long, slender bole; short crown.

LEAVES. In fascicles of 2; 1-3 inches long; stout; often twisted; bright yellow-green; persistent 4-6 years; margins with minute teeth; basal sheath persistent.

FLOWERS. Orange-red; male in spikes; female clustered.

FRUIT. Subsessile; 3/4-2 inches long; ovoid; frequently remaining closed and on the tree for many years; scales at base knoblike, armed with long prickle. Seed: 1/6 inch long; thin, dark red-brown shell; wings 1/2 inch long.

TWIGS. Stout; light orange-brown, becoming black. Winter buds: 1/4 inch long; ovoid; dark chestnut-brown; resinous.

BARK. Very thin, rarely over 2/5 inch thick; orange-brown to gray; covered by thin, loosely appressed scales.

WOOD. Moderately important; soft; fine-textured; sapwood thick; heartwood pale brown; tangential surface with many indentations; used for lumber, ties, poles, and mine timbers.

SILVICAL CHARACTERS. Intolerant; growth slow; maturity reached in about 200 years; shallow root system; reproduction vigorous, typically forming dense stands following fires; fire, bark beetles, and mistletoe cause damage.

HABITAT. Canadian zone; altitudinal range 6,000-11,000 feet; adapted to variety of soil types; in pure, dense, even-aged stands, or in mixture with various conifers.

GENERAL. Some authors distinguish two varieties: shore pine, *P. contorta* var. *contorta*, a stunted, short-leaved, twisted cone tree of the Pacific Coast; and lodgepole pine, *P. contorta* var. *latifolia* Engelm., the taller inland tree described above.

* * *

Chihuahua Pine

Pinus leiophylla var. *chihuahuana* (Engelm.) Shaw

This small essentially Mexican tree is unique in that it matures its cones in 3 years and is a hard pine with a deciduous basal sheath. The needles are in clusters of 3, 2-4 inches long, and persistent 4 years. The cones are 1 1/2-2 inches long, long-stalked, ovoid, and often remaining closed. It is characteristic of poor, dry sites.

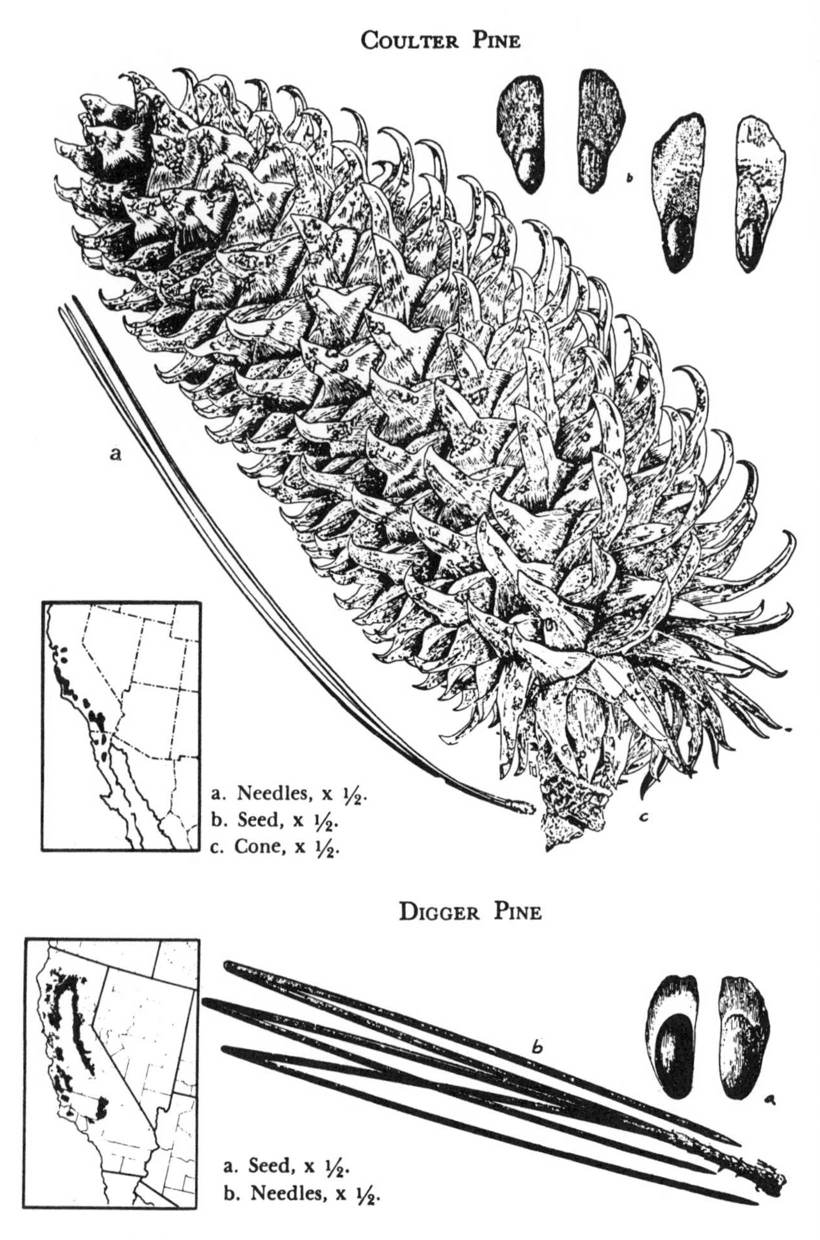

Coulter Pine

a. Needles, x ½.
b. Seed, x ½.
c. Cone, x ½.

Digger Pine

a. Seed, x ½.
b. Needles, x ½.

Coulter Pine

Pinus coulteri D. Don

HABIT. A small tree 40-50 feet high and 1-2 1/2 feet in diameter (max. 80 by 3 1/2 feet); large, open, unsymmetrical crown.

LEAVES. In fascicles of 3; 6-12 inches long; dark blue-green; rigid; persistent 3-4 years; basal sheath persistent.

FLOWERS. Male yellow; female dark red-brown.

FRUIT. Short-stalked; 10-14 inches long; oblong-ovoid; opening at maturity and persistent several years; light yellow-brown; thick scales terminating in large claw; the largest American pine cone, weighing up to 5 pounds. Seed: 1/2 inch long; thick-shelled; wing longer than seed (about 1 inch).

TWIGS. Very stout; rough; orange-brown. Winter buds: 1/2 inch; brown.

BARK. Nearly black; 1 1/2-2 inches thick; broad, scaly ridges.

WOOD. Unimportant; soft and weak; used for fuel.

SILVICAL CHARACTERS. Intolerant; maturity reached in about 150 years; growth slow; wide, deep roots; reproduction scanty.

HABITAT. Dry, rocky slopes 3,000-7,000 feet; pure or mixed stands.

* * *

Digger Pine

Pinus sabiniana Dougl.

HABIT. A small tree 40-50 feet high and 1-2 feet in diameter; bole often forked; open, sparsely foliaged crown.

LEAVES. In fascicles of 3; 7-12 inches long; pale blue-green; flexible; pendent; persistent 3-4 years; basal sheath persistent.

FLOWERS. Male yellow; female dark purple.

FRUIT. Similar to Coulter pine, but smaller (6-10 inches long) and chocolate-brown. Seed: 3/4 inch long; thick-shelled; edible; wing about 1/2 as long as seed.

TWIGS, BARK, WOOD, and SILVICAL CHARACTERS. Similar to Coulter pine.

HABITAT. Dry foothills 100-5,000 feet; able to survive on exceedingly dry sites; in pure stands or mixed with oaks.

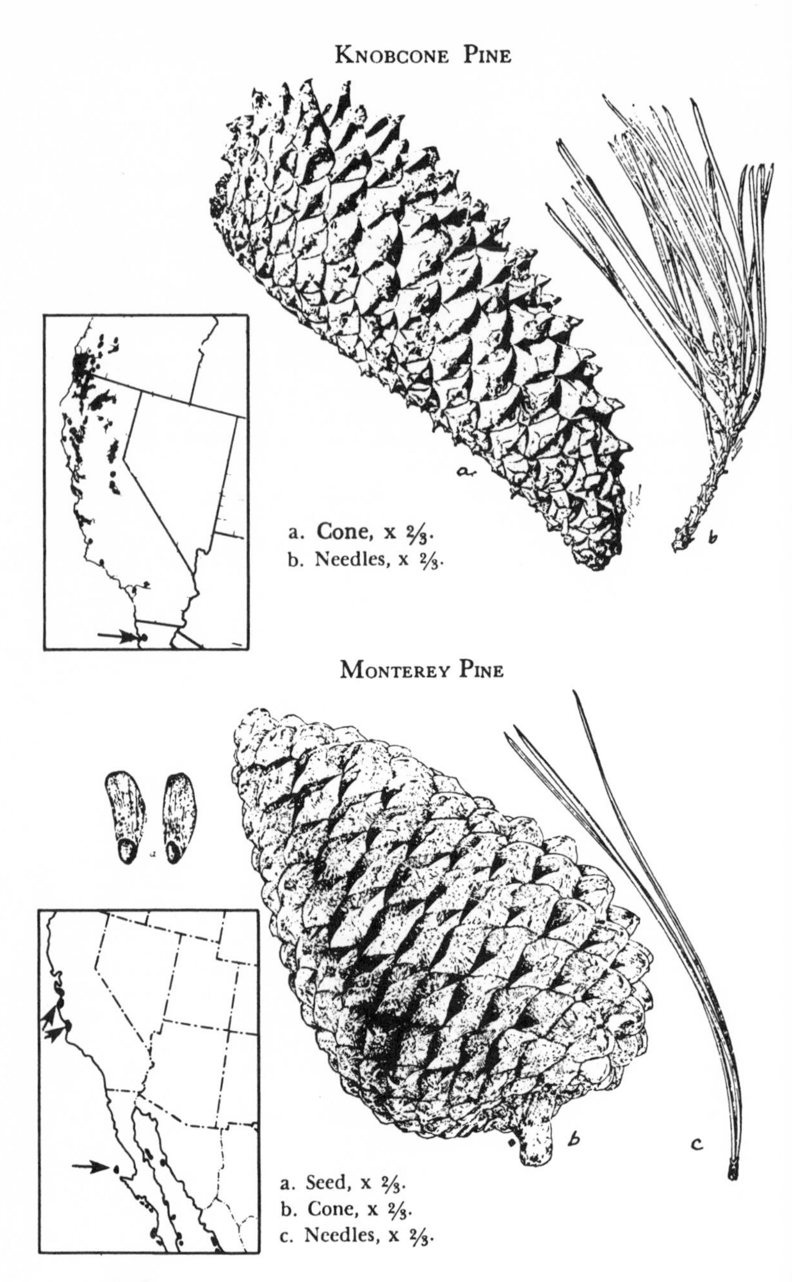

KNOBCONE PINE

a. Cone, x ⅔.
b. Needles, x ⅔.

MONTEREY PINE

a. Seed, x ⅔.
b. Cone, x ⅔.
c. Needles, x ⅔.

PINACEAE

Knobcone Pine

Pinus attenuata Lemm.

HABIT. A small tree 20-60 feet high and 1-2 feet in diameter (max. 100 by 3 feet); open, sparse crown.

LEAVES. In fascicles of 3; 3-7 inches long; yellow-green; slender; persistent 4-5 years; basal sheath persistent.

FLOWERS. Male orange-brown; female fascicled, purple.

FRUIT. Short-stalked; 3-6 inches long; ovoid-conic; light yellow-brown; unsymmetrical with outer basal scales pyramidally knoblike; remaining closed and persistent many years, with clusters characteristically encircling stems of even small trees. Seed: 1/4 inch long; wing 1 1/4 inches long.

TWIGS. Slender; smooth; orange-brown. Winter buds: 1/2 inch long; brown.

BARK. Thin; gray-brown; low, scaly ridges.

WOOD. Unimportant; soft and weak; used locally for fuel.

SILVICAL CHARACTERS. Intolerant; short-lived; rapid growth; wide, deep roots; reproduction aggressive, especially following fire.

HABITAT. Very dry mountain slopes to 5,000 feet; pure or mixed stands.

* * *

Monterey Pine

Pinus radiata D. Don (*Pinus insignis* Dougl.)

HABIT. A handsome tree 40-100 feet high and 2-5 feet in diameter; rounded, open crown; common ornamental.

LEAVES. In fascicles of 3; 4-6 inches long; bright green; slender; flexible; persistent 3 years; basal sheath persistent.

FLOWERS. Male yellow in spikes; female dark purple.

FRUIT. Short-stalked; 3-7 inches long; ovoid; yellow-brown; unsymmetrical, with outer basal scales round, thickened, and domelike; remaining closed and persistent. Seed: 1/4 inch; wings 3/4 inch.

TWIGS. Slender; dark orange. Winter buds: 1/2 inch; brown.

BARK. Thick; red-brown to nearly black; furrowed.

WOOD. Unimportant in the United States; soft and weak.

SILVICAL CHARACTERS. Moderately tolerant; short-lived; rapid growth; prolific seed production. An important forest tree in Australia, New Zealand, and South Africa.

HABITAT. Dry soil (17 inches rain annually) but humid air; in pure stands or mixed with bishop pine and cypress.

Bishop Pine

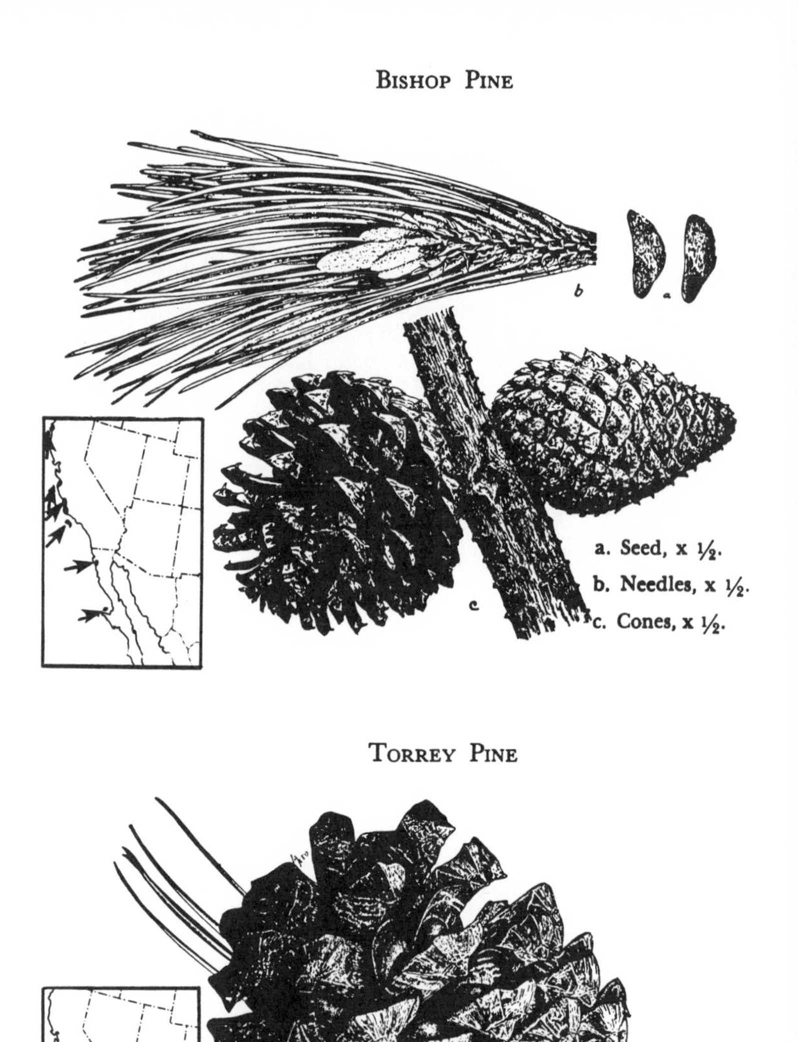

a. Seed, x ½.
b. Needles, x ½.
c. Cones, x ½.

Torrey Pine

a. Cone and needles, x ½.

PINACEAE

Bishop Pine

Pinus muricata D. Don

HABIT. A tree 30-80 feet high and 1-3 feet in diameter; handsome, rounded crown of dense, tufted foliage.

LEAVES. In fascicles of 2; 4-6 inches long; dark yellow-green; thick; rigid; persistent 2-3 years; basal sheath persistent.

FLOWERS. Male in orange-colored spikes; female in whorls.

FRUIT. Sessile; in clusters of 3-7; 2-4 inches long; ovoid; yellow-brown; unsymmetrical, with outer basal scales knoblike and extended into spurlike spines; often remaining closed and persistent for years. Seed: 1/4 inch long; black.

TWIGS. Stout; rough; orange-brown. Winter buds: 1/2 inch long; brown.

BARK. Thick; purple-brown; rough; furrowed.

WOOD. Unimportant; moderately light and strong; lumber and fuel.

SILVICAL CHARACTERS. Moderately tolerant; short-lived; rapid growth; reproduction aggressive, especially following fire.

HABITAT. Variety of sites from wet clays and peat bogs to dry, sandy ridges; pure stands or mixed with lodgepole, madrone, and oaks; commonly planted ornamental.

* * *

Torrey Pine

Pinus torreyana Parry

HABIT. A rare, local tree 25-60 feet high and 8-14 inches in diameter; few, stout, spreading branches; rounded crown.

LEAVES. In fascicles of 5; 7-13 inches long; dark yellow-green; rigid; persistent 3-4 years; basal sheath persistent.

FLOWERS. Male yellow in heads; female on long stalks.

FRUIT. Long-stalked; 4-6 inches long; broad-ovoid; chocolate-brown; thick-scaled; persistent 3-4 years. Seed: 3/4 inch; wing 1/2 inch; edible.

TWIGS. Very stout; smooth; green first year, becoming black.

BARK. About 1 inch thick; red-brown; rough; furrowed.

WOOD. Unimportant; too rare to be of any use.

SILVICAL CHARACTERS. Intolerant; short-lived; reproduction abundant.

HABITAT. Dry soil (15 inches rain annually) but very humid air; in pure, open groves; smallest range of any American pine, on Santa Rosa Island and north of San Diego.

PINACEAE

The Larches

Characteristics of the Genus *Larix* Mill.

HABIT. Tall, pyramidal trees; open crowns, with slender, irregularly disposed, horizontal or pendulous branches; dwarfed, short, spurlike lateral branchlets.

LEAVES. Deciduous; needle-shaped or linear; produced in dense false whorls or clusters on spurlike lateral branches; solitary and spirally arranged on new shoots; numerous lines of stomata on all surfaces; 2 resin canals in cross section.

FLOWERS. Monoecious; terminal; single; appearing with leaves; male naked, globose to oblong, consisting of several yellow, spirally arranged scales, each bearing two pollen sacs; female erect, consisting of few or many rounded, red-purple scales in the axes of much longer scarlet bracts, each scale bearing 2 small inverted ovules.

FRUIT. Woody, erect, short-stalked cones; maturing in one season; cone scales thin, persistent, concave, longer or shorter than their long-pointed bracts. Seed: 2 under each scale; triangular and light brown; large terminal wing.

TWIGS. Smooth and glaucous, or hairy. Winter buds: subglobose; small; nonresinous; accrescent inner scales that mark lateral spur branches with prominent ringlike scars.

WOOD. Rather strong and durable; small, scattered resin ducts; thin, white sapwood and sharply defined, red to russet-brown heartwood.

SILVICAL CHARACTERS. Intolerant; slow-growing; found on a variety of habitats; extensive forests occasionally destroyed by larch sawfly (*Nematus erichsonii*).

GENERAL. This genus contains about 10 species scattered through the Northern Hemisphere. In North America 3 species are native. The European larch (*Larix decidua* Mill.), an important tree, has been planted in the United States for ornamental purposes. It can be identified by its puberulous cones, 3/4-1 1/2 inches long, with inserted bracts and 40-50 suborbicular scales.

KEY TO THE SPECIES OF LARCHES

1. Cones with bracts shorter than scales and usually concealed; twigs slender, yellowish, glabrous.
   2. Cones 1/2-3/4 inch long, composed of 12-15 glabrous scales; twigs glabrous, covered at first with glaucous bloom; eastern, far northern . . . . *L. laricina*, Tamarack, p. 45
   2. Cones 3/4-1 1/2 inches long, composed of 40-50 hairy scales; twigs without bloom; ornamental exotic . . . . *L. decidua*, European Larch, p. 42
1. Cones with bracts conspicuous and exserted beyond the 40 or more scales; twigs stout, brownish, hairy; western.
   3. Twigs with pale hairs, brittle; bark thick, red-brown, plated and deeply furrowed; cones 1-1 1/2 inches long; needles flatly 3-angled; western . . . . *L. occidentalis*, Western Larch, p. 47
   3. Twigs densely woolly, tough; bark thin, gray, scaly; cones 1 1/2-2 inches long; needles 4-angled; western alpine . . . *L. lyallii*, Subalpine Larch, p. 47

* * *

PINACEAE

The Cedars

Characteristics of the Genus *Cedrus* Trew.

The name cedar, though improperly applied to trees in several native genera, is correctly used for four closely related species in the genus *Cedrus*, none of which are native to the Americas. All have linear needles arranged spirally and scattered on new shoots and in dense false whorls on lateral spur branches, as found in the larches; however, the needles are persistent. The upright cones resemble those of *Abies* but require 2-3 years to mature.

The deodar or Himalayan cedar, *Cedrus deodara* (Roxb.) Loud., with pendulous, densely pubescent leading shoots and branchlets, is widely planted in warmer sections of the United States. The biblical cedar of Lebanon, *Cedrus libani* Loud., with stiff, nearly glabrous shoots, is less commonly planted as an ornamental, as are Atlas cedar, *C. atlantica* Man., and Cyprus cedar, *C. brevifolia* Henry, both of which differ from the two species above in having needles less than 1 inch long and cones less than 3 inches long.

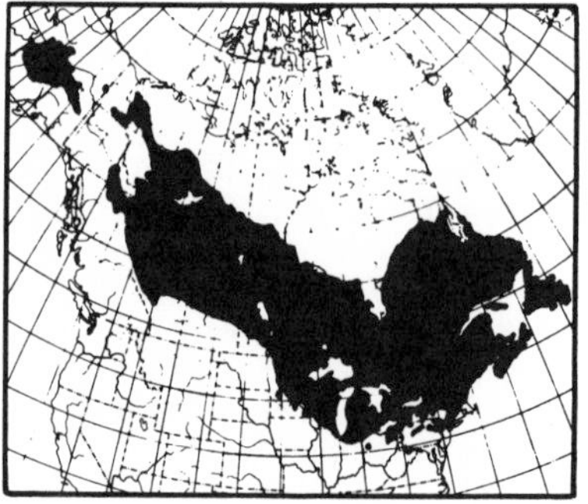

a. Foliage, x 1.

b. Seed, x 1.

c. Branchlet with cones, x 1.

PINACEAE

Tamarack. Eastern Larch

Larix laricina (Du Roi) K. Koch (*Larix alaskensis* Wight)

HABIT. A small to medium-sized tree seldom over 60 feet high and 1-2 feet in diameter (max. 100 by 2 1/2 feet); long, clear, cylindrical bole; open, pyramidal, irregular crown with slender horizontal branches.

LEAVES. Linear; triangular in cross section; 3/4-1 1/4 inches long; bright blue-green; falling in September or October; in clusters of 12-20.

FLOWERS. Male subglobose and sessile; female oblong and short-stalked.

FRUIT. 1/2-3/4 inch long; short-stalked; oblong to subglobose; chestnut-brown; falling during second year; cone scales less than 20, slightly longer than broad, erose at margin, glabrous and lustrous and twice as long as their bracts except at base of cone. Seed: 1/8 inch long; light chestnut-brown wings about 1/4 inch long.

TWIGS. Slender; smooth; glaucous at first, becoming orange-brown during first year. Winter buds: conspicuous; globose; small; lustrous; dark red.

BARK. Thin and smooth on young stems; 1/2-3/4 inch thick, red-brown, scaly on mature trunks.

WOOD. Heartwood yellow-brown; medium texture; strong, hard; heavy; durable; not widely used, chief uses being poles, railroad ties, and rough lumber.

SILVICAL CHARACTERS. Intolerant; growth moderate, maturity being reached in 100-200 years; reproduction vigorous on favorable sites; shallow root system; larch sawfly and fire cause serious damage.

HABITAT. Hudsonian and Canadian zones; restricted to sphagnum bogs or swamps in southern part of range and making best growth on moist beaches and well-drained uplands farther north; chiefly with black spruce, also balsam fir, aspen, birch, and jack pine; extending northward to limits of tree growth.

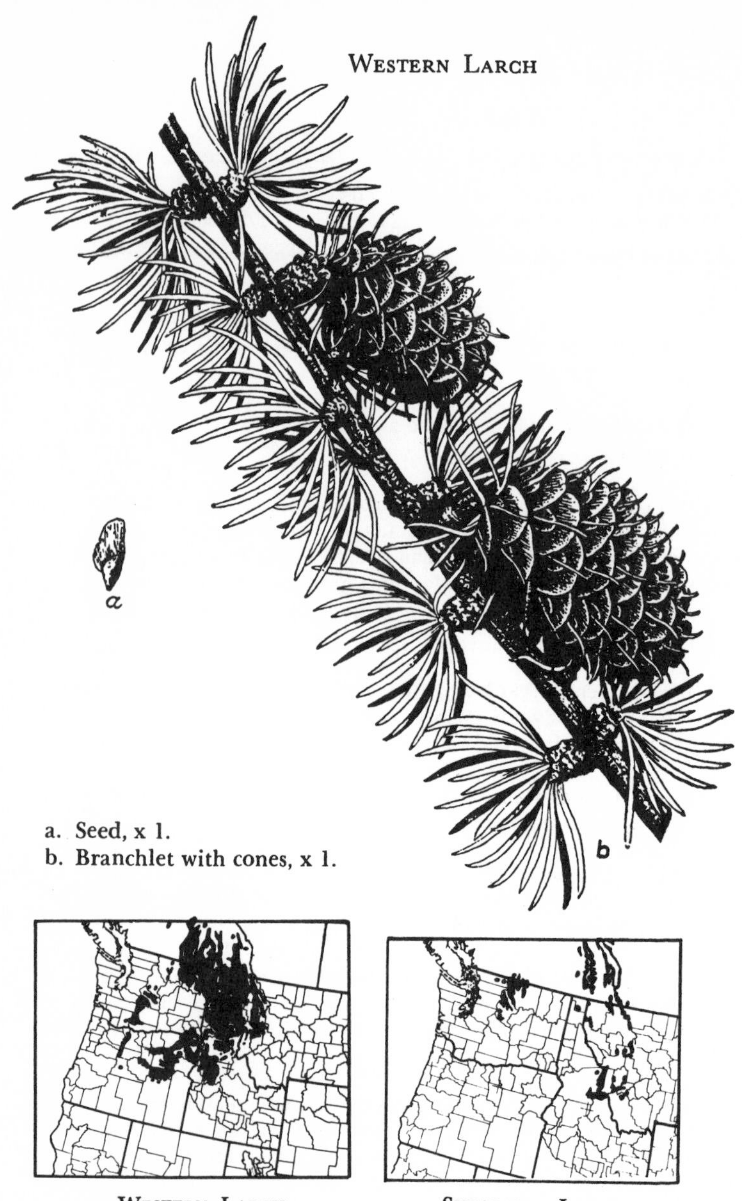

a. Seed, x 1.
b. Branchlet with cones, x 1.

PINACEAE

Western Larch
Larix occidentalis Nutt.

HABIT. A tree 140-180 feet in height and 3-4 feet in diameter; long, clear, cylindrical bole, often with a swollen butt; short, open crown of small, horizontal branches.

LEAVES. Linear; flatly triangular in cross section; 1-1 3/4 inches long; light pale green, becoming yellow; falling in early autumn; in clusters of 14-30.

FLOWERS. Male short, oblong; female oblong, subsessile.

FRUIT. 1-1 1/2 inches long; short-stalked; oblong; purple-red to red-brown; falling first year; cone scales 40 or more, broader than long, sometimes toothed at reflexed apex, usually white-woolly on the outside, and shorter than exserted long-tipped bracts. Seed: 1/4 inch long; wings 1/2 inch long.

TWIGS. Stout; brittle; at first with a pale pubescence, soon becoming glabrous and orange-brown. Winter buds: 1/8 inch in diameter; subglobose; chestnut-brown.

BARK. Scaly on young stems; 4-6 inches thick, plated, deeply furrowed on old trunks; red-brown to cinnamon-red.

WOOD. Heartwood red-brown; heavy; strong; durable; similar to and sold as Douglas-fir for lumber, poles, ties, etc.

SILVICAL CHARACTERS. Intolerant; growth rather slow, maturity being reached in 300-400 years (trees over 700 years reported); reproduction vigorous, competing with lodgepole pine on burned areas; windfirm with deep, widespreading root system; fungi and mistletoe often cause damage.

HABITAT. Transition zone; altitudinal range 2,000-7,000 feet; best development on deep, moist soils but does well on dry, gravelly slopes; in nearly pure stands or with Douglas-fir; western white, ponderosa, and lodgepole pines; western hemlock; Engelmann spruce; and alpine and grand fir.

* * *

Subalpine larch, *Larix lyallii* Parl., is a small timberline tree in regions of heavy snows that fall early and remain late. Its range lies within that shown for western larch. It is characterized by densely woolly twigs and cone scales; cones 1 1/2-2 inches long with exserted bracts; 4-angled needles; and thin, furrowed, scaly bark.

PINACEAE

The Spruces

Characteristics of the Genus *Picea* Dietr.

HABIT. Evergreen trees with sharp-pointed, pyramidal crowns, and straight, tapering trunks; branches in regular whorls.

LEAVES. Spirally arranged; linear; sessile; stiff; single; extending from all sides of twigs; persistent 7-10 years, but deciduous when dried; 4-angled or flattened; in falling, leave basal peglike projections (sterigmata) on the twig.

FLOWERS. Monoecious; catkinlike; solitary; male, or pollen-bearing, axillary, yellow to red or purple, 3/4-1 inch long, consisting of numerous spirally arranged scales, each bearing two pollen sacs; female, or cone- and seed-producing, terminal, erect, yellow-green or red, 3/4-1 1/4 inches long, consisting of numerous 2-ovuled, bracted scales.

FRUIT. Woody, pendent cone; matures in one season; borne mostly near top of crown; scales numerous, thin, unarmed, persistent, much longer than bracts. Seed: 2 under each fertile scale; small; compressed; highly buoyant with thin wing.

TWIGS. Roughened by sterigmata. Winter buds: ovoid or conical; of overlapping scales; usually not resinous.

BARK. Thin and scaly (furrowed on old trunks in blue spruce).

WOOD. Light; soft; resilient; fine-textured; long-fibered; straight-grained; small, scattered resin ducts; not resinous; high satiny luster; strong for weight; highly important for paper pulp, lumber, boxes, etc.

SILVICAL CHARACTERS. Tolerant; no taproot; generally shallow-rooted. Natural enemies: fire; leaf aphid (*Adelges abietis*), which causes conelike gall; spruce budworm (*Harmologa fumiferana*), which often destroys young stands; white pine weevil (*Pissodes strobi*).

HABITAT. Cool, moist sites; typically in swampy areas or along the margins of streams and lakes.

GENERAL. This genus contains about 40 species, largely restricted to cooler regions in the Northern Hemisphere. In North America there are 7 indigenous species. The Norway spruce, *Picea abies* (L.) Karst., characterized by cones 4-7 inches long, is commonly planted throughout the United States and has become naturalized in the East.

KEY TO THE SPECIES OF SPRUCES

1. Cone scales rounded at tip, smooth or wavy (erose) on margin.
   2. Cones under 2 inches long; needles 4-sided; eastern or northwestern.
      3. Cones 1/2-1 1/2 inches long, ovoid, purple, persistent many years; scales stiff, rigid, brittle, with wavy margins; twigs hairy; needles blunt-pointed; transcontinental and northern . . . . *P. mariana*, Black Spruce, p. 53
      3. Cones 1 1/4-2 inches long, oblong, brown, falling in 1 year; scales smooth on margin or nearly so; needles pointed.
         4. Cone scales stiff and rigid at maturity; twigs more or less hairy; eastern United States and Canada . . . . . *P. rubens*, Red Spruce, p. 55
         4. Cone scales soft and flexible at maturity; twigs not hairy; transcontinental and northern . . . . *P. glauca*, White Spruce, p. 51
   2. Cones 2-4 inches long; scales with entire margins; needles flattened; alpine in California and Oregon . . . . *P. breweriana*, Brewer Spruce, p. 59
1. Cone scales wedge-shaped at tip, margin wavy (erose).
   5. Cones 4-7 inches long; drooping foliage; needles 4-sided, not prickly; common European ornamental . . . . *P. abies*, Norway Spruce, p. 48
   5. Cones rarely 4 inches long; native and western.
      6. Needles flattened, not prickly, yellow-green; cone 2 1/2-4 inches long; Pacific Coast area . . . . *P. sitchensis*, Sitka Spruce, p. 59
      6. Needles 4-sided, blue-green; western mountain region.
         7. Cone 1-2 1/2 inches long, persistent 1 year; needles flexible, acute but not prickly to touch; twigs minutely hairy; buds 1/8-1/4 inch long with scales usually appressed; bark on mature trees thin and scaly . . . . *P. engelmannii*, Engelmann Spruce, p. 57
         7. Cone 2 1/4-4 1/2 inches (mostly about 3 1/2 inches) long, persistent 2 years; needles stiff, bristle-pointed; twigs glabrous; buds 1/4-1/2 inch long with scales usually reflexed, bark on mature trees thick and furrowed; central Rocky Mountains . . . *P. pungens*, Blue Spruce, p. 61

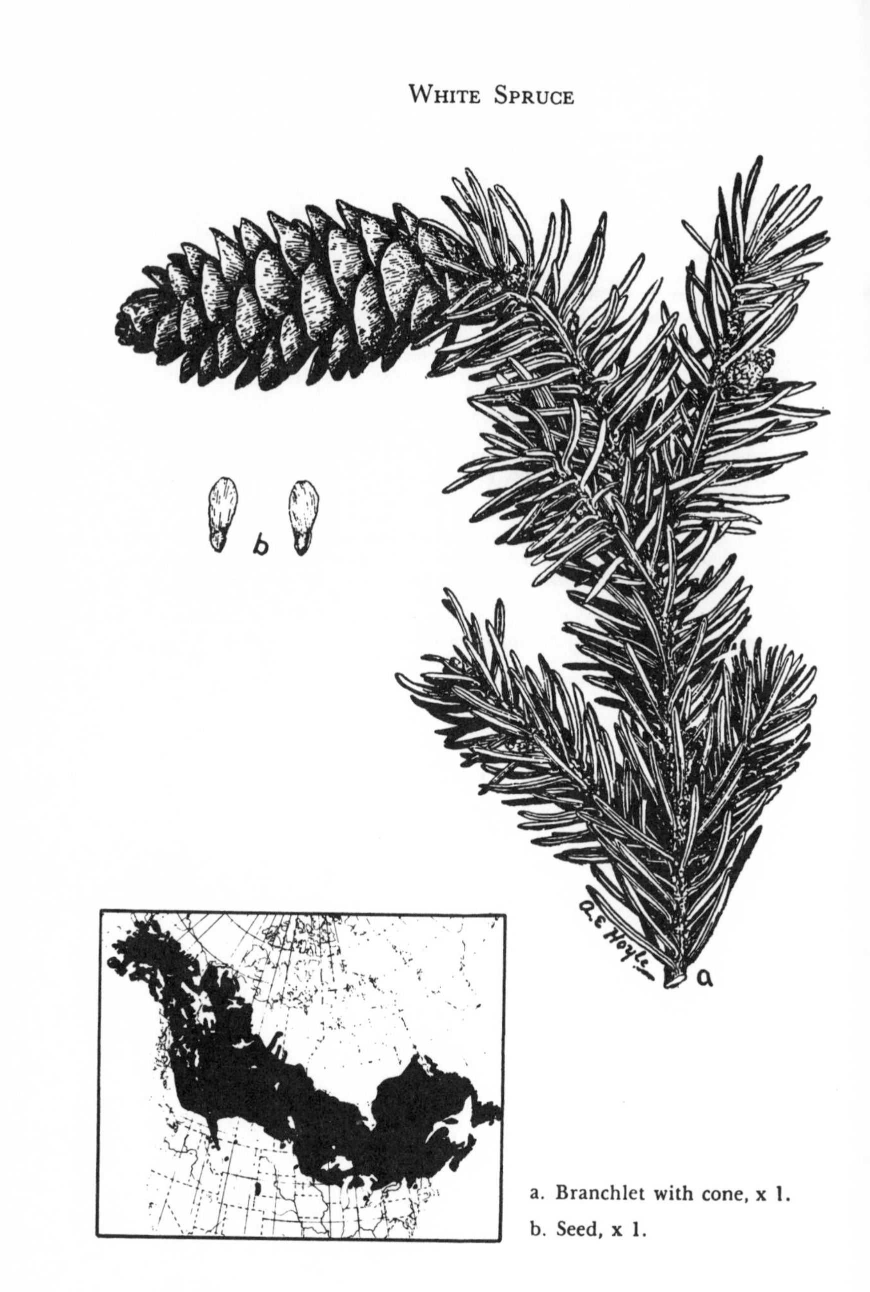

a. Branchlet with cone, x 1.

b. Seed, x 1.

PINACEAE

White Spruce

Picea glauca (Moench) Voss (*Picea canadensis* B.S.P.)

HABIT. A tree 60-70 feet high and 1 1/2-2 feet in diameter (max. 120 by 4 feet); slender symmetrical bole; crown narrowly to broadly pyramidal with long, thick branches.

LEAVES. Tending to be crowded on upper side of branch by twisting of those on lower side; 1/3-3/4 inch long; 4-angled; blue-green, occasionally with whitish tinge; rigid acute tips; odor pungent when crushed; 2 resin ducts in cross section.

FLOWERS. Male pale red to yellow; female with red or yellow-green scales.

FRUIT. 1-2 1/2 inches long; oblong-cylindrical; nearly sessile; cone scales flexible, rounded, and smooth at apex; light green or reddish before shedding seed and becoming light brown and falling soon after. Seed: 1/8 inch long; pale brown; oblique wing 1/4-3/8 inch long.

TWIGS. Glabrous, or in far Northwest downy; rather slender; orange-brown; skunklike odor when bruised. Winter buds: 1/8-1/4 inch long; ovoid; obtuse; and chestnut-brown.

BARK. Thin, 1/4-1/2 inch thick; ash-brown; silvery inner bark; separated into irregular thin plates or scales.

WOOD. Important in Northeast; used for pulp, construction lumber, boxes and crates, sounding boards, etc.

SILVICAL CHARACTERS. Tolerant of considerable shade, recovers from suppression well; growth slow but faster than black spruce; tree attains age of 250-300 years; reproduction abundant on moist sites; shallow, spreading root system.

HABITAT. Hudsonian and Canadian zones; typical of low, damp woods and banks of streams and lakes; altitudinal range from sea level to 5,000 feet; frequently gives way to black spruce or tamarack on wet sites and to lodgepole pine on dry sites; often forms pure, dense forests, but also in mixture with black spruce, fir, birch, poplars, and willows.

GENERAL. The western white spruce, *Picea glauca* var. *albertiana* (S. Brown) Sarg., is the form found in the Rocky Mountain region and Black Hills. It is characterized by somewhat shorter and broader cones and a narrow crown.

Black Spruce

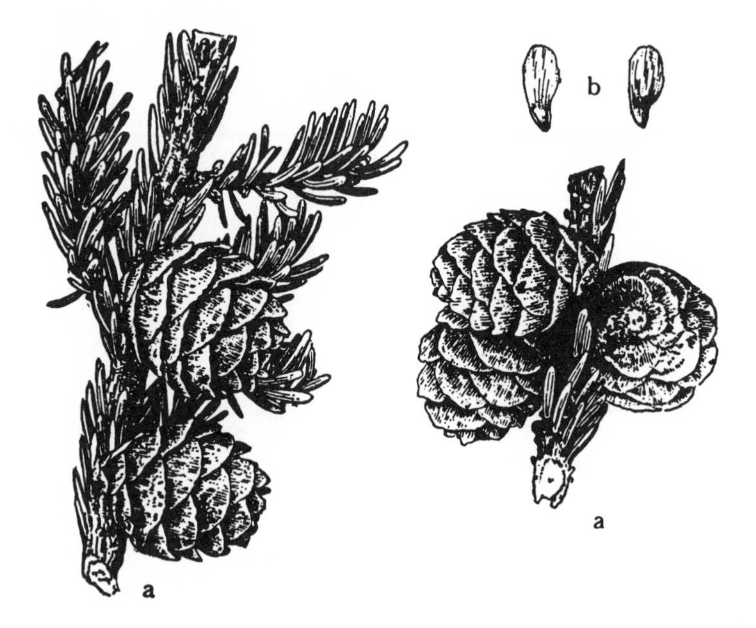

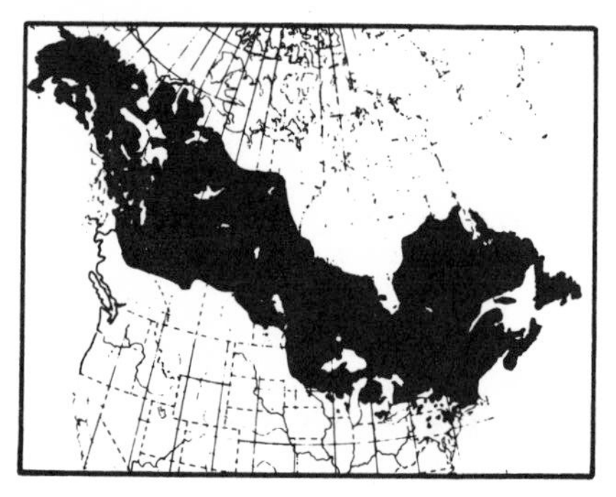

a. Branchlets with cones, x 1.

b. Seed, x 1.

PINACEAE

Black Spruce

Picea mariana (Mill.) B.S.P.

HABIT. A tree sometimes 40-80 feet high and 2/3-3 feet in diameter, but commonly much smaller; short, slender bole, usually pruning poorly; crown open, conical, more or less irregular.

LEAVES. Spreading in all directions; 1/4-3/4 inch long; 4-angled; pale blue-green and glaucous; blunt at apex; more or less incurved; hoary on upper surface from broad bands of stomata; lustrous and slightly stomatiferous below; 2 resin ducts in cross section.

FLOWERS. About 1/2 inch long; male red; female purple.

FRUIT. 1/2-1 1/2 inches long; ovoid; on strongly incurved, short stalks; cone scales stiff, brittle, rigid, rounded, smooth or erose at apex, puberulous, and dull gray-brown; persistent many years. Seed: 1/8 inch long; dark brown; pale brown oblique wing 1/4-3/8 inch long.

TWIGS. Rusty-pubescent; rather slender; at first green, becoming dull red-brown. Winter buds: 1/8 inch long; ovoid; acute; light red-brown; puberulous.

BARK. Thin, 1/4-1/2 inch; gray-brown; separated into thin, closely appressed scales or flakes; inner bark often olive-green.

WOOD. Not important except for pulp; used interchangeably with white spruce.

SILVICAL CHARACTERS. Very tolerant, recovering from suppression at an advanced age; growth slow; tree rather short-lived, attaining an age of 200 years; shallow, spreading root system; reproduction good on moist sites; lower branches often take root, forming clusters of small trees.

HABITAT. Hudsonian and Canadian zones; typical of cold sphagnum bogs and swamps, but also found on dry slopes in Northwest; altitudinal range 100-3,500 feet; with white spruce and tamarack, reaches northern limit of tree growth; in dense, pure stands, or in mixture with tamarack, balsam fir, white spruce, white birch, aspen, etc.

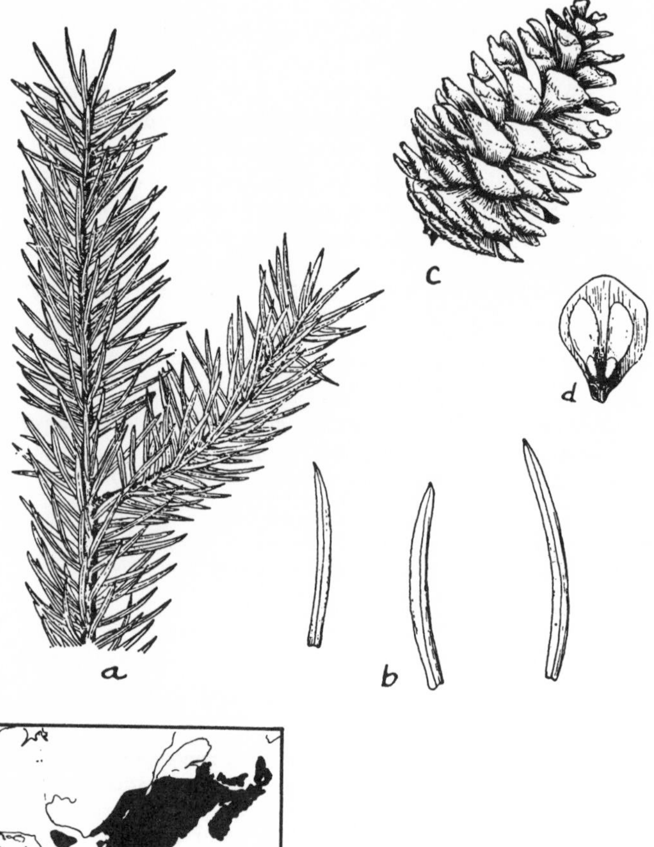

a. Twig, x 1.

b. Needles, x 1½.

c. Cone, x 1.

d. Scale with seed, x 1.

Red Spruce

Picea rubens Sarg. (*Picea rubra* Link)

HABIT. A tree 70-80 feet high and 1 1/2-2 feet in diameter (max. 120 by 3 1/2 feet); slender, symmetrical bole; crown narrow and conical with long branches.

LEAVES. Extending at nearly right angles from all sides of twig; 1/2-5/8 inch long; 4-angled; dark yellow-green; blunt or pointed at apex.

FLOWERS. Male bright red at maturity; female on different branches, red-green.

FRUIT. 1 1/4-2 inches long; ovoid-oblong; nearly sessile; cone scales rigid, rounded, and smooth at apex; light green or purplish before shedding seed, becoming red-brown and falling the first winter or following spring. Seed: 1/8 inch long; dark brown; wing 1/4 inch long.

TWIGS. More or less pubescent at first, becoming smooth second year; orange-brown. Winter buds: 1/4-1/3 inch long; ovoid; acute; red-brown.

BARK. Thin, 1/4-1/2 inch thick; gray-brown to red-brown, with red-brown inner bark; separating into irregular, close scales.

WOOD. Important; light; soft; even-grained; lustrous; used for lumber, pulp, musical instruments, containers, etc.

SILVICAL CHARACTERS. Very tolerant, though growing slowly under heavy shade; long-lived, attaining age of 400 years; good reproduction; spreading root system; susceptible to damage by fire and windthrow.

HABITAT. Varied, ranging from swamps and bogs where growth is slow to mountaintops; best growth on well-drained uplands and mountain slopes; in pure stands or mixed with yellow birch, beech, maple, white pine, and hemlock on better sites and with black spruce, balsam fir, tamarack, and red maple on swampy sites. The common spruce in the mountains of New York and New England and with white pine the most important timber species in this region.

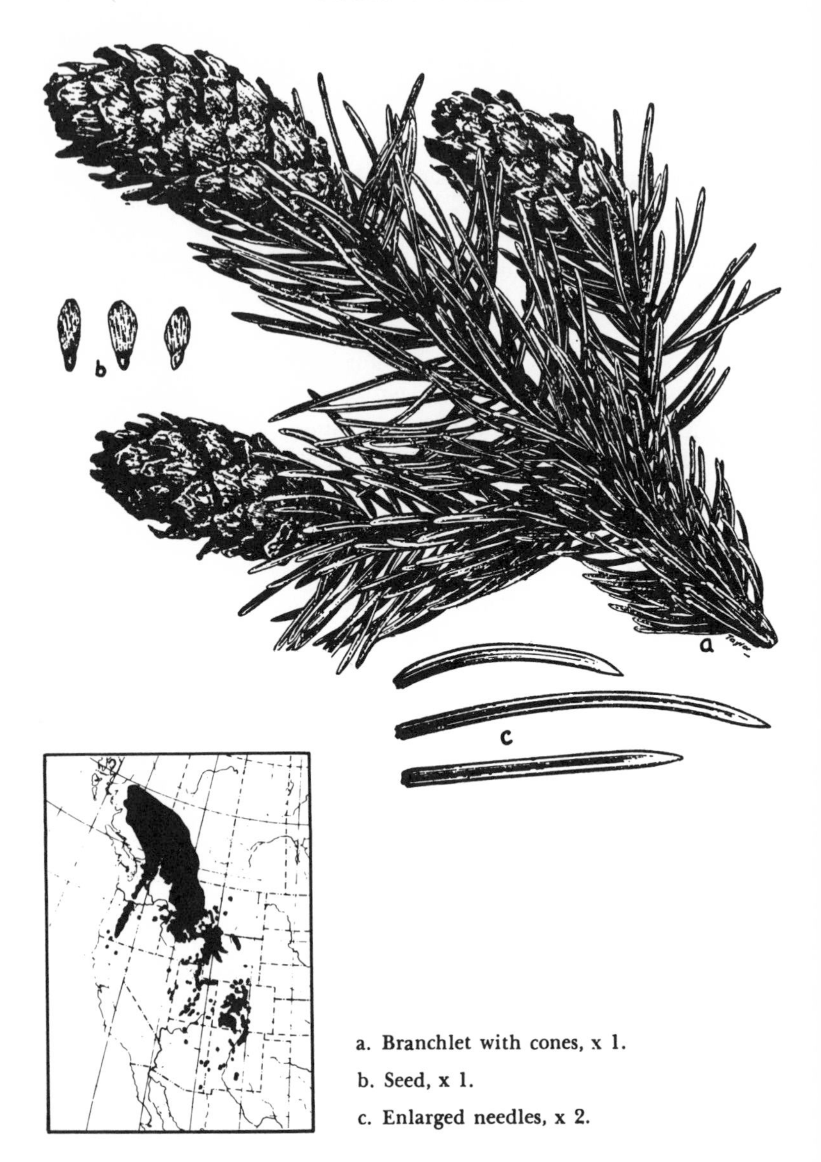

a. Branchlet with cones, x 1.

b. Seed, x 1.

c. Enlarged needles, x 2.

PINACEAE

Engelmann Spruce

Picea engelmannii Parry

HABIT. A tree 60-120 feet high and 1 1/2-3 feet in diameter (max. 165 by 5 feet); bole long but limby, cylindrical; crown compact, somewhat scraggly, narrowly pyramidal, with short, whorled branches. A prostrate shrub at high elevations.

LEAVES. Tending to be crowded on the upper side of the branch by the curving of those on the lower side; 1-1 1/8 inches long; 4-angled; blue-green, occasionally with whitish, glaucous bloom; blunt or acute tips (not very sharp to touch); flexible; no resin ducts in cross section.

FLOWERS. Male dark purple; female bright scarlet.

FRUIT. 1-2 1/2 inches long; oblong-cylindrical; sessile or short-stalked; cone scales flexible, variable in outline and erose-dentate at apex; light chestnut-brown; falling during autumn or winter of first season. Seed: 1/8 inch long; nearly black; broad, oblique wing 1/2 inch long.

TWIGS. Minutely pubescent (visible with hand lens); rather stout; orange-brown to gray-brown. Winter buds: 1/8-1/4 inch long; broadly ovoid to conic; pale chestnut-brown; scales usually appressed.

BARK. Thin, 1/4-1/2 inch thick; cinnamon-red to purple-brown; broken into large, thin, loosely attached scales.

WOOD. Properties similar to white spruce; this is the longest-fibered and lightest weight spruce, but at present not widely used because inaccessible; lumber, telephone poles, railroad ties, mine timbers, and fuel.

SILVICAL CHARACTERS. Tolerant and recovering well from prolonged suppression; growth generally rather slow because of short summer season; a long-lived tree, reaching ages of 350-500 or more years; reproduction abundant and vigorous; shallow, spreading root system.

HABITAT. Hudsonian and Canadian zones; varying from 1,500-5,000 feet in the northern Rockies to 10,000-12,000 feet in the southern Rockies; rich, loamy soils with abundance of moisture; in pure stands or in mixture with alpine fir, lodgepole pine, and other conifers growing at high elevations.

SITKA SPRUCE

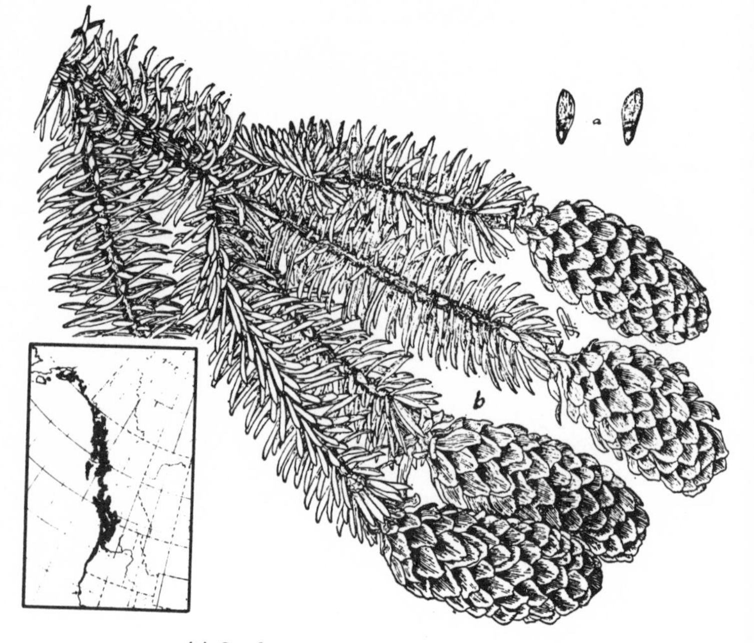

(a) Seeds, x ⅔; (b) Branch with cones, x ⅔.

BREWER SPRUCE

Twig with cone, x ⅔.

PINACEAE

Sitka Spruce

Picea sitchensis (Bong.) Carr.

HABIT. A large tree 180-200 feet high and 3-4 1/2 feet in diameter (max. 300 by 16 feet); bole often clear for 100 feet, cylindrical; crown pyramidal, short, open.

LEAVES. Extending at nearly right angles from all sides of twig; 1/2-1 1/8 inches long; flattened; bright yellow-green above, bluish-white, glaucous below; very sharp-pointed.

FLOWERS. Male dark red; female covered by elongated bracts.

FRUIT. 2-4 inches long; oblong-cylindrical; short-stalked; cone scales thin, rough, papery, stiff, wedge-shaped at apex and erose-dentate at margin; chestnut-brown; falling first season. Seed: 1/8 inch long; red-brown; wings 3/8 inch long.

TWIGS. Glabrous; slender; orange-brown to gray-brown. Winter buds: 1/4-1/2 inch long; ovoid; red-brown.

BARK. Thin, 1/2-1 inch thick; red-brown to purple; broken into thin, loose, concave scales.

WOOD. Very important and much in demand for specialized uses such as wood aircraft; during World War II aircraft quality lumber reported to have sold for $600 per mbf; soft, light, uniform grain.

SILVICAL CHARACTERS. Tolerant; growth fairly rapid; long-lived, ages of 700-800 years; reproduction vigorous; shallow roots.

HABITAT. A tidewater species; best development on deep, moist loams; in pure stands or mixed with hemlock, Douglas-fir, western redcedar, alder, and maple.

* * *

Brewer spruce, *Picea breweriana* S. Wats., is an uncommon timberline species restricted to the Siskiyou Mountains of southern Oregon and northern California. It is characterized by long, pendulous, weeping branches; flattened, blunt, dark green needles; rounded, entire-margined, purplish cone scales; and cones 2-4 inches long.

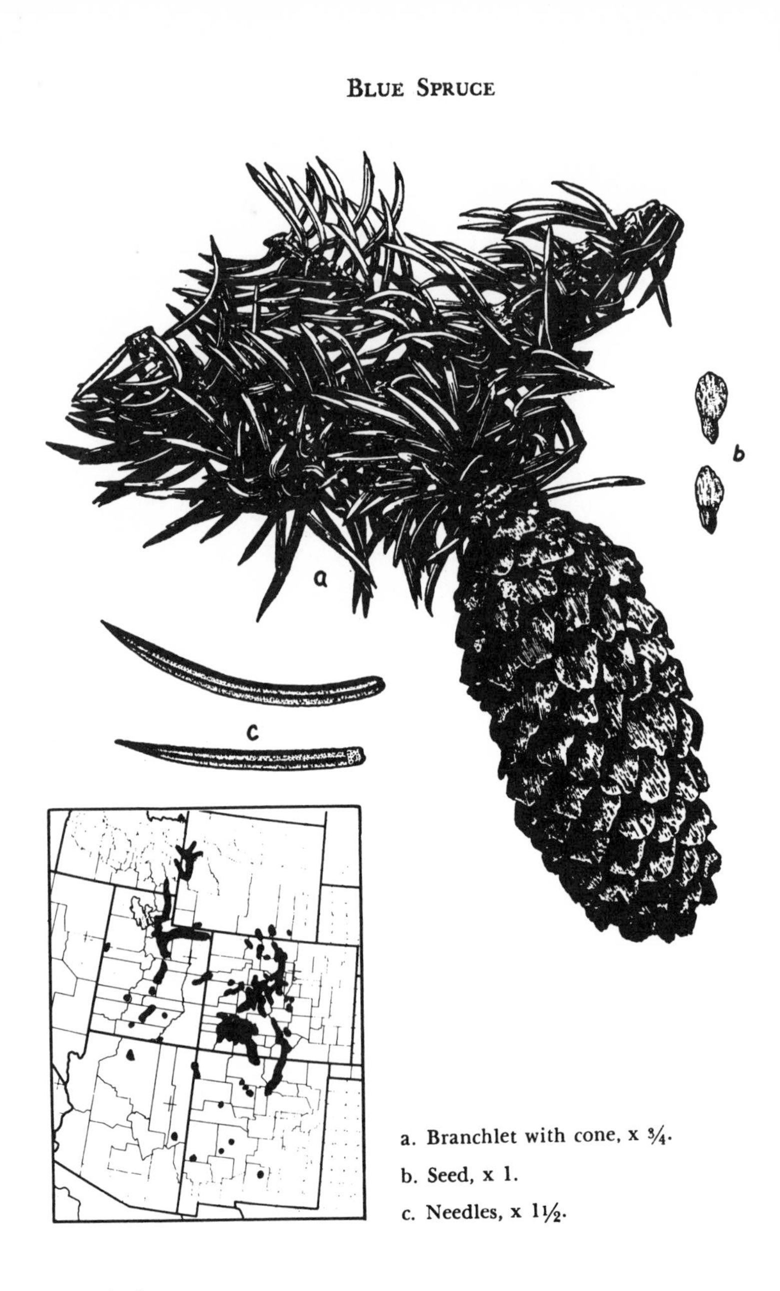

a. Branchlet with cone, x ¾.

b. Seed, x 1.

c. Needles, x 1½.

Blue Spruce. Colorado Blue Spruce

Picea pungens Engelm. (*Picea parryana* Parry)

HABIT. A tree 80-100 feet high and 1-2 feet in diameter (max. 150 by 4 feet); bole symmetrical, tapering, knotty; crown typically dense and conical when young, becoming thin, ragged and pyramidal in age, and extending to the ground on open-grown species. The state tree of Colorado.

LEAVES. Extending at nearly right angles of all sides of twig; 1-1 1/4 inches long; 4-angled; blue-green, frequently with a silvery, glaucous bloom that persists for 3-4 years on young trees; rigid, tipped with long, bristle-sharp point; 1 resin duct in an angle of leaf in cross section.

FLOWERS. Male yellow, tinged with red; female pale green.

FRUIT. 2 1/4-4 1/2 (mostly 3 1/2) inches long, oblong-cylindrical; sessile or short-stalked; cone scales tough, stiff, spreading, with erose margins; shiny, light chestnut-brown; not falling until fall of second season. Seed: 1/8 inch long; dark chestnut-brown; broad oblique wing about 1/2 inch long.

TWIGS. Glabrous; stout and rigid; orange-brown to gray-brown. Winter buds: 1/4-1/2 inch long; broadly ovoid and obtuse; light chocolate-brown; bud scales usually reflexed.

BARK. Pale to dark gray; thin and scaly on young trunks, becoming 3/4-1 1/2 inches thick and deeply furrowed with rounded ridges on old trunks.

WOOD. Rather similar to white spruce but brittle, knotty, and of little value. The chief use of this tree is for ornamental planting.

SILVICAL CHARACTERS. Moderately tolerant, but least so of spruces; slow-growing; long-lived; reproduction generally scanty because of dense ground cover; widespread, moderately deep root system, and decidedly windfirm.

HABITAT. Transition and Canadian zones, but mostly below the Engelmann spruce belt; varying from 6,000-9,000 feet in the north to 8,000-11,000 feet in the south; rich, moist soils, typically on stream banks; never abundant; in scattered pure groves or singly in mixture with ponderosa pine, Douglas-fir, alpine fir, Engelmann spruce, and hardwoods.

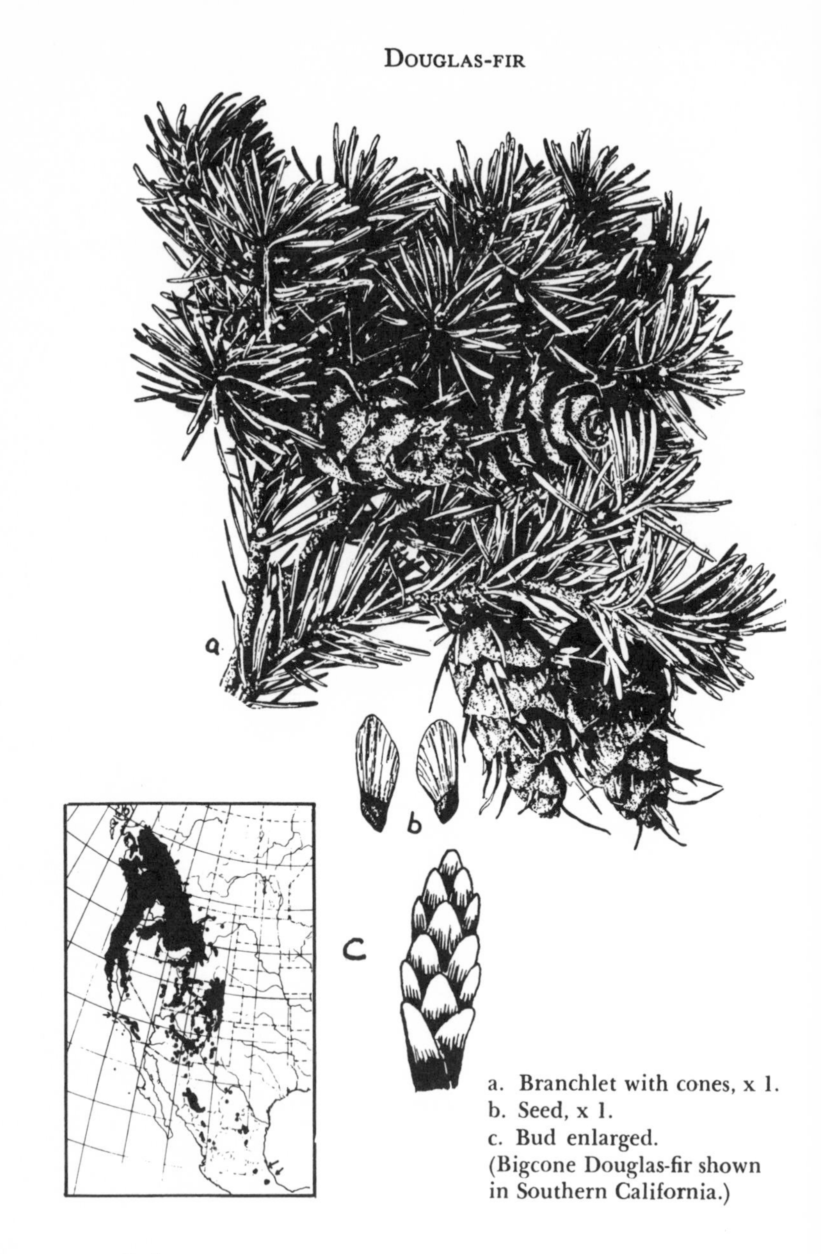

a. Branchlet with cones, x 1.
b. Seed, x 1.
c. Bud enlarged.
(Bigcone Douglas-fir shown in Southern California.)

PINACEAE

Douglas-fir

Pseudotsuga menziesii Franco
[*Pseudotsuga taxifolia* (Poir.) Britt.]

HABIT. A large evergreen tree attaining a height of over 300 feet on the West Coast and 130 feet in the Rocky Mountains; compact, pyramidal crown, with irregularly disposed branches.

LEAVES. Linear; single; more or less flattened; blunt to pointed; spirally arranged; petiolate; 3/4-1 1/4 inches long; grooved above and stomatiferous below; persistent 5-8 years or longer.

FLOWERS. Monoecious; male orange-red; female red-green.

FRUIT. Pendent, woody cones; 2-4 1/2 inches long; oblong-ovoid; maturing in one season; scales thin, rigid, rounded, much shorter than their long, exserted, 3-lobed bracts. Seed: 1/4 inch long; large, rounded, terminal wing.

TWIGS. Slender; pubescent; orange-brown, becoming gray-brown. Winter buds: 1/4 inch long, characteristically long; conical; sharp-pointed; lustrous; brown.

BARK. Smooth; gray-brown; resin blisters on young trees; becoming very thick (6-24 inches); rough; red-brown ridges separated by deep furrows.

WOOD. Highly variable from yellowish, narrow-ringed, moderately light, and soft to red-brown, wide-ringed, with weak spring wood and very dense summer wood. This tree produces more timber than any other species.

SILVICAL CHARACTERS. Intermediate in tolerance; reproduction abundant and vigorous; well-developed, widespreading lateral root system; attaining great age.

HABITAT. Sea level to 11,000 feet; adapted to variety of soils but best on moist, deep, porous soils of northern exposure; will endure considerable drought; in pure stands or mixed with Rocky Mountain conifers.

GENERAL. European taxonomists have separated this species into 11 species; however, this view is not commonly accepted in North America. The Rocky Mountain form is recognized as the separate geographical variety *glauca* (Mayr) Sudw.

* * *

The bigcone Douglas-fir *Pseudotsuga macrocarpa* (Vasey) Mayr, of southern California is distinguished by its larger fruit (4-6 1/2 inches long), with bracts only slightly longer than the thick, stiff cone scales. Its distribution is in solid color on the map.

PINACEAE

The Hemlocks

Characteristics of the Genus *Tsuga* (Endl.) Carr.

HABIT. Tall, broadly pyramidal, evergreen trees; long, slightly tapering trunks; pyramidal or conical crown with scattered, slender, horizontal, and often pendulous branches; leading shoots characteristically drooping.

LEAVES. Spirally arranged, often appearing 2-ranked by a twist of the petioles; linear; single; abruptly petiolate; flattened or rounded; persistent 3-6 years and leaving conspicuous, woody, persistent bases (sterigmata) when they fall; deciduous in drying; usually grooved above, with 2 conspicuous bands of stomata below; 1 centrally located resin duct in cross section.

FLOWERS. Monoecious; single; on twigs of previous season; male, or pollen-bearing, axillary, globose, of numerous short stamens; female, or cone- and seed-bearing, terminal, erect, or numerous, circular scales of nearly the same length as their membranous bracts.

FRUIT. Woody, pendent cones; maturing in one season; scales thin, rounded, entire-margined, several times longer than bracts. Seed: 2 under each scale; small, light, and widely disseminated; long, terminal, obovate wing; dotted with small resin vesicles.

TWIGS. Slender; round; usually roughened by persistent leaf bases. Winter buds: small; nonresinous; ovoid to globose.

BARK. Rough; hard; ridged; deeply furrowed; clear chocolate-red color when broken; containing tannin.

WOOD. Moderately soft; moderately strong; resin ducts normally absent; light to red-brown; considered inferior to pine and Douglas-fir.

SILVICAL CHARACTERS. Tolerant trees; requiring abundant moisture; seldom attaining ages of more than 500 years; shallow, wide-spreading root system.

GENERAL. This genus contains 10 or more species widely scattered through North America and Asia. In North America there are 4 native species, 2 western and 2 eastern.

KEY TO THE SPECIES OF HEMLOCKS

1. Needles flat, obtuse, grooved above, with 2 whitish bands below; cones light brown, small (1/2-1 1/2 inches long), ovoid.
   2. Needle margin finely toothed; needles mostly 2-ranked; cones 1/2-1 inch long.
      3. Needles tapering from base to apex, with well-defined. narrow white bands below; cones stalked. 1/2-3/4 inch long, oblong-ovoid, with scale margins smooth; buds ovoid, pointed; eastern . . . . *T. canadensis*, Eastern Hemlock, p. 67
      3. Needles of uniform width from base to apex, with poorly defined bands below; cones sessile, 3/4-1 inch long, ovoid, with scale margins undulate; buds globose; western . . *P. heterophylla*, Western Hemlock, p. 69
   2. Needle margin entire; needles spreading in all directions; cones 1-1 1/2 inches long; Appalachian Mountains . *P. caroliniana*, Carolina Hemlock, p. 67
1. Needles rounded or keeled above, stomatiferous on all surfaces, spreading in all directions; bluntly pointed; cones yellow-green to purple, 3/4-3 1/2 inches long (mostly about 2 inches long), oblong-cylindric . . . . *T. mertensiana*, Mountain Hemlock, p. 71

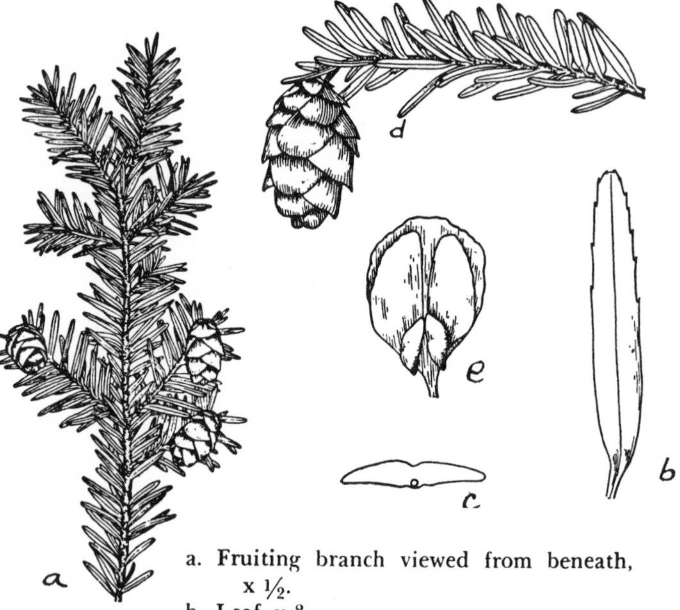

a. Fruiting branch viewed from beneath, x ½.
b. Leaf, x 3.
c. Cross section of leaf, enlarged.
d. Branchlet with partly opened cone, x 1.
e. Cone scale with seeds, x 3.

EASTERN HEMLOCK

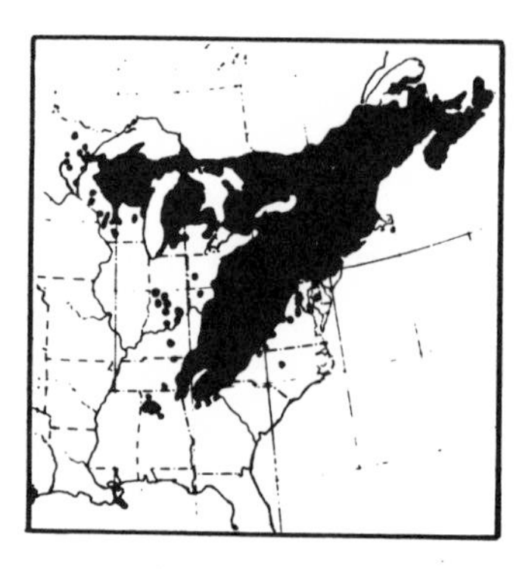

CAROLINA HEMLOCK

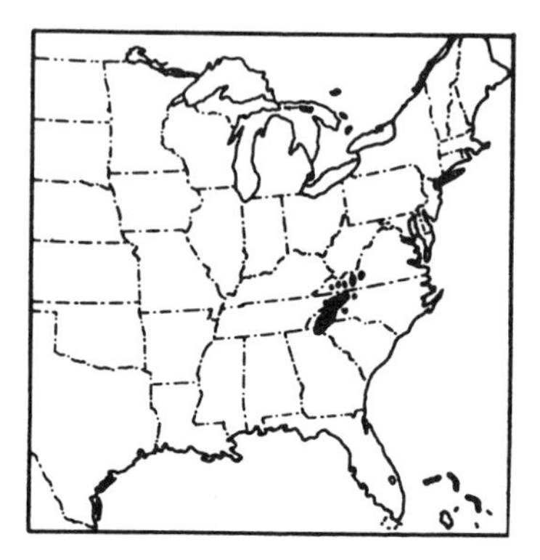

Eastern Hemlock

Tsuga canadensis (L.) Carr.

HABIT. A tree usually 60-70 feet high and 2-3 feet in diameter (max. 160 by 6 feet); dense, pyramidal crown with horizontal branches extending nearly to the ground in open-grown trees; typically flexible, drooping terminal leader.

LEAVES. Flattened; 1/3-2/3 inch long; tapering from base to apex; dark yellow-green and grooved above, 2 narrow, well-defined bands of stomata below; abrupt, slender petiole; appearing 2-ranked; rounded or notched at apex.

FLOWERS. Male yellow; female pale green, bracts shorter than scales.

FRUIT. 1/2-3/4 inch long; oblong-ovoid; light brown; scales suborbicular, smooth-margined. Seed: 1/16 inch long; wing about 1/3 inch long; light brown.

TWIGS. Slender; light brown, and pubescent during first year, becoming gray-brown and glabrous. Winter buds: ovoid; 1/16 inch long.

BARK. Scaly on young trees, becoming deeply furrowed and ridged; red to gray in color with purple streaks on freshly cut surfaces; important source of tannin.

WOOD. Light, brash, coarse-grained, splintery; used for poorer grades of lumber and pulp; knotty as a result of persistent branches.

SILVICAL CHARACTERS. Tolerant; growth rapid; maximum age about 600 years; reproduction abundant and vigorous; shallow, wide-spreading root system.

HABITAT. Cool, moist sites; in small pure groves, or more commonly in mixed stands with white pine, red spruce, or hardwoods.

* * *

Carolina hemlock, *Tsuga caroliniana* Engelm., is a handsome and somewhat rare tree found on the upper slopes of the Appalachian Mountains from Virginia to northern Georgia. This tree, commonly planted as an ornamental, differs from eastern hemlock in having the needles extending from all sides of the twigs and larger cones that are 1-1 1/2 inches long.

Western Hemlock

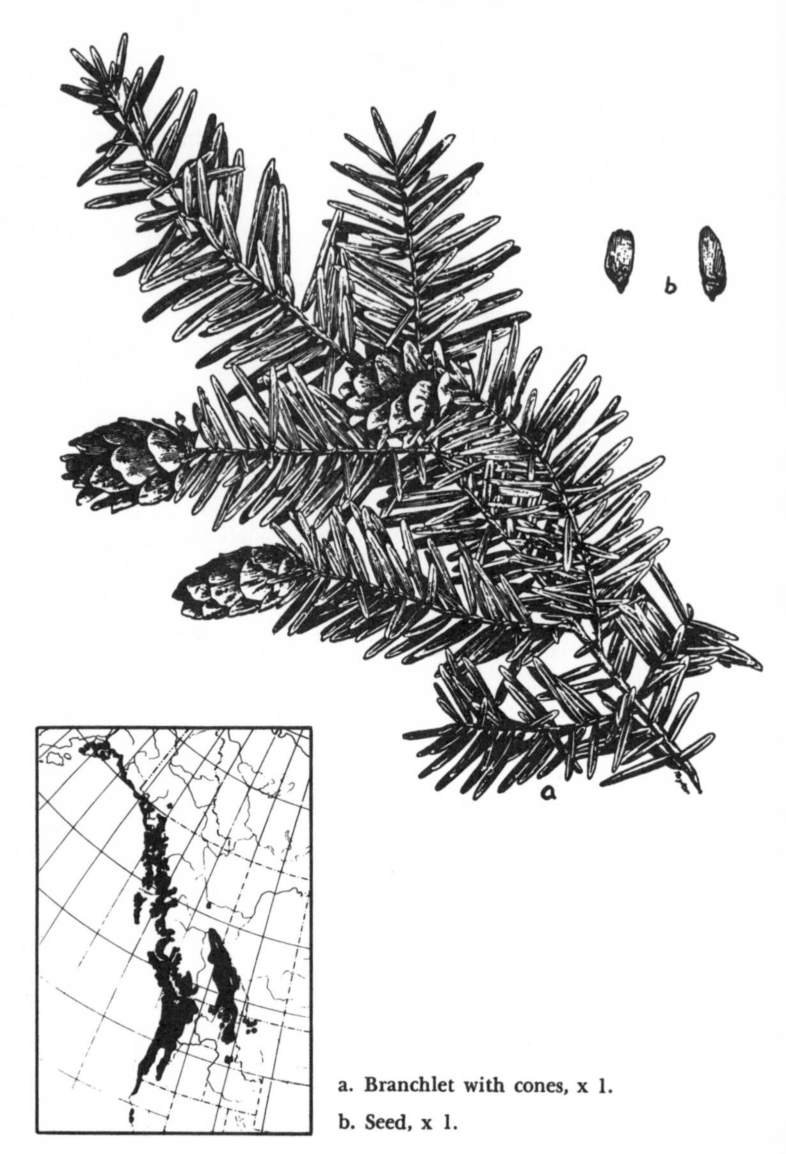

a. Branchlet with cones, x 1.

b. Seed, x 1.

PINACEAE

Western Hemlock

Tsuga heterophylla (Rafn.) Sarg.

HABIT. A tree 125-175 feet high and 2-4 feet in diameter (max. 259 by 9 feet); tall, clear trunk; short, open, pyramidal crown with typically flexible, drooping terminal leader.

LEAVES. Flattened; 1/4-3/4 inch long; dark, shiny green and grooved above, 2 broad bands of stomata below; abrupt, slender petiole; mostly 2-ranked; rounded or blunt at apex.

FLOWERS. Male yellow; female red or purple with rounded bracts shorter than scales.

FRUIT. 3/4-1 inch long; ovoid; light brown; scales suborbicular; wavy-margined. Seed: 1/16 inch long; ovoid; 1/3 as long as narrow; straw-colored wing.

TWIGS. Slender; pubescent for 5-6 years; pale yellow-brown becoming dark red-brown; drooping. Winter buds: ovoid; 1/16 inch long; blunt; bright chestnut-brown.

BARK. Thin (1-1 1/2 inches) even on largest trees; young bark scaly, russet-brown; on old trunks hard, dark russet-brown, furrows separating wide flat ridges; inner bark dark red streaked with purple; used for its tannin content.

WOOD. Superior in quality to eastern hemlock; one of 4 major timber-producing species of Pacific Northwest; uniform texture, not very harsh or splintery; suitable for all uses but heavy construction; most important pulpwood species of region.

SILVICAL CHARACTERS. Tolerant throughout life; growth rapid, comparing favorably with Douglas-fir, seldom attaining age of over 500 years; reproduction very abundant and vigorous; shallow, wide-spreading root system; susceptible to fire injury; butt rot common in old trees.

HABITAT. Transition and Canadian zones; altitudinal range, sea level to 7,000 feet; prefers deep, moist, porous soils, but hardy in drier situations; in pure, dense stands or mixed at lower levels with Douglas-fir, silver and grand firs, giant arborvitae, redwood, and hardwoods; at higher levels with noble fir, Alaska cedar, mountain hemlock, western white and lodgepole pines.

Mountain Hemlock

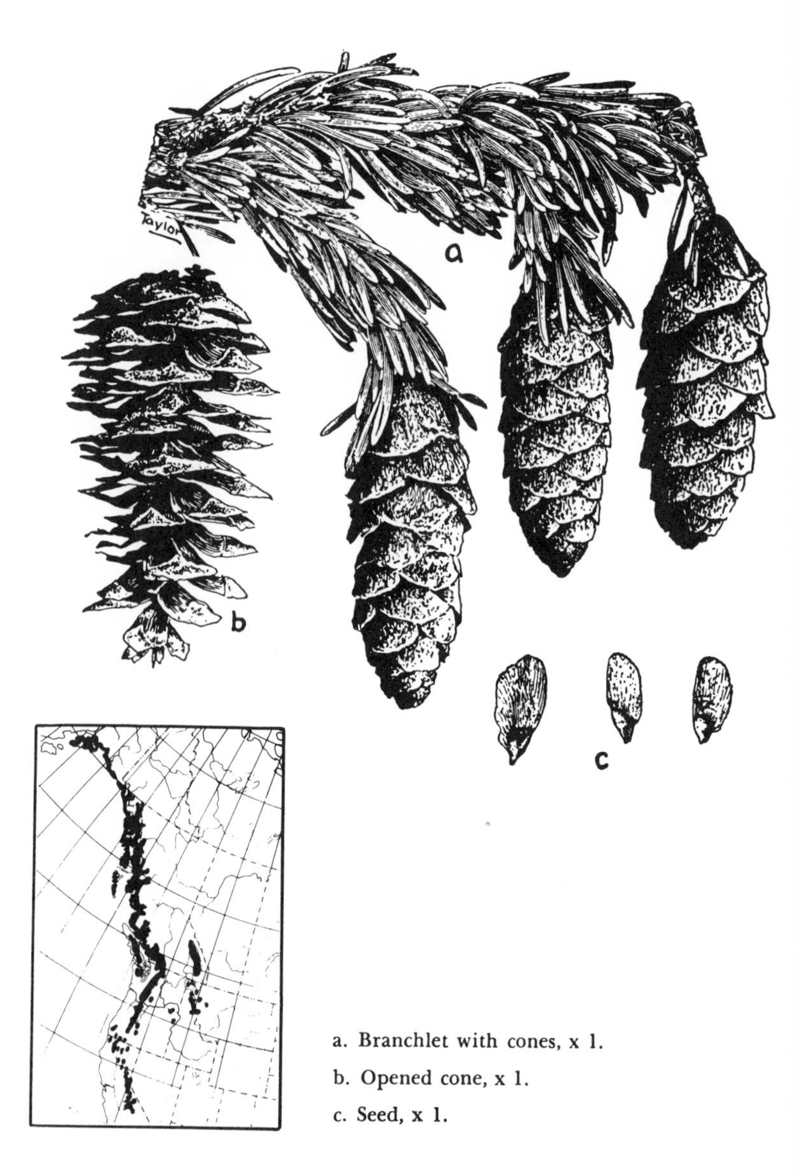

a. Branchlet with cones, x 1.

b. Opened cone, x 1.

c. Seed, x 1.

Mountain Hemlock. Black Hemlock

Tsuga mertensiana (Bong.) Carr.

HABIT. An alpine tree 75-100 feet high and 2 1/2-3 1/2 feet in diameter; trunk tapering, long and clear or knotty and malformed; crown open, pyramidal, with slender, drooping branches and drooping terminal leader; a sprawling shrub at timberline.

LEAVES. Semicircular in cross section; 1/2-1 inch long; pale bluish-green and stomatiferous on all surfaces; upper surface often keeled or grooved; abruptly narrowed into straight or twisted petiole; extending from all sides of twig or crowded toward upper side; bluntly pointed.

FLOWERS. Male purple on slender, drooping stems; female purple or green, slender-tipped bracts longer than scales.

FRUIT. 1/2-3 1/2 inches long, (mostly 2 inches); oblong-cylindric; yellow-green to purple; scales oblong-obovate and spreading at right angles or reflexed when mature. Seed: 1/8 inch long; 1/4 as long as wing.

TWIGS. Thin or stout; dense; short; pale; pubescent for 2-3 years; light red-brown, becoming gray-brown and scaly. Winter buds: conical; 1/8 inch long; acute; red-brown; outer scales with awllike tip.

BARK. Rather thin on large trees (1-1 1/2 inches), early broken and rough on young trees; on old trunks hard, purplish to red-brown, with deep, narrow furrows separating narrow, rounded ridges; containing large quantities of tannin.

WOOD. Little used and inferior in quality to western hemlock; light; soft; not strong; close-grained.

SILVICAL CHARACTERS. Tolerant; growth slow; trees over 500 years of age seldom found; reproduction generally abundant; shallow, wide-spreading root system.

HABITAT. Hudsonian and Canadian zones; altitudinal range, sea level (Alaska) to 11,000 feet, but mostly near timberline; at its best on cool, moist, deep soils of northern exposure, moisture being essential; in pure stands or in mixture with alpine fir; alpine larch; Engelmann spruce; and whitebark, lodgepole, and western white pines.

PINACEAE

The Firs

Characteristics of the Genus *Abies* Mill.

HABIT. Tall, pyramidal, evergreen trees; dense, spirelike crowns; slender, horizontal, whorled branches; straight, gradually tapering trunks.

LEAVES. Spirally arranged; linear; sessile; single; usually flat and blunt; extending from all sides of twig but mostly appearing 2-ranked by a twist near their base; persistent for 7-10 years; usually grooved above with stomatiferous lines below; 2 resin canals in cross section; when falling, leave a conspicuous, smooth, circular scar on twig.

FLOWERS. Monoecious, axillary, single; male, or pollen-bearing, numerous on lower sides of lower crown branches, oval or cylindrical, yellow to scarlet anthers; female, or cone- and seed-bearing, on upper side of topmost branches, erect, globose to oblong, consisting of numerous, imbricated, 2-ovuled scales much shorter than their bracts.

FRUIT. Woody, erect cones; maturing in one season; scales thin, fan-shaped, and falling at maturity from the central, spike-like axis that persists many years. Seed: 2 under each scale; large, thin wing; peculiar, conspicuous resin vesicles.

TWIGS. Smooth, glabrous, or pubescent. Winter buds: small; mostly subglobose or ovoid; thin, loosely imbricated scales; usually thickly covered with resin.

BARK. Young bark thin, smooth, with numerous blisterlike resin pockets; old bark smooth or furrowed; Canada balsam obtained from the resin blisters of balsam fir.

WOOD. Light; soft; weak; color whitish to light brown, summer wood frequently with a purplish tinge; resin ducts absent; used for pulp, containers, and general construction.

SILVICAL CHARACTERS. Tolerant, moisture-loving trees of cool sites; slow- to fast-growing; moderately long-lived.

GENERAL. This genus contains about 35 species widely scattered through North and Central America, Europe, Asia, and northern Africa. There are 9 species native to North America.

KEY TO THE SPECIES OF FIRS

1. Winter buds subglobose, 1/8-1/3 inch long; needles not bristle-tipped; cones without spiny-tipped bracts.
   2. Needles crowded toward the upper side of the twigs; cones large and broad (3 1/2-9 inches long); Pacific Coast region.
      3. Needles 4-sided, stomatiferous (covered with minute white dots) on all sides and usually glaucous; bark furrowed or scaly on mature trunks.
         4. Cones 4-6 inches long; bracts much longer than, and covering, cone scales; needles with sharply defined groove on upper surface; Washington, Oregon, northern California. . . . . *A. procera*, Noble Fir, p. 87
         4. Cones 6-9 inches long; bracts never covering scales; needles ribbed above and below.
            5. Cones with bracts inserted . . . . *A. magnifica*, California Red Fir, p. 85
            5. Cones with bracts exserted . . . . *A. magnifica shastensis*, Shasta Red Fir, p. 85
      3. Needles flattened, dark green, and lustrous above, with 2 distinct white bands (stomata) on lower surface; bark smooth and ashy gray on mature trunks; cones 3 1/2-6 inches long with inserted bracts; British Columbia and Alaska to Washington and Oregon . . . . *A. amabilis*, Pacific Silver Fir, p. 83
   2. Needles flattened, on sterile branches spreading and not crowded, often 2-ranked; cones narrow, 2-5 inches long.
      5. Cones with bracts inserted.
         6. Needles on lower branches 3/4-1 3/4 inches long; cones dark purple, scales slightly longer than broad.
            7. Needles dark green and lustrous above with 2 distinct white bands (stomata) below, 2-ranked on lower branches; northeastern United States to northwestern Canada . . . *A. balsamea*, Balsam Fir, p. 75
            7. Needles stomatiferous on both surfaces, nearly erect on lower branches; western.
               8. Mature bark hard, smooth, gray; high western mountains . . . . *A. lasiocarpa*, Subalpine Fir, p. 77
               8. Mature bark distinctly soft and corky, yellow-white; Colorado, Arizona, New Mexico . . . . *A. lasiocarpa arizonica*, Corkbark Fir, p. 77
         6. Needles 1 1/2-3 inches long; cones yellow to green-purple, scales slightly broader than long; western.
            9. Needles distinctly 2-ranked, 1 1/2-2 inches long, dark green above with 2 white bands of stomata below; Montana to British Columbia and California . . . . *A. grandis*, Grand Fir, p. 81
            9. Needles spreading or obscurely 2-ranked, 2-3 inches long, stomatiferous and pale green on both surfaces; western United States . . . . *A. concolor*, White Fir, p. 79
      5. Cones with bracts exserted; needles 1/2-1 inch long; Appalachian Mountains . . . . *A. fraseri*, Fraser Fir, p. 75
1. Winter buds ovoid, pointed, 3/4-1 inch long; needles 1 1/2-2 1/4 inches long, flat, rigid, bristle-tipped; cones 3-4 inches long with spiny bracts exserted 3/4-1 3/4 inches; California (Monterey County) . . . . *A. venusta*, Bristlecone Fir, p. 87

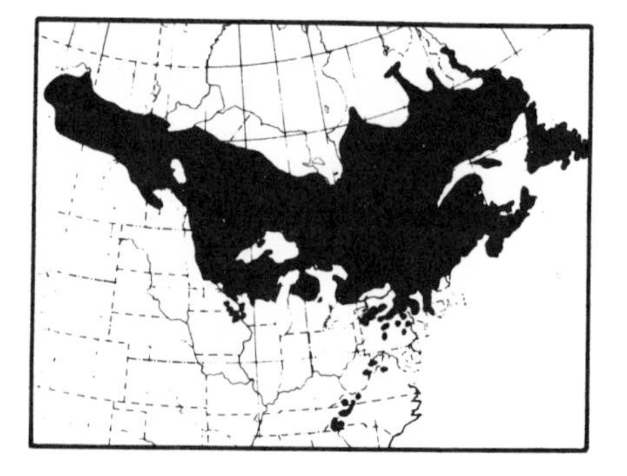

a. Lower crown branchlet.

b. Branchlet with cone, x 1.

c. Upper and lower sides of cone scale, x 1.

d. Seed, x 1.

(Fraser fir shown in Southeast.)

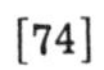

Balsam Fir

Abies balsamea (L.) Mill.

HABIT. A medium-sized tree 40-60 feet high and 1-1 1/2 feet in diameter (max. 85 by 2 feet); a dense, dark green, narrowly pyramidal crown with a slender spirelike tip.

LEAVES. On lower branches 2-ranked, 3/4-1 1/2 inches long, scattered, flattened, blunt, notched; on upper branches shorter, spreading, crowded; dark green above, two silvery bands of stomata below.

FLOWERS. Male of yellow anthers; female purple.

FRUIT. 2-4 inches long; oblong-cylindric; dark purple; scales longer than broad and twice as long as short, pointed bracts. Seed: 1/4 inch long; broad purple-brown wings.

TWIGS. Slender; finely pubescent; yellow-green; becoming smooth and gray to purple. Winter buds: subglobose; resin-covered; 1/8-1/4 inch long; orange-green scales.

BARK. Thin; ash-gray; smooth except for numerous resin blisters on young trees; becoming 1/2 inch thick, red-brown, and broken into thin scales.

WOOD. Soft and brittle; used for pulp, boxes, etc.; resin in bark blisters is source of Canada balsam.

SILVICAL CHARACTERS. Tolerant (less so than spruce), recovering well from suppression; a short-lived tree reaching an age of 150 years, but generally defective before 90 years; reproduction plentiful and aggressive; shallow root system; lower branches sometimes take root, producing new trees.

HABITAT. Canadian and Hudsonian zones; demands abundant soil moisture and humid atmosphere; forms pure stands in swamps; on higher sites in mixture with spruce, hemlock, broadleafed species; var. *phanerolephis* Fern. ranging north from northern Virginia has slightly exposed fruit bracts.

* * *

Fraser fir, *Abies fraseri* (Pursh) Poir., is a tree of the high mountains of Tennessee, North Carolina, and southwestern Virginia (shown in these areas on map). Very similar to balsam fir except for the long fruit bracts that extend beyond the scales and are strongly reflexed.

Subalpine Fir

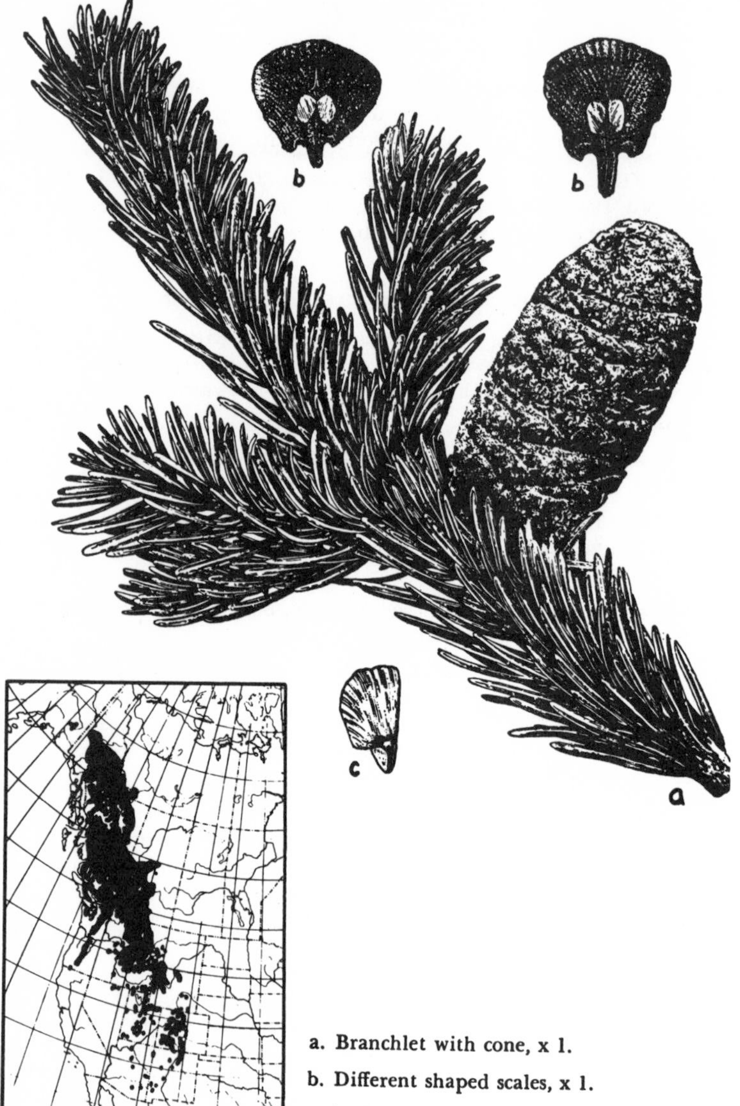

a. Branchlet with cone, x 1.

b. Different shaped scales, x 1.

c. Seed, x 1.

Subalpine Fir

Abies lasiocarpa (Hook.) Nutt.

HABIT. A tree 60-100 feet high and 1 1/2-2 feet in diameter (max. 160 by 3 feet); a dense, narrowly pyramidal, spirelike crown often extending to the ground, with short, thick branches; a prostrate shrub at timberline.

LEAVES. On lower branches, 1-1 3/4 inches long (mostly about 1 inch), flattened, blunt, or notched; on upper branches 1/2 inch long and pointed; deep blue-green; crowded and nearly erect by a twist at their base; stomatiferous on both surfaces (less conspicuous above).

FLOWERS. Male dark indigo-blue; female dark purple.

FRUIT. 2-4 inches long; oblong-cylindric; dark purple; scales mostly longer than broad and 3 times longer than long-tipped bracts. Seed: 1/4 inch long; dark lustrous wings.

TWIGS. Stout, pubescent; pale orange-brown; becoming smooth and gray or silver-white. Winter buds: subglobose; resinous; 1/8-1/4 inch long; light orange-brown scales.

BARK. Thin; gray; smooth except for numerous resin blisters on young trees; becoming shallowly fissured.

WOOD. Similar to balsam fir but little used except for fuel.

SILVICAL CHARACTERS. Tolerant (of its associates, only Engelmann spruce and mountain hemlock are more so); growth not rapid; reproduction abundant and vigorous; shallow root system; lower branches sometimes taking root.

HABITAT. Canadian and Hudsonian zones; growing from 3,500 feet to timberline in the north and from 10,500 feet to timberline in the south; in cool, moist sites. Commonly with Engelmann spruce; lodgepole, whitebark, limber, or bristle cone pines; alpine larch; cork fir; and aspen.

GENERAL. Corkbark fir *A. lasiocarpa* var. *arizonica* (Merr.) Lemm. of New Mexico, Arizona, and southern Colorado, differs from alpine fir in having soft, corky, yellow-white to ash-gray trunk bark.

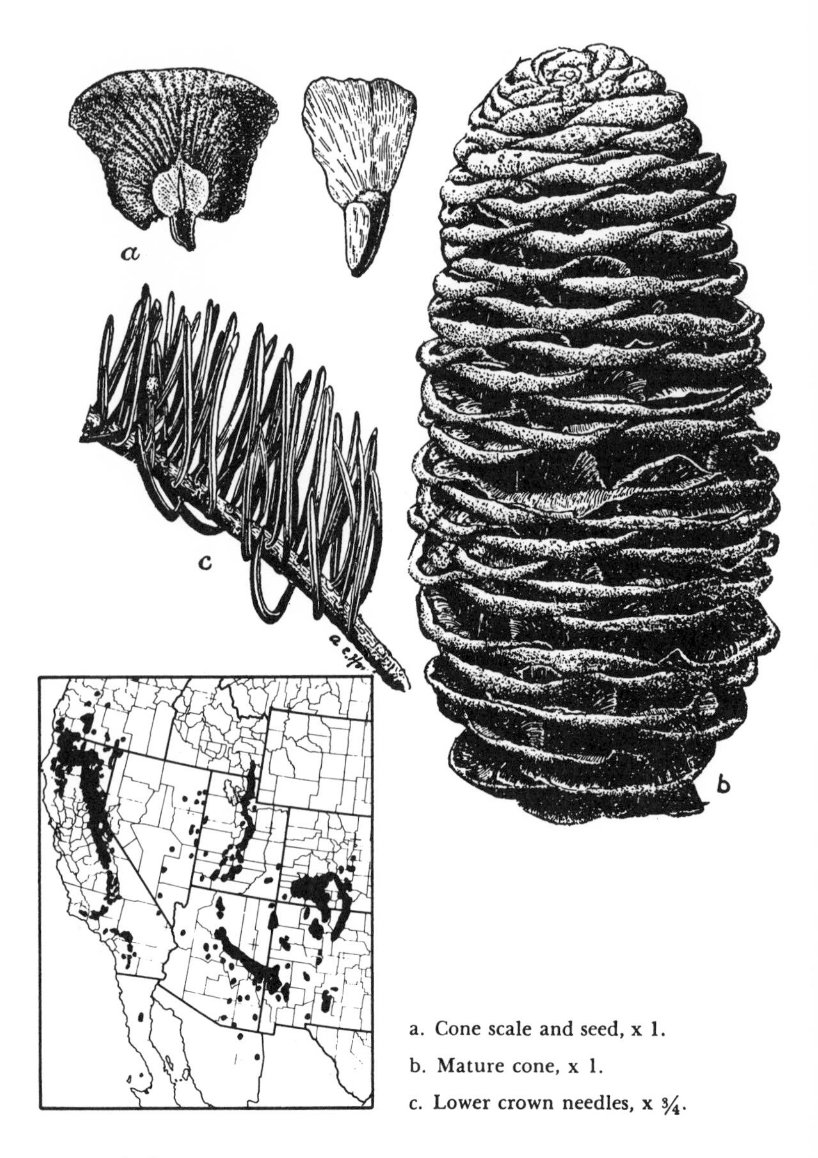

a. Cone scale and seed, x 1.

b. Mature cone, x 1.

c. Lower crown needles, x ¾.

White Fir

Abies concolor (Gord. & Glend.) Lindl.

HABIT. A tree 120-150 feet high and 3-4 feet in diameter (max. 200 by 6 feet); a dense conelike crown with heavily foliaged, long-persisting, short branches.

LEAVES. On lower branches 2-3 inches long, flat, straight, and acute at apex; on fertile branches or on old trees, 3/4-1 1/2 inches long, thick, keeled above, usually curved, acute or rarely notched at apex; silver-blue to silver-green; crowded; more or less obscurely 2-ranked or extending from all sides of twig; stomatiferous above and below.

FLOWERS. Male rose to dark red; female greenish.

FRUIT. 3-5 inches long; oblong; bright yellow to olive-green or purple; scales much broader than long and twice as long as short-tipped bracts. Seed: 1/3-1/2 inch long; yellow-brown; rose-tinted broad wing.

TWIGS. Moderately stout; smooth; yellow-green to brown-green, ultimately gray-brown. Winter buds: subglobose; resin-covered; 1/8-1/4 inch long; yellow-brown.

BARK. Thin, gray, smooth except for numerous resin blisters on young trees; becoming 4-7 inches thick, ash-gray, hard and horny, with deep furrows and wide ridges.

WOOD. Similar to balsam fir; used for lumber, pulp, boxes and novelties.

SILVICAL CHARACTERS. Tolerant, although less so than alpine fir; growth moderately rapid, maturity being reached in about 300 years; reproduction generally abundant and aggressive; root system normally shallow.

HABITAT. Transition and Canadian zones; altitudinal range 6,000-11,000 feet; requires less moisture than other western firs, existing surprisingly well on poor, dry sites; seldom in pure stands, usually with ponderosa and limber pine, Douglas-fir, alpine fir, Engelmann spruce, and aspen.

GENERAL. The Pacific white fir, considered a variety by some authors, [*A. concolor* var. *lowiana* (Gord.) Lemm.] of Oregon and California differs from the species in having smaller buds and somewhat longer needles that are more pectinately arranged.

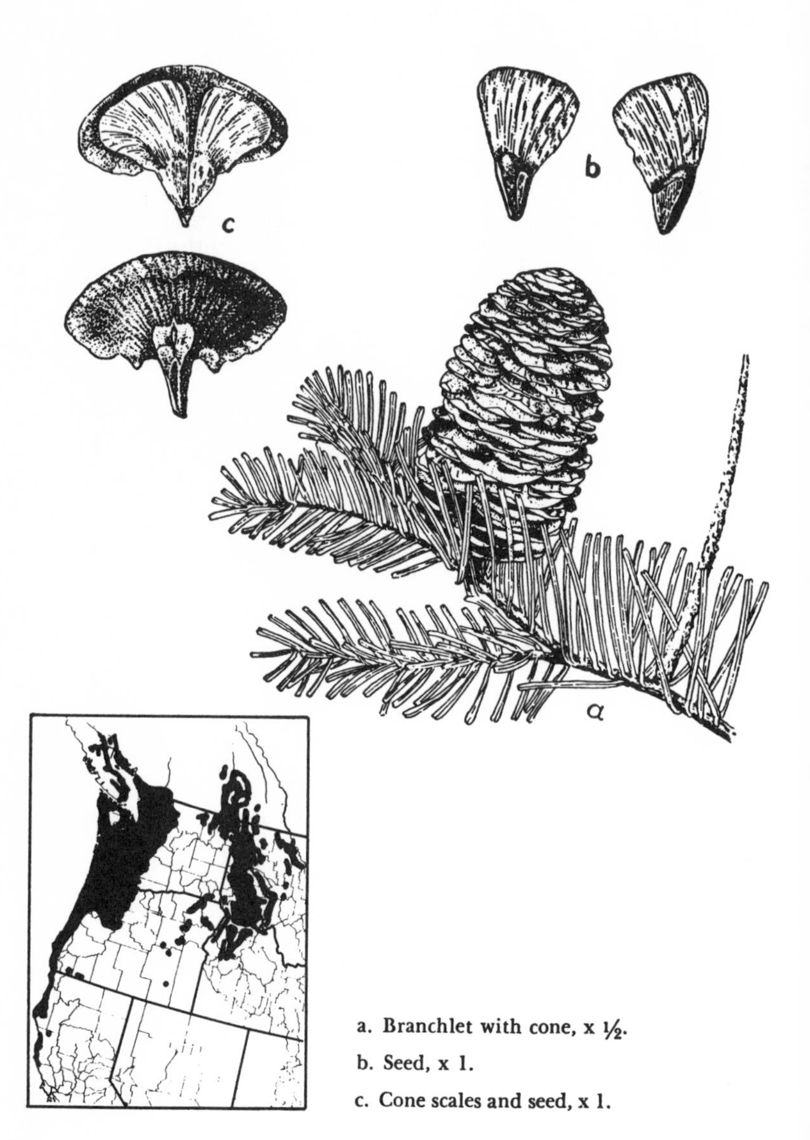

a. Branchlet with cone, x ½.

b. Seed, x 1.

c. Cone scales and seed, x 1.

Grand Fir. Lowland White Fir

Abies grandis (Dougl.) Lindl.

HABIT. A tree 140-160 feet in height and 2-4 feet in diameter (max. 250 by 6 feet); rather open domelike crown, on old trees appearing wider in the middle because of the drooping of the lower branches.

LEAVES. On lower branches 1 1/2-2 inches long, scattered, distinctly 2-ranked; on fertile branches 1-1 1/2 inches long, more crowded, obscurely 2-ranked or nearly erect; blunt, flat, lustrous, dark yellow-green, and grooved above; white with 2 bands of stomata below.

FLOWERS. Male pale yellow; female light yellow-green.

FRUIT. 2-4 1/2 inches long; cylindrical; yellow-green to green-purple; scales 1/3 broader than long, 3-4 times longer than short-tipped bracts. Seed: 3/8 inch long; light brown; straw-colored wing about 3/4 inch long.

TWIG. Slender; yellow-green to orange-brown; puberulous, becoming glabrous in second year. Winter buds: subglobose; 1/8-1/4 inch long; resinous.

BARK. Thin, gray-brown, smooth except for resin blisters and chalky white blotches on young trees; becoming 2-3 inches thick, red-brown, plated or divided into flat ridges separated by deep furrows.

WOOD. Similar to balsam fir, but with disagreeable odor, and so known as "stinking fir"; lumber, pulp, and boxes.

SILVICAL CHARACTERS. Moderately tolerant, but less so than associated firs; growth moderate, maturity being reached in about 200 years; reproduction abundant if sufficient moisture and protection against frost present; windfirm with deep, spreading root system; subject to attack by spruce budworm and stringy brown-rot fungus.

HABITAT. Transition and Canadian zones; altitudinal range, sea level to 5,000 feet; on deep, moist, alluvial soils along streams or on mountain slopes. In limited pure stands, or more frequently in mixed hardwood and coniferous forests with ponderosa, western white, and lodgepole pines; Douglas-fir; western larch; alpine fir; Engelmann spruce; etc.

Pacific Silver Fir

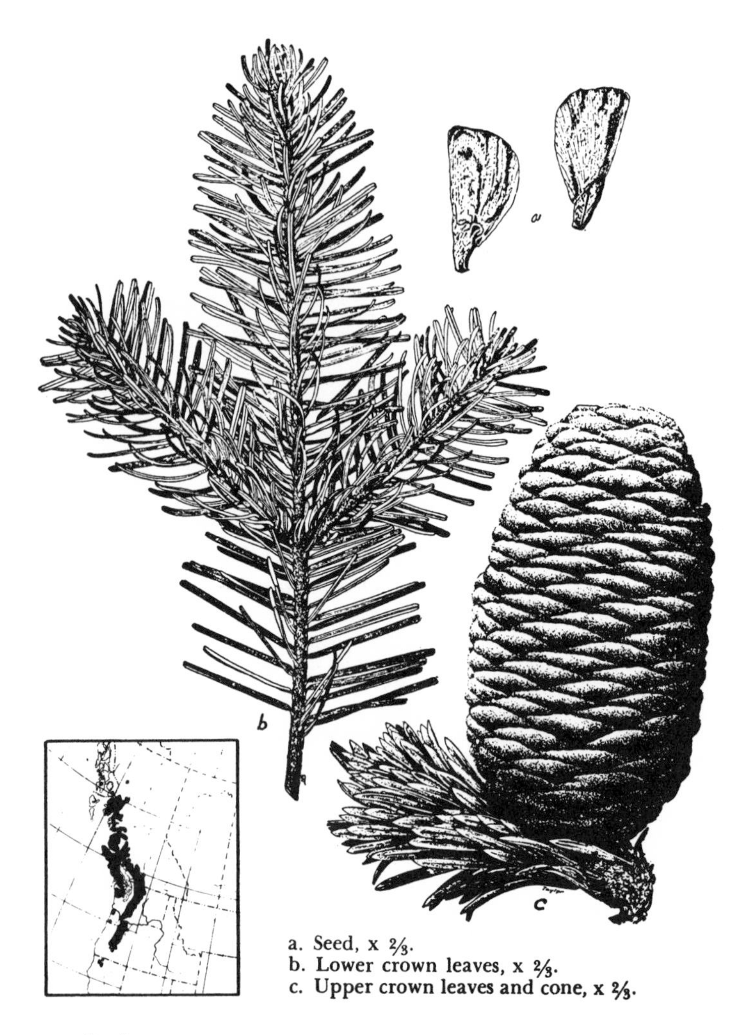

a. Seed, x ⅔.
b. Lower crown leaves, x ⅔.
c. Upper crown leaves and cone, x ⅔.

Pacific Silver Fir

Abies amabilis (Dougl.) Forb.

HABIT. A tree 140-160 feet in height and 2-4 feet in diameter (max. 200 by 6 feet); crown spirelike or pyramidal; bole clear in dense forest, but clothed to the ground with rather short branches in open.

LEAVES. On lower branches 3/4-1 1/4 inches long, crowded toward the upper side of the twig, flat, lustrous dark green and grooved above, silvery white with stomata below, notched or pointed at apex; on fertile branches often somewhat thickened and stomatiferous above at apex.

FLOWERS. Male red; female with broad scales and lustrous purple bracts.

FRUIT. 3 1/2-6 inches long; cylindrical to barrel-shaped; deep purple; scales slightly broader than long, longer than the spiny-tipped, inserted bracts. Seed: 1/2 inch long; light yellow-brown; pale brown wings 3/4 inch long.

TWIG. Stout; orange-brown; puberulous the first year, becoming red-brown. Winter buds: subglobose; 1/4 inch long; dark purple; resin-covered.

BARK. Thin; silver-white to ash-gray; chalk-colored blotches and resin blisters on trees at maturity; on overmature trunks, becoming scaly at the base.

WOOD. Similar to balsam fir; used in limited quantities for lumber and pulp.

SILVICAL CHARACTERS. Moderately tolerant; growth moderate, maturity being reached in about 250 years; reproduction abundant; often planted as ornamental because of handsome crown shape and dense foliage; the most abundant fir of the Northwest.

HABITAT. Transition and Canadian zones; altitudinal range from sea level to 6,000 feet; best on deep, moist soils with southern or western exposure; in pure stands or mixed with Sitka spruce, Douglas-fir, grand fir, western hemlock, and western redcedar.

California Red Fir

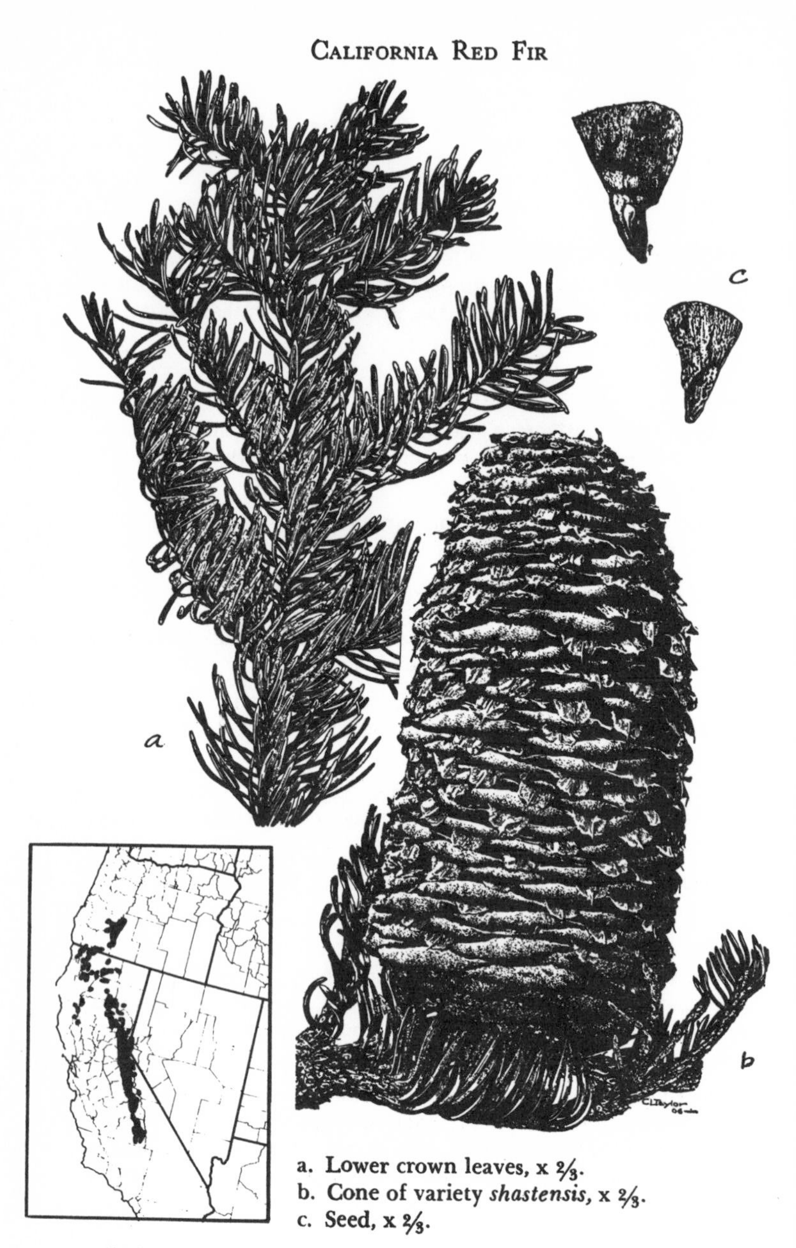

a. Lower crown leaves, x ⅔.
b. Cone of variety *shastensis*, x ⅔.
c. Seed, x ⅔.

California Red Fir

Abies magnifica A. Murr.

HABIT. The largest of the firs, varying from 150-180 feet in height and 4-5 feet in diameter (max. 230 by 10 feet); crown narrow but round-topped; bole clear for much of its height, then with short, small branches.

LEAVES. On lower branches 3/4-1 1/2 inches long, somewhat flattened, rounded apex, and somewhat 2-ranked; on upper, fertile branches almost equally 4-sided, erect, crowded, and with short callous tips; glaucous during first year, becoming blue-green; stomatiferous on all sides.

FLOWERS. Male red-purple; female with green bracts much longer than rounded scales.

FRUIT. 6-9 inches long; cylindrical to barrel-shaped; dark purplish brown; scales longer than broad and longer than the spiny-tipped, inserted bracts. Seed: 1/2-3/4 inch long; dark brown; large, broad, rose-colored wings.

TWIG. Stout; yellow-green and slightly rusty-pubescent the first year; becoming smooth and light red-brown. Winter buds: ovoid; 1/4 inch long; slightly resinous at the tip; dark brown.

BARK. Distinctly reddish-colored, thick (4-6 inches), furrowed on old trunks; smooth and chalky-gray on young trunks.

WOOD. Similar to and substituted for noble fir; light and soft; used for lumber.

SILVICAL CHARACTERS. Moderately intolerant at maturity; growth rather slow, with maturity being reached in about 300 years; reproduction abundant on moist mineral soil; deep, spreading roots.

HABITAT. Transition zone to timberline in some places; altitudinal range 5,000-10,000 feet; in pure stands or mixed with Douglas-fir, sugar and ponderosa pine, and numerous other species.

GENERAL. A distinctive variety, the Shasta red fir, *Abies magnifica* var. *shastensis* Lemm., is occasionally found throughout the range of the species. It is characterized by cone bracts that are exserted and strongly reflexed and by winter buds almost free of resin.

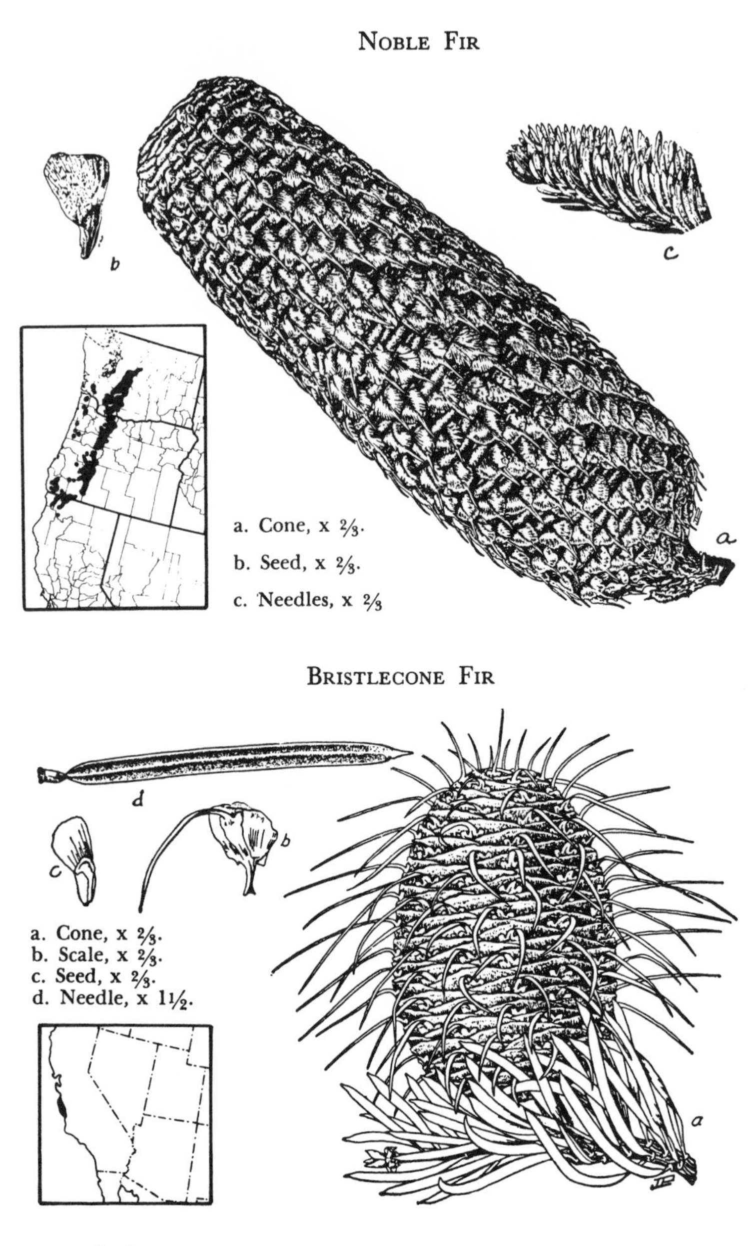

Noble Fir

a. Cone, x ⅔.
b. Seed, x ⅔.
c. Needles, x ⅔

Bristlecone Fir

a. Cone, x ⅔.
b. Scale, x ⅔.
c. Seed, x ⅔.
d. Needle, x 1½.

Noble Fir

Abies procera Rehd. [*Abies nobilis* (Dougl.) Lindl.]

HABIT. A large tree 150-200 feet high and 4-6 feet in diameter (max. 250 by 8 feet); crown broad and rounded; bole clear for much of its length.

LEAVES. On lower branches 1-1 1/2 inches long, flattened, rounded, and often notched at the apex; on upper, fertile branches almost equally 4-sided, with long, rigid, callous tips; marked on upper surface by sharply defined groove; glaucous during first year, becoming blue-green; stomatiferous on all sides.

FLOWERS. Male red-purple; female with orbicular bracts much longer than rounded scales.

FRUIT. Very distinctive; 4-6 inches long; cylindrical; purple or olive-brown; scales wider than long, nearly or entirely covered by strongly reflexed, greenish bracts. Seed: 1/2 inch long; red-brown; short, lustrous, light brown wings.

TWIG. Slender; red-brown; rusty-pubescent. Winter buds: oblong-conic; 1/8 inch long; slightly resinous at tip; red-brown; blunt.

BARK. 1-2 inches thick, red-brown, and fissured on old trunks; smooth and gray for many years on younger trunks, with prominent resin blisters.

WOOD. Best quality of native true firs; used in aircraft; light; straight-grained; not durable.

SILVICAL CHARACTERS. Seedlings and trees both intolerant; growth rather rapid; maturity reached in about 350 years; reproduction sparse; deep, spreading roots.

HABITAT. Transition zone; in stands of mixed conifers; altitudinal range 2,000-5,000 feet; on cool, deep, moist sites.

* * *

Bristlecone Fir

Abies bracteata D. Don [*Abies venusta* (Dougl.) K. Koch]

This curious species is native only to Monterey County, California. It is easily distinguished by its flat, bristle-pointed needles 1-2 inches long, its ovoid, nonresinous bud 3/4-1 inch long, and its 3-4 inch cone with its spiny, long-exserted bracts.

Redwood

a. Fruiting branch, x 2/3.
b. Seed, x 2/3.
c. Cone, x 1.

Redwood

Sequoia sempervirens (D. Don) Endl.

HABIT. This tree reaches greater heights than any in the world, commonly rising 200-275 feet, and is 8-12 feet in diameter (max. 372 by 20 feet); crown short and narrowly conical; bole clear, straight; often tight circles of young trees found around old stumps due to sprouts.

LEAVES. Spirally arranged; flattened; decurrent; on lower branches 1/2-1 inch long, linear, appearing 2-ranked, acute tips, short petioles; on leaders and fertile branches 1/4-1/2 inch long, awl- to needle-shaped, in several ranks; dark yellow-green above, two bands of stomata below; persistent several years, falling with twigs.

FLOWERS. Monoecious; solitary; male oblong, of several spirally arranged bracts containing stamens; female of 15-20 spirally arranged, peltate scales, each bearing 3-12 erect ovules; ovuliferous bract and scale partially fused.

FRUIT. Woody, ovoid, pendent cone; 3/4-1 inch long; red-brown; wrinkled, peltate scales; maturing in one season. Seed: 1/16 inch long; light brown; laterally 2-winged with wings as broad as seed; 2-9 seeds on each scale.

TWIG. Slender; greenish; smooth. Winter buds: small; globose; covered by numerous, imbricated, acute scales.

BARK. 6-12 inches thick; fibrous; red-brown to cinnamon-brown; deeply furrowed.

WOOD. Important; light clear red weathering to dark red; soft and weak; very durable; used for lumber, dimension stock, tanks. A single tree has yielded 480,000 bd ft.

SILVICAL CHARACTERS. Very tolerant; growth rapid; maturity reached in 400-1,800 years; reproduces vigorously by stump sprouts as well as seeds; deep, wide-spreading, lateral root system; remarkably free from enemies.

HABITAT. Restricted to the fog belt; the dominant species in this narrow belt; in pure stands or mixed with Douglas-fir, Sitka spruce, grand fir, western hemlock, and hardwoods.

* * *

Dawn Redwood, *Metasequoia glyptostroboides* Hu & Cheng, long thought extinct, was found in China in 1941 and is now widely planted. The foliage resembles redwood but is deciduous; cones and needles are longer and narrower.

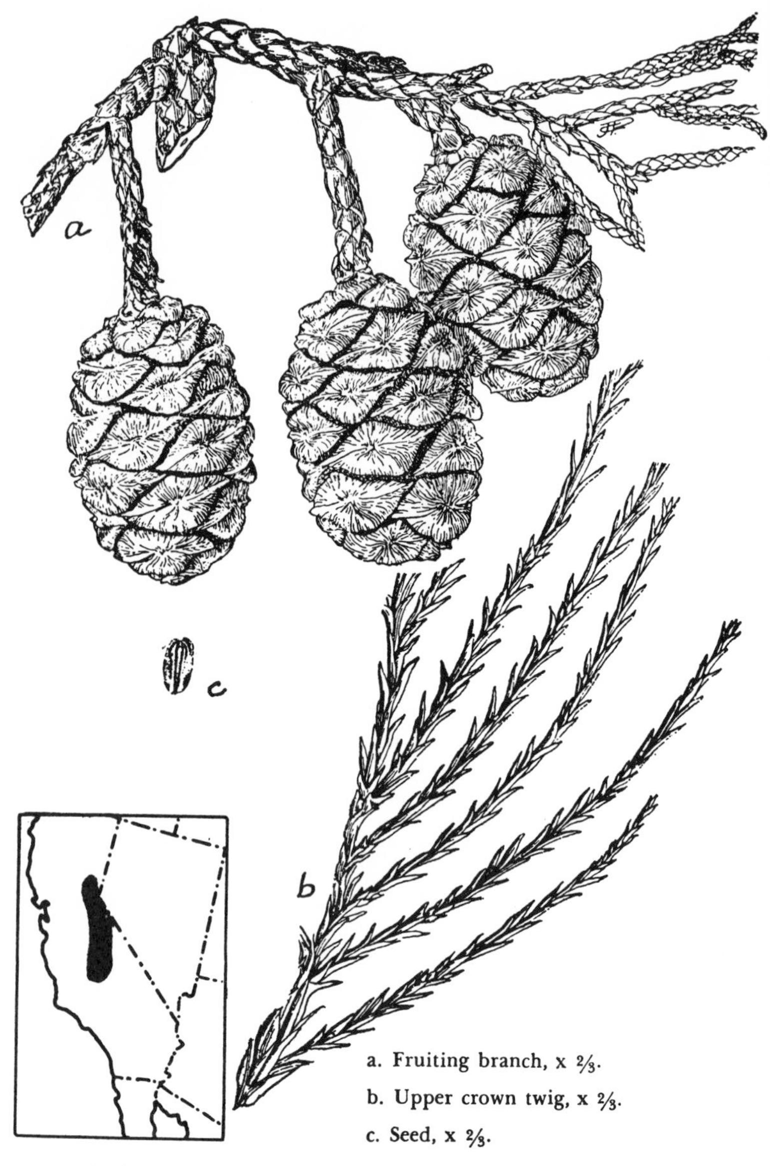

a. Fruiting branch, x ⅔.

b. Upper crown twig, x ⅔.

c. Seed, x ⅔.

TAXODIACEAE

Giant Sequoia. Bigtree

Sequoiadendron giganteum (Lindl.) Buckholz
[*Sequoia gigantea* (Lindl.) Decne.]

HABIT. The most massive and possibly the oldest of all living things; commonly reaching a height of 250-280 feet and a diameter of 10-15 feet (max. 293 by 37 feet); crown scraggly and open; bole clear with short thick branches.

LEAVES. Spirally arranged; 1/4-1/2 inch long; ovate to lanceolate; appressed or spreading, but thickly clothing the twig; rigid; sharp-pointed; decurrent at base; blue-green; turning brown in 2 to 3 years, but persisting for several years.

FLOWERS. Similar to redwood, but female with 25-40 scales.

FRUIT. Woody, ovoid-oblong, pendent cones; 2-3 1/2 inches long; red-brown; peltate, wrinkled scales; reaching full size first year, but not maturing until second year; unique in that seed may be retained in cone up to 20 years, while peduncle lives and grows and cone looks like young cone. Seed: 1/4 inch long; light brown; laterally 2-winged, with wings broader than seed; 2-9 seeds on each scale.

TWIG. Slender; leaf-covered. Winter buds: small; naked.

BARK. 12-24 inches thick; fibrous cinnamon-red; furrowed between broad, rounded ridges.

WOOD. Unimportant; similar to redwood, but brittle; impractical to log because of size.

SILVICAL CHARACTERS. Intermediate in tolerance; growth rapid; attaining ages of 4,000-5,000 years; reproduction sparse, mineral soil needed for seed germination; tree does not sprout; wide-spreading lateral root system; no natural enemies aside from lightning and fire; no tree is known to have died from old age, insects, or fungal attack.

HABITAT. Native only in some 32 groves of varying extent found along the western middle slopes of the Sierra Nevada of California; in former geologic periods this genus was widely scattered through the forests of the Northern Hemisphere. Generally associated with sugar, ponderosa, and Jeffrey pines; white and red fir; and incense-cedar.

Baldcypress

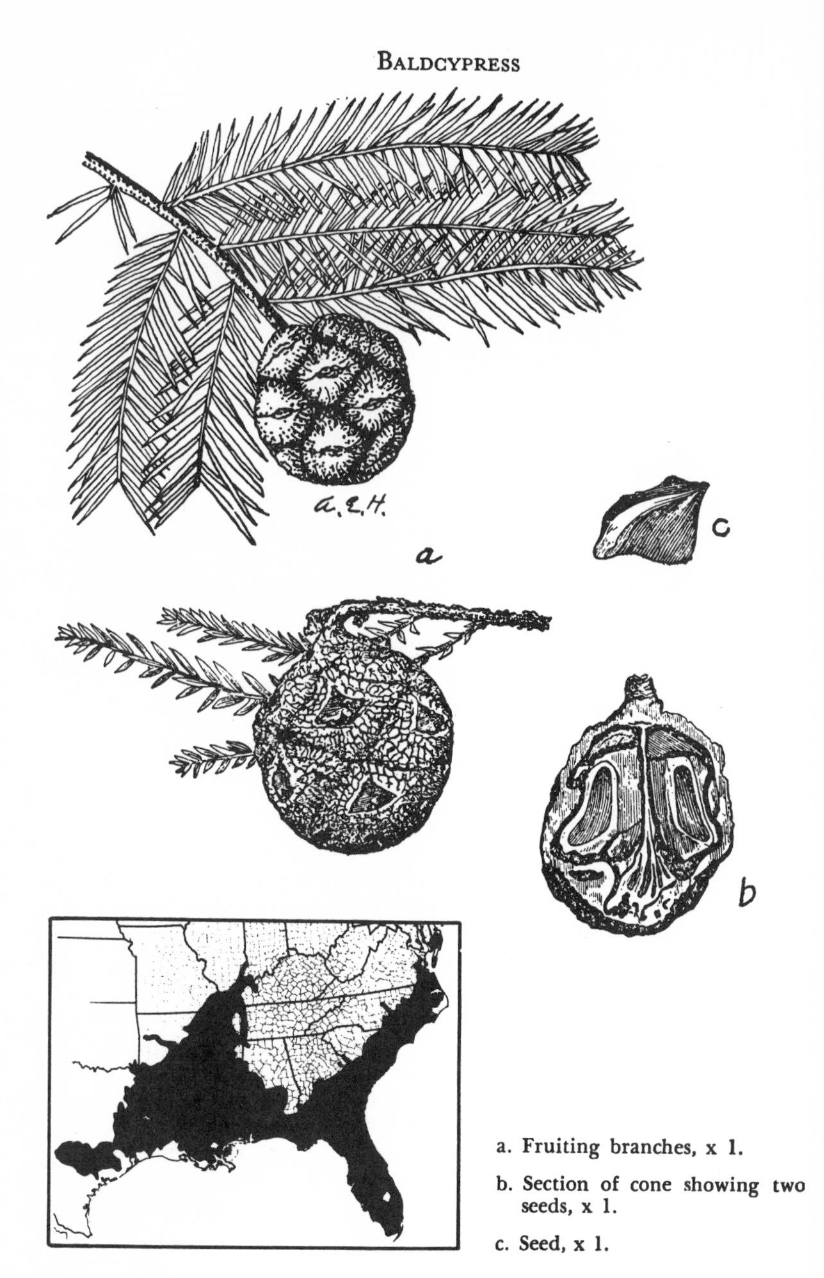

a. Fruiting branches, x 1.

b. Section of cone showing two seeds, x 1.

c. Seed, x 1.

TAXODIACEAE

Baldcypress

Taxodium distichum (L.) Rich.

HABIT. A large and important tree, 100-120 feet high and 3-5 feet in diameter (max. 150 by 17 feet); young crown narrowly pyramidal, becoming irregular and flattened with age; bole tapered with fluted butt; peculiar conical structures known as knees arise from roots on wetter sites.

LEAVES. Spirally arranged; deciduous; lateral twigs falling, with attached leaves; on lower branches 1/2-3/4 inch long, linear, appearing 2-ranked; on fertile branches about 1/2 inch long, appressed, and nearly scalelike; yellow-green.

FLOWERS. Monoecious; male cones in drooping panicles, each composed of 6-8 stamens; female subglobose, of several spirally arranged, peltate scales each bearing 2 erect ovules.

FRUIT. Woody, subglobose, pendent cone; 3/4-1 inch in diameter; brown; of several wrinkled, peltate scales; maturing in one year and usually disintegrating at maturity. Seed: 1/4 inch long; brown; irregularly 3-angled and 3-winged; 2 seeds to each scale.

TWIG. Terminal twigs bearing axillary buds and persistent; lateral twigs deciduous, with needles still attached. Winter buds: small; subglobose; several imbricated scales.

BARK. Variable; thin and scaly to fibrous and 1 1/2 inches thick; red-brown to ash-gray.

WOOD. Important; variegated from light to dark brown; moderately heavy, hard, and strong; very durable; used for construction lumber, siding, caskets, shingles, etc.

SILVICAL CHARACTERS. Intolerant; growth rather slow on wet sites; attains age of 1,000-2,000 years; reproduces well from seed, and stump sprouts vigorously; shallow, wide-spreading roots; surprisingly windfirm even on wet, unstable soils; wood of old trees riddled with pecky rot.

HABITAT. Typical in swamps; in pure stands or with water tupelo, sweetgum, and other bottomland hardwoods.

* * *

A closely related, unimportant, smaller tree, *Taxodium distichum* var. *nutans* (Ait.) Sweet, the pondcypress, differs from baldcypress in having smaller scalelike to needle-shaped leaves that are closely appressed. This is found in bogs and along ponds from southern Virginia to Louisiana. *Taxodium mucronatum* Ten., the Montezuma baldcypress of Mexico, which enters southern Texas, differs from other species in having persistent needles.

California Incense-cedar

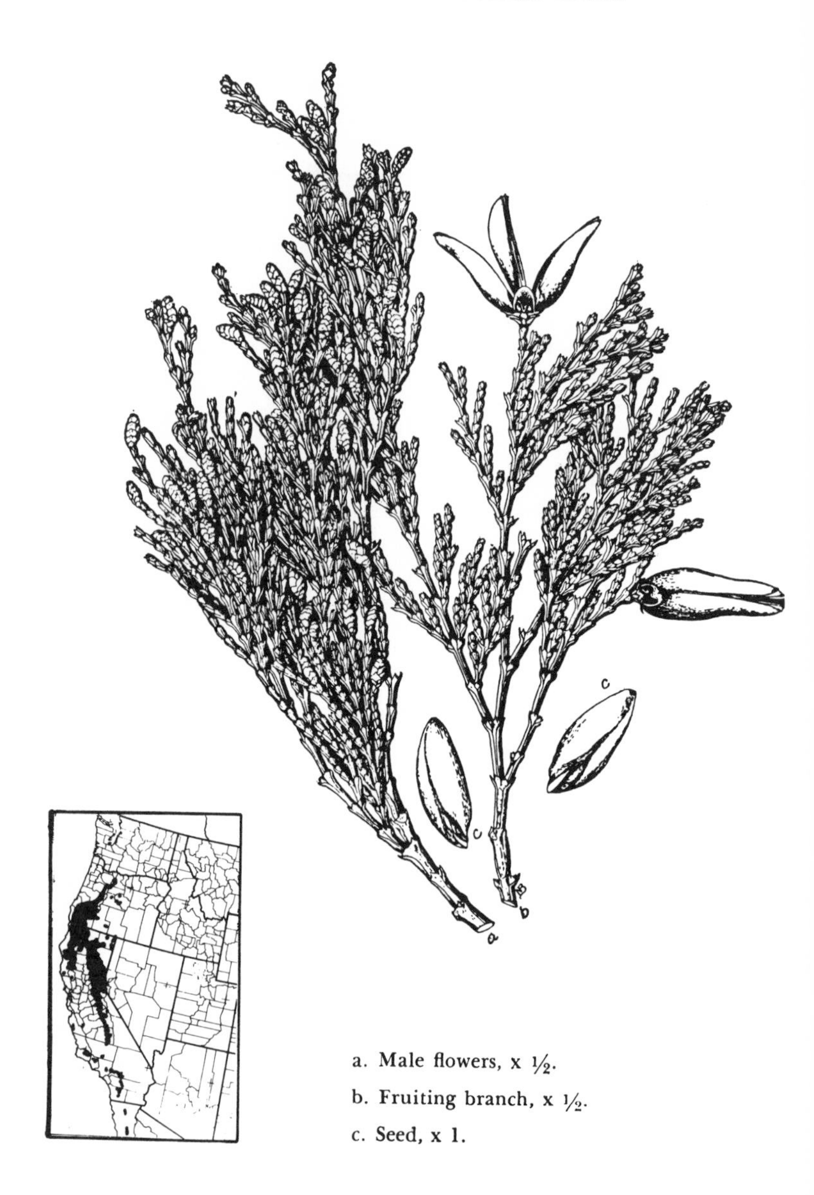

a. Male flowers, x ½.

b. Fruiting branch, x ½.

c. Seed, x 1.

CUPRESSACEAE

Incense-cedar

Libocedrus decurrens Torr.

HABIT. A medium-sized tree 80-120 feet high and 3-4 feet in diameter (max. 186 by 8 feet); tapering, often fluted trunk covered for up to half its length with lustrous, irregular foliage.

LEAVES. Small; scalelike; whorled in 4's; facial leaves flattened; the lateral keeled and almost ensheathing facial leaves; oblong-ovate; 1/8-1/2 inch long; persistent 3-5 years; glandular; aromatic when crushed; fronds usually in a vertical plane.

FLOWERS. Monoecious; terminal; male oblong, golden, with 12-16 decussate 4-celled anthers; female oblong, yellow-green, of 6 scales, the inner 2 each bearing 2 erect ovules.

FRUIT. Leathery, pendent cones; 3/4-1 1/2 inches long; oblong; 6-scaled, with 2 becoming greatly enlarged and spreading at maturity; maturing in one season. Seed: 1/3-1/2 inch long; in pairs on fertile scales; unequally laterally winged; straw-colored.

TWIGS. Slender; flattened; leaf-covered; in long sprays that are more often in a vertical plane than in a horizontal plane. Winter buds: minute; naked; inconspicuous.

BARK. Thin; smooth to scaly; gray-green on young stems; becoming 3-8 inches thick, yellow-brown to cinnamon-red, fibrous, and deeply furrowed on old trees.

WOOD. Chief source of pencil stock; would be highly important for many uses except for prevalence of pecky rot; sapwood white; heartwood reddish, soft, fragrant.

SILVICAL CHARACTERS. Tolerant; growth rather slow; maturity reached in about 300 years; reproduction abundant and vigorous; moderately deep, lateral root system; fire and pecky rot caused by *Polyporus amarus* cause extensive damage.

HABITAT. Transition zone; altitudinal range 1,000-9,000 feet; on cool, moist soils. In mixed stands with sugar, ponderosa, Jeffrey, and white pine; white fir; and Douglas-fir.

Western Redcedar
and
Northern White-cedar

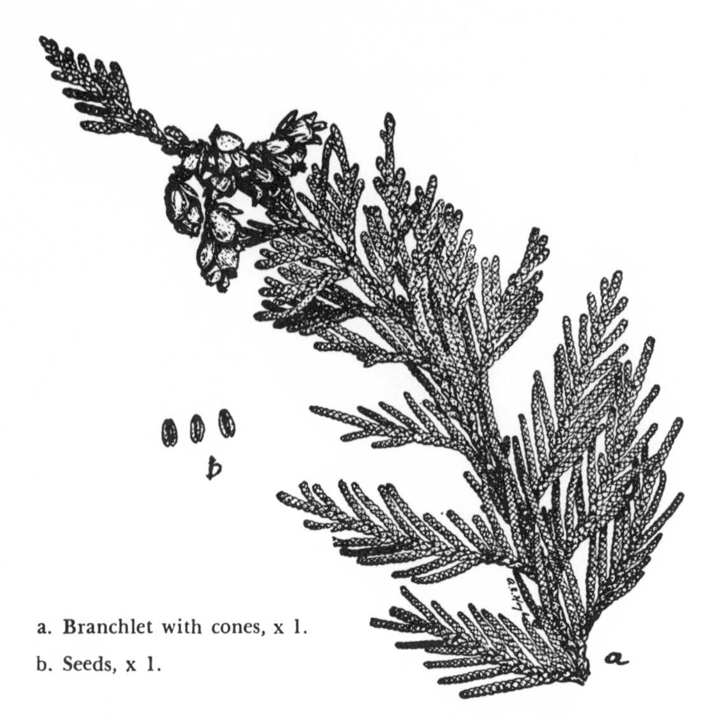

a. Branchlet with cones, x 1.

b. Seeds, x 1.

Western Redcedar

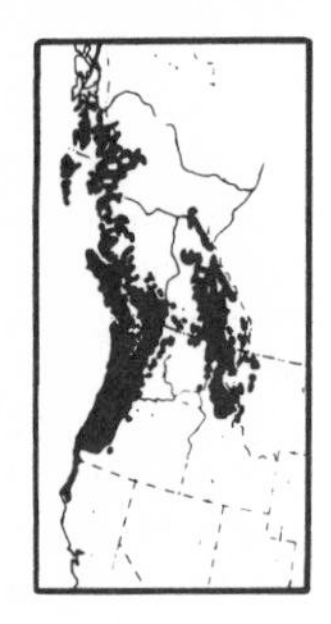

Northern White-cedar

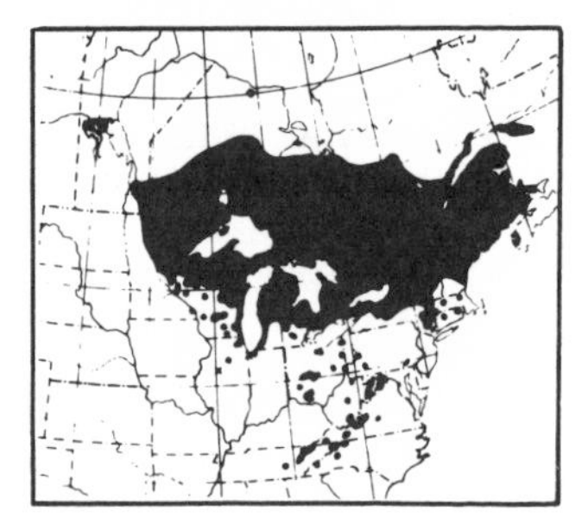

CUPRESSACEAE

Western Redcedar. Giant Arborvitae

Thuja plicata Donn

HABIT. A large tree, 150-200 feet high and 4-8 feet in diameter (max. 250 by 16 feet); tapering, fluted trunk; irregular crown of horizontal or drooping branches.

LEAVES. Small, scalelike; persistent 2-5 years; decussate; the facial leaves flattened, grooved; the lateral leaves rounded or keeled; lustrous, dark yellow-green.

FLOWERS. Monoecious; terminal; dark brown; male with 3-6 pairs of decussate stamens; female 8-12-scaled.

FRUIT. Leathery or subwoody erect cones; 1/2 inch long; ovoid-oblong; maturing in one season; scales 8-12 (only 6 being fertile), thin, spine-tipped. Seed: 1/8 inch long; brown; lateral wings each about as wide as seed.

TWIGS. Slender; flattened; leaf-covered in long drooping sprays. Winter buds: minute; naked; inconspicuous.

BARK. Thin (1/2-1 inch); fibrous; narrow interlacing ridges; cinnamon-red to gray-brown on old trunks.

WOOD. Widely used where durability rather than strength is required; sapwood white; heartwood reddish, soft, fragrant; shingles, siding, interior trim, boats, poles, etc.

SILVICAL CHARACTERS. Tolerant; growth rather rapid; trees over 1,000 years old reported; reproduction generally plentiful; shallow, wide-spreading root system; fire and pecky heart rot cause serious damage.

HABITAT. Transition and Canadian zones; altitudinal range 2,000-7,000 feet in Rocky Mountains; on rich soils with abundant moisture; in mixed coniferous stands.

* * *

Northern White-cedar. Eastern Arborvitae

Thuja occidentalis L.

Aside from its eastern distribution, this species differs from *Thuja plicata* by only minor characters, the cones having but 4 fertile scales and the leaves being dull green and usually glandular-pitted. This small tree grows on a variety of sites and is typical of swampy areas. It is not important, although used to some extent for posts and poles. The oriental arborvitae, *T. orientalis* L., is widely planted and perhaps naturalized. It is easily distinguished from the native species by the vertical disposition of the leaf sprays, the thick cone scales, and the wingless seed.

CUPRESSACEAE

The Cypresses and White-cedars

Characteristics of the Genera *Cupressus* L. and *Chamaecyparis* Spach

HABIT. Evergreen, resinous, pyramidal trees (sometimes shrubs); clear tapering trunks; dense crowns of stout, erect, or horizontal branches.

LEAVES. Small; scalelike (often awl-shaped on leading shoots); ovate; slender, spreading, or appressed tip; thickened, rounded, often glandular on the back; persistent 3-6 years; decussate (in pairs alternately crossing at right angles); becoming brown and woody before falling; margins smooth to finely serrate.

FLOWERS. Monoecious; minute; terminal; male oblong of numerous decussate stamens each bearing 2-6 subglobose anthers; female of 4-14 peltate decussate scales, each with 2-20 erect, basal ovules; bract and ovuliferous scale wholly fused.

FRUIT. Subglobose, leathery to woody, erect cone; maturing in 1-2 seasons; cone scales peltate, thick, often with a central boss or mucro that is a remnant of flower scales. Seed: 1-20 in several rows on each scale; erect; lateral wings.

TWIGS. Slender; quadrangular or flattened; leaf-covered. Winter buds: minute; naked; inconspicuous.

WOOD. Strongly aromatic; light brown to yellow heartwood; pale yellow sapwood; very durable.

SILVICAL CHARACTERS. Tolerant; usually slow-growing.

GENERAL. The genus *Chamaecyparis* is considered by many authors to be but a section of the genus *Cupressus*, and because of their similarity they are described together here. The chief differences are described in the key on the next page. Six species of *Cupressus* and three of *Chamaecyparis* are native to the United States. Italian cypress and two Asiatic species of *Chamaecyparis* are commonly planted ornamentals and are included in the key.

KEY TO THE SPECIES OF CUPRESSUS AND CHAMAECYPARIS

1. Cone woody, 6-16-scaled, 1/2-1 1/2 inches diameter, ripening second year, 6-20 seeds per scale; branchlets round or 4-angled; leaves fringed, narrow translucent border visible with lens . . . . CUPRESSUS, The CYPRESS
   2. Leaves glaucous, pale blue to silvery gray-green; cone 6-8-scaled, 3/4-1 1/4 inches diameter.
      3. Leaves acute, glandular; twigs gray; Arizona to Texas and Mexico. . . . . *Cupressus arizonica* Greene, Arizona Cypress, p. 101
      3. Leaves obtuse; twigs bright red; southern and lower California, Mexico . . . . *Cupressus guadalupensis* S. Wats., Tecate Cypress
   2. Leaves dark or bright green.
      4. Leaves conspicuously glandular-pitted on back; cone 6-8-scaled, 1/2-1 inch diameter.
         5. Cones with prominent hornlike umbos; rare; Oregon, California . . . . *Cupressus macnabiana* A. Murr., McNab Cypress
         5. Cones with short, conical umbos; southern Oregon, northern California . . . . *C. bakeri* Jeps., Modoc Cypress
      4. Leaves without glands or obscurely glandular.
         6. Cones 6-8-scaled, 1/2-3/4 inch diameter; California (Mendocino County) . . . . *Cupressus goveniana* Gord., Gowen Cypress
         6. Cones 8-14-scaled, 7/8-1 1/2 inches diameter.
            7. Branchlets and boss of cone scales thick; California . . . . *Cupressus macrocarpa* Hartw., Monterey Cypress
            7. Branchlets and boss of cone scales thin; planted ornamental . . . . *Cupressus sempervirens* L., Italian Cypress
1. Cone leathery to semifleshy, 4-10-scaled, 1/4-1/2 inch diameter, ripening in 1 or 2 years, 2 seeds (rarely 5) per scale; twigs somewhat flattened; leaves entire . . . . CHAMAECYPARIS, The WHITE-CEDAR
   8. Branchlets not arranged in conspicuous horizontal planes; leaves green on both sides; bark on mature trees thin; cones with 4-6 scales; lateral and facial leaves nearly equal in size.
      9. Branchlets slender; leaves about 1/16 inch long, usually glandular; cones 1/4 inch diameter, maturing in 1 year, 1-2 seeds per scale; heartwood light brown; Atlantic Coast from Maine to Mississippi . . . . *Chamaecyparis thyoides* (L.) B.S.P., Atlantic White-cedar, p. 105
      9. Branchlets stout; leaves about 1/8 inch long, often without glands; cones 1/4-1/2 inch diameter, maturing in 2 years, 2-4 seeds per scale; heartwood clear yellow; Pacific Northwest . . . . *Chamaecyparis nootkatensis* (D. Don) Spach, Alaska-cedar, p. 105
   8. Branchlets arranged in conspicuous horizontal planes; cones with 10 scales; leaves with glaucous or whitish marks below; heartwood light brown.
      10. Bark on mature trees thick and furrowed; leaves conspicuously glandular on the back and indistinctly marked with white streaks below, lateral much larger than facial; Oregon and California, also widely planted . . . *Chamaecyparis lawsoniana* (A. Murr.) Parl., Port Orford-cedar, p. 103
      10. Bark on mature trees thin; leaves without glands or glands indistinct, but with distinct white markings below; Japanese ornamentals widely planted in the United States.
          11. Leaves closely appressed, obtuse to acute, the lateral much larger than the facial, conspicuous Y-shaped white mark below; cone 1/3-1/2 inch diameter . . . . *Chamaecyparis obtusa* (S. & Z.) Endl., Hinoke White-cedar, p. 103
          11. Leaves loosely appressed, acuminate, the lateral slightly larger than the facial, conspicuous white patches below; cone 1/4 inch diameter *Chamaecyparis pisifera* (S. & Z.) Endl., Sawara White-cedar, p. 103

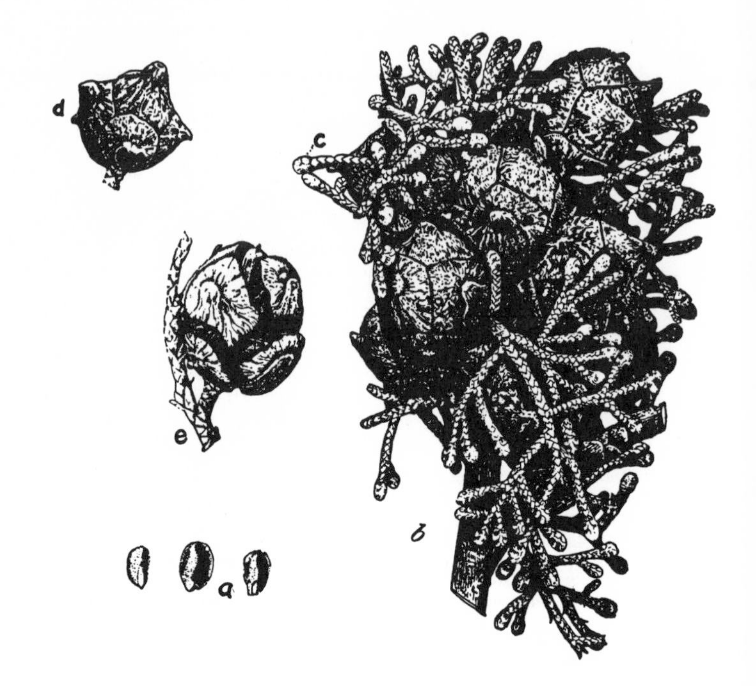

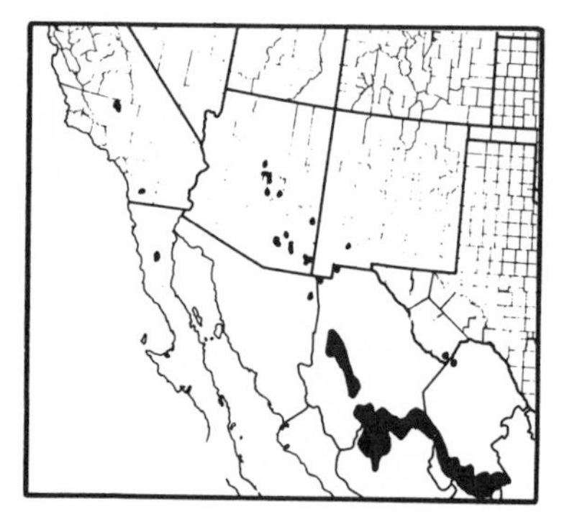

a. Seeds, x ¾.
b. Branchlet with cones, x ¾.
c. Male flower bud in autumn.
d. Different form of cone, x ¾.
e. Opened cone, x ¾.

CUPRESSACEAE

Arizona Cypress

Cupressus arizonica Greene

HABIT. Under favorable conditions a tree 50-60 feet high and 1-2 1/2 feet in diameter (max. 80 by 4 feet); trunk short, limby, sharply tapering; crown dense, sharply conical.

LEAVES. Scalelike, pointed, 1/16 inch long; silvery gray-green; dying and turning red-brown the second year and falling about 4 years later; commonly without glands or pits on back; giving off a skunklike odor when bruised.

FLOWERS. Male oblong, yellow; female subglobose.

FRUIT. 3/4-1 inch in diameter; subglobose; dark red-brown on stout stalks; of 6-8 peltate scales, each with stout, incurved, prominent boss; maturing during second summer and remaining on tree many years. Seed: 6-20 on each scale; 1/16-1/8 inch long; oblong to triangular; deep red-brown; thin, narrow, lateral wings.

TWIGS. 4-angled; dark gray; loose-scaly bark, with smooth, reddish inner bark visible below. Winter buds: minute; inconspicuous; naked.

BARK. Loose-scaly on young trunks and branches, showing smooth reddish inner bark below; on old trunks 1 1/4 inches thick, fibrous, deeply furrowed, ridged, dark red-brown.

WOOD. Slightly aromatic; durable; heartwood light brown; sapwood straw-colored; soft and light, splitting easily; used locally for fence posts, mine timbers, etc.

SILVICAL CHARACTERS. Tolerant throughout life; growth slow; trees seldom over 400 years old; reproduction generally scanty, although seed produced every year.

HABITAT. Transition zone; altitudinal range 4,500-8,000 feet; best growth on moist, gravelly, north slopes and benches but hardy on dry, sterile, rocky sites.

* * *

Five similar but unimportant species are native to the Pacific Coast. These are separated in the key on page 99. The distinctions between these are minor and it is often difficult to identify the species. These species as well as the Italian cypress are commonly planted as ornamentals in warmer sections of the United States.

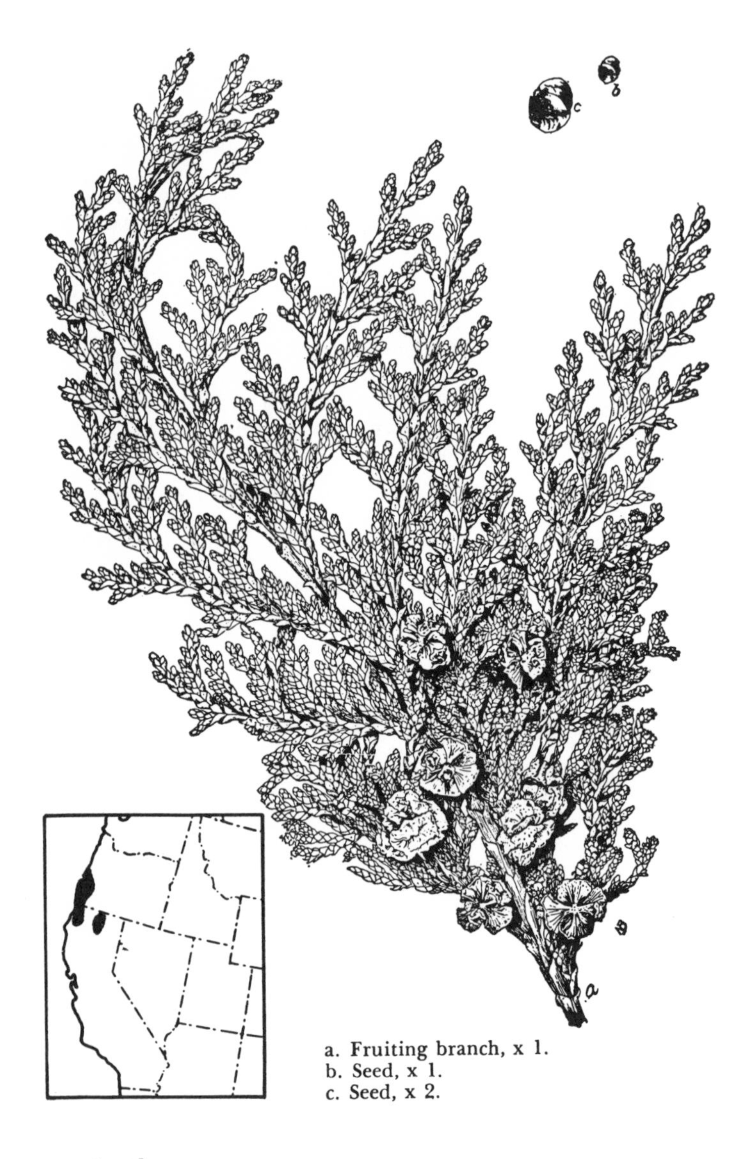

a. Fruiting branch, x 1.
b. Seed, x 1.
c. Seed, x 2.

CUPRESSACEAE

Port Orford-cedar

Chamaecyparis lawsoniana (A. Murr.) Parl.

HABIT. A large, handsome tree 140-180 feet high and 4-6 feet in diameter (max. 225 by 16 feet); bole clear and often buttressed; crown short-conical, with many branches.

LEAVES. Scalelike; blunt; 1/16 inch long; entire; yellow-green to blue-green; turning brown after 2-3 years, but persisting for several years; glandular; forming flat, horizontal, feathery sprays.

FLOWERS. Male oblong, bright red; female subglobose.

FRUIT. 1/3 inch in diameter; globose; red-brown; abundant; of 6-8 decussate, peltate scales, each with thin, acute, reflexed boss; maturing in 1-2 years; persistent on twig after seed released. Seed: 2-5 on each scale; 1/8 inch long; ovoid; chestnut-brown; 2 broad lateral wings.

TWIGS. Slender; flattened; leaf-covered. Winter buds: minute; naked; inconspicuous.

BARK. 6-10 inches thick; red-brown to silver-brown; fibrous and furrowed on old trunks; young bark thin and scaly.

WOOD. Important; aromatic; very durable; pale brown; light; easily worked; used for battery separators, venetian blinds, boats, aircraft; cut far exceeds growth.

SILVICAL CHARACTERS. Moderately tolerant; growth moderate; maturity reached in 300-350 years, but often living for more than 500 years; reproduction aggressive; free from insect and fungal enemies, though damaged by fire.

HABITAT. Altitudinal range from sea level to 5,000 feet; requires abundant soil and atmospheric moisture; in pure stands or mixed with numerous conifers.

* * *

Two Japanese species are widely planted as ornamentals:

1. The Sawara white-cedar, *Chamaecyparis pisifera* (S. & Z.) Endl., is a narrow, pyramidal tree of fairly rapid growth, easily identified by its acuminate leaves that are obscurely glandular and dark green above and have conspicuous white lines below. The cones are 1/4 inch or less in diameter and consist of 10-12 scales.

2. The Hinoki white-cedar, *Chamaecyparis obtusa* (S. & Z.) Endl., is a tree with broad, pyramidal crown, characterized by obtuse leaves that are dark green above and not glandular and with conspicuous white lines below. The cones are 1/3-1/2 inch in diameter and consist of 8-10 scales.

Alaska Yellow-cedar

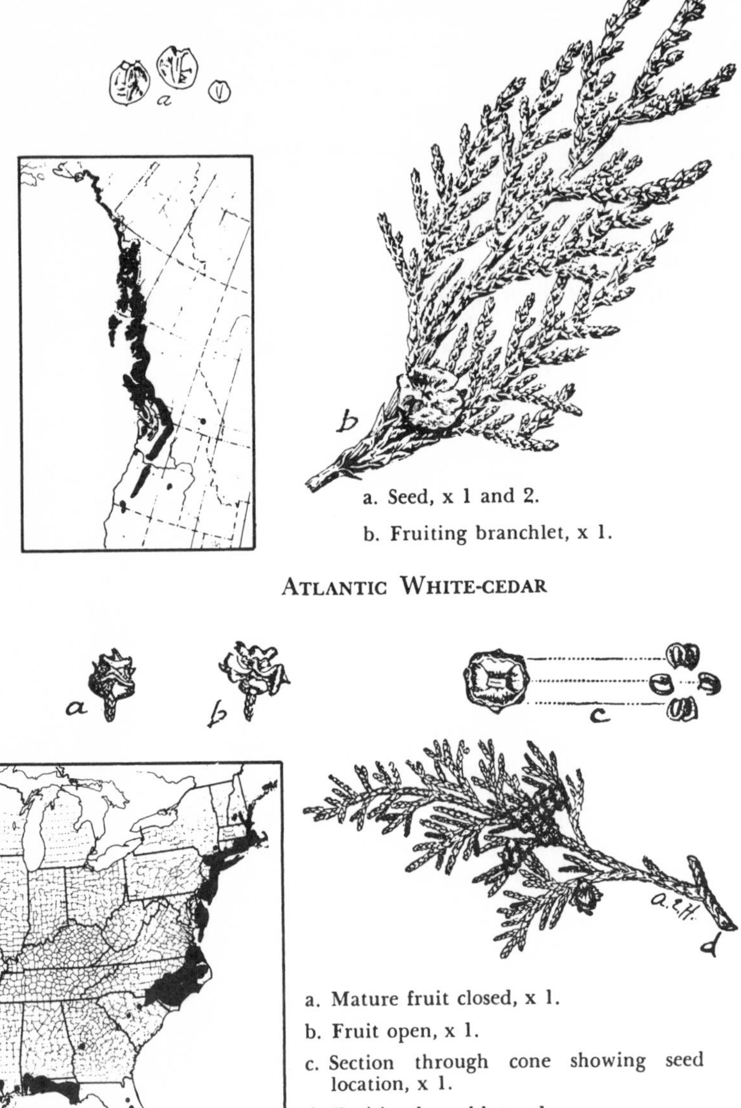

a. Seed, x 1 and 2.

b. Fruiting branchlet, x 1.

Atlantic White-cedar

a. Mature fruit closed, x 1.

b. Fruit open, x 1.

c. Section through cone showing seed location, x 1.

d. Fruiting branchlet, x 1.

Alaska-cedar. Yellow-cedar

Chamaecyparis nootkatensis (D. Don) Spach

HABIT. A medium-sized tree 60-90 feet high and 2-3 feet in diameter (max. 130 by 6 feet); clear, buttressed, often fluted bole; crown conical, of drooping branches.

LEAVES. Scalelike; acute; 1/8 inch long; blue-green to gray-green; persistent 2-3 years; usually without glands; sprays appearing limp and wilted.

FLOWERS. Male bright yellow; female subglobose, brown.

FRUIT. 1/4-1/2 inch in diameter; globose; purple to red-brown; of 4-6, decussate, peltate, bossed scales; maturing in 2 years. Seed: 2-4 on each scale; 1/4 inch long; broad wings.

TWIGS. Rather stout; pendulous. Winter buds: minute.

BARK. Thin; gray-brown; fibrous; furrowed on old trunks; young bark thin and scaly.

WOOD. Moderately important; durable; heartwood clear yellow; used for boats, trim, blinds, etc.

SILVICAL CHARACTERS. Intermediate in tolerance; growth slow; ages over 600 years attained; reproduction sparse; lateral roots.

HABITAT. Sea level (Alaska) to timberline; moist sites; pure stands or mixed with many conifers.

* * *

Atlantic White-cedar. Southern White-cedar

Chamaecyparis thyoides (L.) B.S.P.

This tree of the Atlantic seaboard differs from other native species by the following characters:

HABIT. Small- to medium-sized (max. 120 by 5 feet); bole clear, cylindrical; crown conical, with drooping branches.

LEAVES. 1/16-1/8 inch long; blue-green; keeled and glandular.

FRUIT. 1/4 inch in diameter; globose; purple to red-brown; reflexed boss; somewhat fleshy; maturing in 1 year. Seed: 1-2 on each scale; 1/8 inch long.

BARK. Thin; gray to red-brown; fibrous; furrowed.

WOOD. Similar to northern white-cedar; used for posts and poles.

SILVICAL CHARACTERS. Moderately tolerant; growth slow; reproduction vigorous; shallow, lateral roots.

HABITAT. Swamps and bogs in the coastal plains; pure stands or mixed with many species.

CUPRESSACEAE

The Junipers

Characteristics of the Genus *Juniperus* L.

HABIT. Evergreen, aromatic small trees or shrubs.

LEAVES. Persistent for several years; sessile; aromatic; needlelike or awl-shaped on young growth; of three types on older growth—(1) ternate, spreading, and entirely needlelike or subulate; (2) decussate, appressed, decurrent, and entirely scalelike; (3) a combination of the preceding types.

FLOWERS. Dioecious (rarely monoecious); minute and inconspicuous; male yellow, solitary, of numerous ternate or decussate stamens; female of 3-8 decussate or ternate pointed scales, some or all bearing 1 or 2 ovules.

FRUIT. A berrylike, succulent, indehiscent cone, formed by coalescence of flower scales; subtended by persistent flower bracts; maturing in 1-3 years. Seed: 1-21 in cone; ovoid; unwinged; marked at base by a scar (hilum).

BUDS. Small; naked; covered by leaves (scaly in dwarf juniper).

BARK. Thin; soft; fibrous; shreddy (brittle, thick, and divided into nearly square plates in alligator juniper).

WOOD. Durable; weak; close-grained; aromatic; heartwood red-purple to brown; sapwood whitish.

SILVICAL CHARACTERS. Exceedingly variable; mostly intolerant; slow-growing; long-lived; seed disseminated by birds or mammals; deep lateral roots.

GENERAL. This genus contains 40-60 species widely scattered through the Northern Hemisphere, with 13 species native to the United States. Only *J. virginiana* is of any commercial importance, although several other species form conspicuous parts of the vegetation. Two appear as small shrubs:

(1) The dwarf juniper, *Juniperus communis* L., while circumpolar and extending through most of the United States and Canada, attains tree size only in New England and southern Illinois. This species has ternate leaves 1/3-1/2 inch long that are all needlelike, scaly buds, and axillary flowers; the prostrate, high-mountain form has been designated *J. communis* var. *montana* Ait. (*J. communis* var. *sibirica* Rydb.).

(2) The prostrate juniper, *J. horizontalis* Moench, entering the United States from the north, differs from dwarf juniper in having closely appressed, decussate leaves, 1/8-1/4 inch long. This species often forms dense creeping mats.

KEY TO THE SPECIES OF JUNIPERS

1. Leaves all needlelike or awl-shaped, 1/3 to 1/2 inch long, ternate; bud scale, 1/8 inch long; usually a small shrub, rarely a small tree . . . . *J. communis*, Common Juniper, p. 106
1. Leaves on mature branches usually scalelike, 1/8 inch long or less, opposite (on young or vigorous shoots often needlelike); buds indistinct, naked.
   2. Trunk bark in thick, squarish plates; fruit red-brown, 1/2 inch diameter, usually 4-seeded, ripening in 2 years; heartwood brown; southwest . . . . *J. deppeana*, Alligator Juniper, p. 113
   2. Trunk bark fibrous and shreddy.
      3. Fruit bright red to red-brown beneath, whitish bloom; western.
         4. Fruit bright red (rarely copper-colored), 1/4 inch diameter, 1-seeded; heartwood brown; western Texas . . . . *J. pinchotii*, Pinchot Juniper, p. 113
         4. Fruit dull red-brown or copper-colored; heartwood brown.
            5. 4-12 seeds per fruit; drooping branches; southern Texas, Mexico . . . . *J. flaccida*, Drooping Juniper, p. 113
            5. 1 (rarely 2) seeds per fruit; branches seldom drooping.
               6. Fruit 1/4-3/4 inch diameter; seed completely enclosed; fruit maturing in 2 years; heartwood brown.
                  7. Leaves in whorls of 3's, rarely opposite; seed marked at base by light-colored hilum; California . . . . *J. californica*, California Juniper, p. 115
                  7. Leaves opposite, rarely in 3's; seed marked to middle by conspicuous hilum; Rocky Mountains . . . . *J. osteosperma*, Utah Juniper, p. 113
               6. Fruit 1/8-1/3 inch diameter; seed marked at base by hilum; fruit maturing in 1 year; Nevada to Colorado and south to Texas and Arizona . . . . . . *J. monosperma*, One-seed Juniper, p. 111
      3. Fruit bluish to blue-black, 1-3-seeded.
         8. Leaf margin smooth; heartwood distinctly reddish.
            9. Fruit maturing in 2 years; western. . . . . *J. scopulorum*, Rocky Mountain Juniper, p. 111
            9. Fruit maturing in 1 year; eastern.
               10. Fruit 1/24-1/12 inch diameter; drooping branches; South Carolina to eastern Texas . . . . *J. silicicola*, Southern Redcedar, p. 109
               10. Fruit 1/4-1/3 inch diameter; branches erect or spreading . . . . *J. virginiana*, Eastern Juniper, p. 109
         8. Leaf margin minutely fringe-toothed (under hand lens); heartwood brownish.
            11. Fruit with 2-3 thick-shelled seeds; maturing in 2 years; Northwest and Pacific Coast . . . . *J. occidentalis*, Western Juniper, p. 115
            11. Fruit and 1-2 thin-shelled seeds, maturing in 1 year.
               12. Seed 1 (rarely 2), pale brown; leaves gray-green; Nevada to California and south . . . . *J. monosperma*, One-seed Juniper, p. 111
               12. Seed 1-2; dark brown; leaves blue-green; Missouri to Texas and Mexico . . . . *J. ashei*, Ashe Juniper, p. 113

EASTERN REDCEDAR

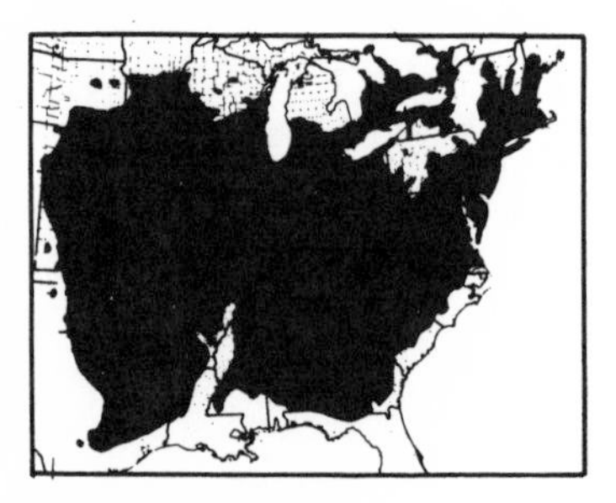

a. Branchlet with awl-shaped leaves, x 1.
b. Tip of branchlet, showing awl-shaped leaves, enlarged.
c. Fruiting branchlet with scalelike leaves, x 1.
d. Tip of branchlet, showing scalelike leaves, enlarged.

CUPRESSACEAE

Eastern Redcedar

Juniperus virginiana L.

HABIT. A small to medium-sized tree 30-40 feet high and 1-2 feet in diameter (max. 100 by 4 feet); crown dense, irregular, pyramidal; numerous ornamental varieties widely planted.

LEAVES. The scalelike leaves decussate, closely appressed, dark green, acute, 1/16 inch long, smooth-margined, glandular; the awl-shaped, juvenile leaves often ternate, sharp-pointed, 1/4-1/2 inch long.

FLOWERS. Dioecious; rarely monoecious.

FRUIT. Glaucous; dark blue; berrylike; 1/4-1/3 inch in diameter; subglobose; maturing in 1 year. Seed: 1-2 (rarely 3-4) in each cone; angled and acute; requiring 2-3 years to germinate.

TWIGS. Slender; 4-angled. Winter buds: naked; minute; leaf-covered.

BARK. Thin (1/8-1/4 inch); red-brown; fibrous, exfoliating into long, narrow strips.

WOOD. Bright red, often streaked with white; aromatic; durable; easily worked; moderately heavy and hard; weak; used for chests, trim, pencils, posts.

SILVICAL CHARACTERS. Intermediate in tolerance; growth slow; maximum age about 300 years; deep, lateral roots; reproduction abundant and vigorous; often disfigured by "cedar apples" caused by *Gymnosporangium juniperi*.

HABITAT. Adapted to a variety of sites but typical on poor, dry soils; in pure stands or mixed with hardwoods such as oaks and hickory.

* * *

Southern Redcedar

Juniperus silicicola (Small) Bailey (*Juniperus lucayana* Britt.)

This species is very similar to eastern redcedar, but can be distinguished by its very small fruit (1/12-1/24 inch in diameter) and very slender, pendulous branches. It also differs from eastern redcedar in that it is found on wet, swampy, coastal plain sites from South Carolina to eastern Texas.

Rocky Mountain Juniper

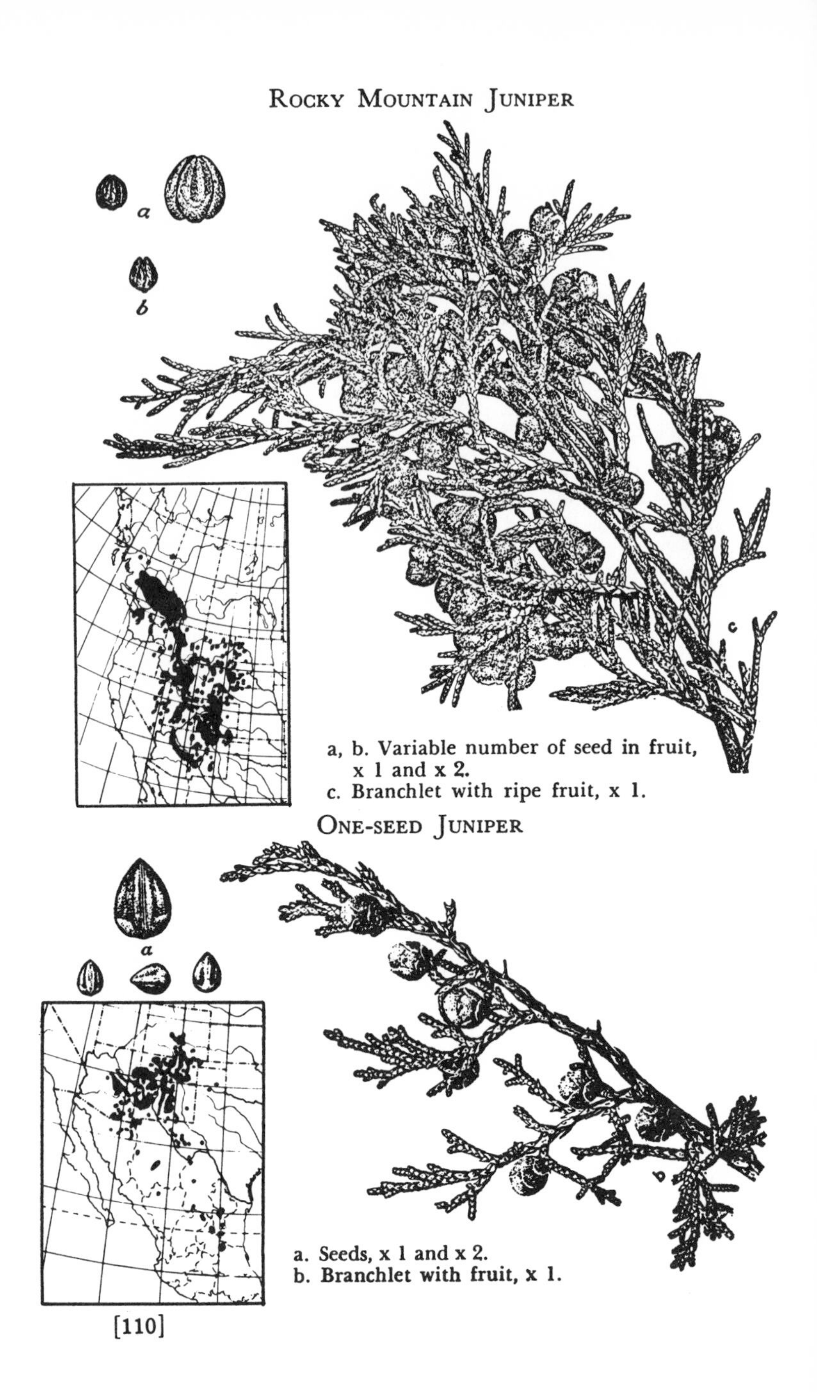

a, b. Variable number of seed in fruit, x 1 and x 2.
c. Branchlet with ripe fruit, x 1.

One-seed Juniper

a. Seeds, x 1 and x 2.
b. Branchlet with fruit, x 1.

Rocky Mountain Juniper

Juniperus scopulorum Sarg.

HABIT. From a bushy shrub on exposed sites to a tree 40-55 feet high and 15-30 inches in diameter; trunk often dividing near the ground; crown typically irregular and rounded.

LEAVES. 1/8 inch long; acute or acuminate; pale to dark green; obscurely glandular on back; smooth margins.

FLOWERS. Dioecious.

FRUIT. Glaucous; blue; 1/4-1/3 inch in diameter; maturing in 2 years. Seed: 1-2 per cone; angled; acute; grooved.

WOOD. Heartwood dull red or bright red and streaked with white; sapwood thick and white; durable.

HABITAT. Upper Sonoran and transition zones; largest distribution of any western juniper; very drouth-resistant; in pure stands or mixed with pinyon, ponderosa pine, etc.

* * *

One-seed Juniper

Juniperus monosperma (Engelm.) Sarg.

HABIT. A small tree, or shrubby as a result of large branches leaving trunk at or below root collar.

LEAVES. 1/8 inch long; acute; gray-green; denticulately fringed; usually glandular.

FLOWERS. Dioecious.

FRUIT. Glaucous; copper-colored (rarely blue); 1/8-1/4 inch long; maturing in 1 year. Seed: 1 (rarely 2) in each cone; ovoid.

WOOD. Heartwood yellow-brown to red-brown; sapwood white; durable; rather heavy and hard; used for posts and fuel.

HABITAT. Upper Sonoran zone; growing on dry, rocky slopes, in pure stands or with pinyon and ponderosa pines.

GENERAL. A related southwestern form with a single seed exposed at the apex and formerly given specific rank as open-seed juniper, *J. gymnocarpa* (Lemm.) Cory, is now considered a form of *J. monosperma*.

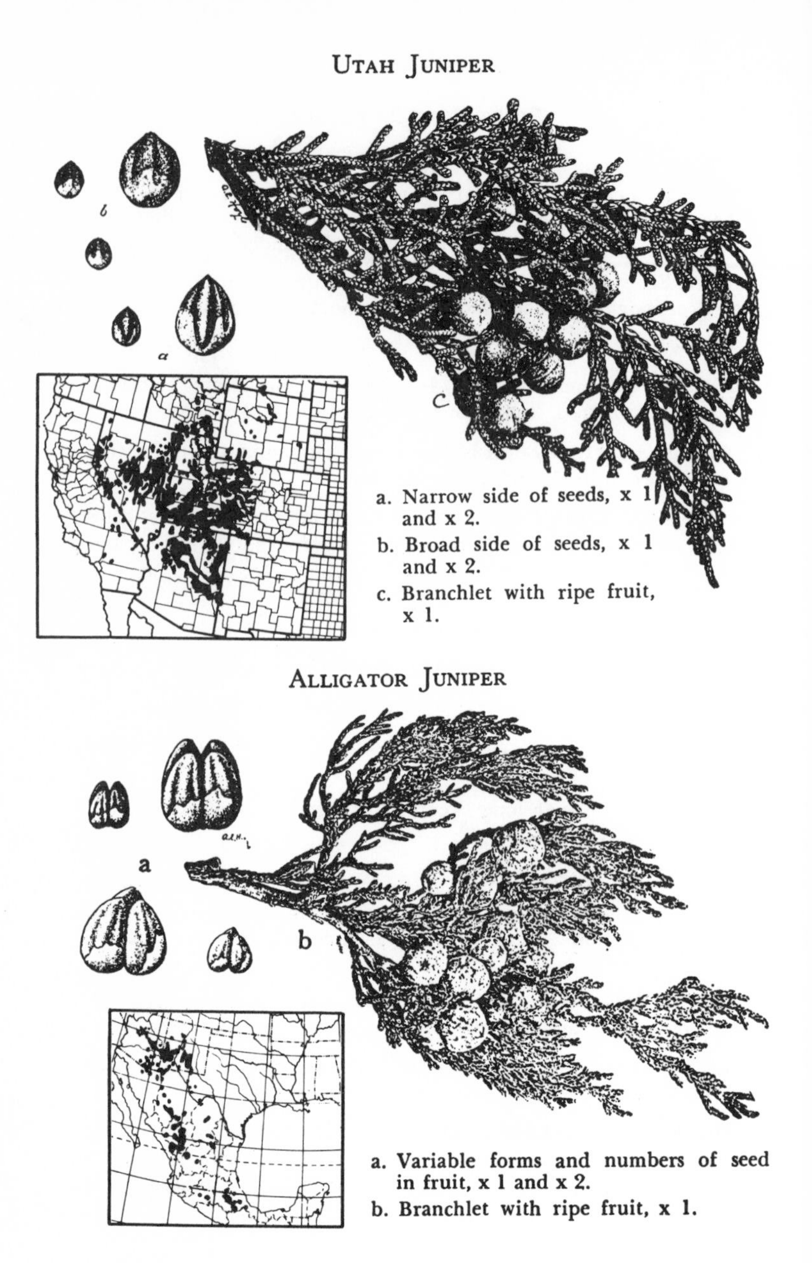

Utah Juniper

a. Narrow side of seeds, x 1 and x 2.
b. Broad side of seeds, x 1 and x 2.
c. Branchlet with ripe fruit, x 1.

Alligator Juniper

a. Variable forms and numbers of seed in fruit, x 1 and x 2.
b. Branchlet with ripe fruit, x 1.

CUPRESSACEAE

Utah Juniper

Juniperus osteosperma (Torr.) Little
[*Juniperus utahensis* (Eng.) Lemm.]

HABIT. A spreading shrub or small tree; trunk single, or many-stemmed just above the ground; crown rounded.

LEAVES. 1/8 inch long; acute; pale yellow-green; minutely toothed; usually glandular.

FLOWERS. Usually monoecious; sometimes dioecious.

FRUIT. Glaucous; red-brown; 1/4-3/4 inch long; maturing in 2 years. Seed: 1 (rarely 2) in each cone; ovoid; sharply angled; acute; marked to the middle by a conspicuous hilum.

WOOD. Heartwood light yellow-brown; sapwood thick and white; very durable; used locally for fuel and fence posts.

HABITAT. Upper Sonoran zone; on dry, rocky slopes; in pure stands or with one-seed juniper, pinyon, and desert shrubs.

* * *

Alligator Juniper

Juniperus deppeana Steud. (*Juniperus pachyphloea* Torr.)

HABIT. A spreading shrub or small tree; crown broad.

LEAVES. 1/8 inch long; acute; blue-green; minutely toothed; conspicuously glandular.

FLOWERS. Monoecious.

FRUIT. Glaucous; dark red-brown; 1/3-1/2 inch long; maturing in 2 years. Seed: 1-4 (usually 4) in each cone; distinctly grooved; conspicuously swollen on back.

BARK. Very characteristic; 1/2-4 inches thick; red-brown; deeply furrowed into square plates 1-2 inches across.

WOOD. Heartwood brown; durable; used locally.

HABITAT. Upper Sonoran and transition zones; drouth-resistant; commonly mixed with nut pines and oaks.

* * *

1. Ashe Juniper, *J. ashei* Buch., is in Missouri, Arkansas, Oklahoma and Texas.
2. Drooping Juniper, *J. flaccida* Schl., is in southwestern Texas.
3. Pinchot Juniper, *J. pinchotii* Sudw., is in Texas and Oklahoma.

Western Juniper

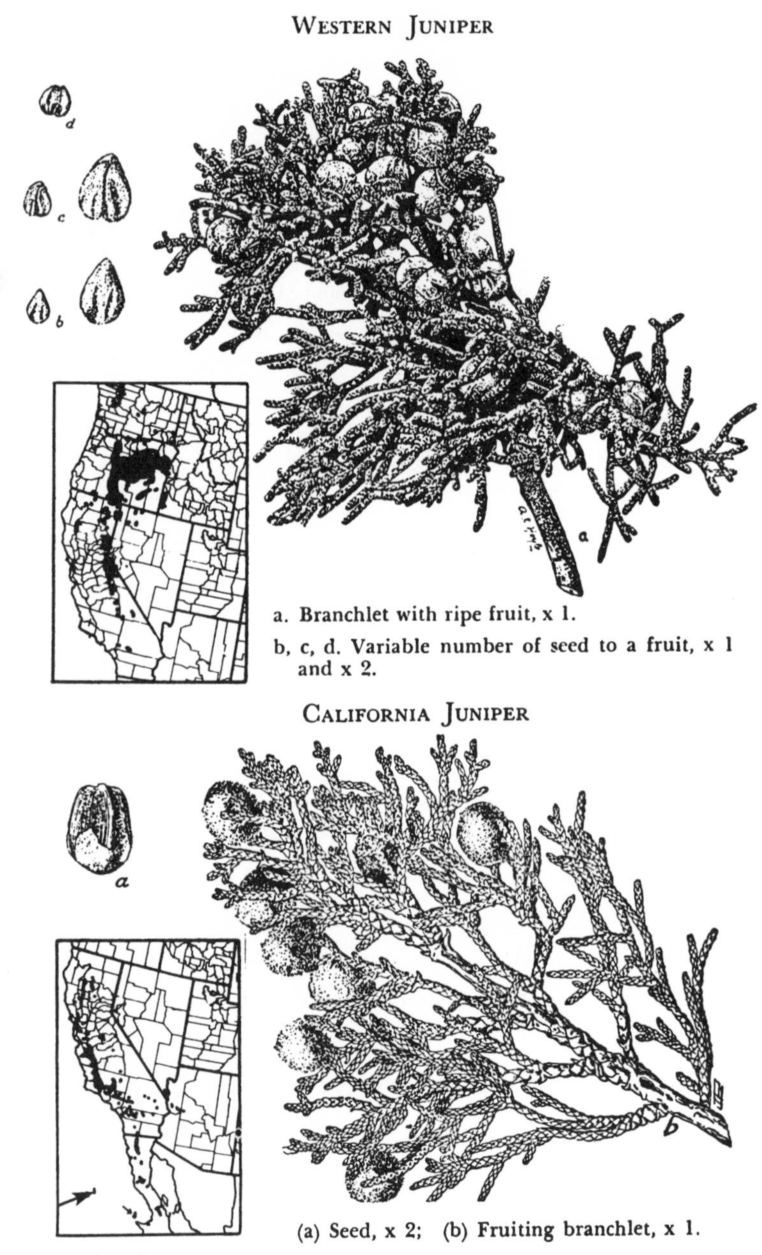

a. Branchlet with ripe fruit, x 1.

b, c, d. Variable number of seed to a fruit, x 1 and x 2.

California Juniper

(a) Seed, x 2; (b) Fruiting branchlet, x 1.

Western Juniper. Sierra Juniper.

Juniperus occidentalis Hook.

HABIT. A bushy shrub or small tree; short, straight trunk; broad, rounded crown, extending nearly to ground.

LEAVES. 1/8 inch long; acute or acuminate; gray-green; conspicuously glandular-pitted on back; denticulately fringed.

FLOWERS. Dioecious.

FRUIT. Glaucous; blue-black; 1/4-1/3 inch in diameter; maturing in 2 years. Seed: 2-3 in each cone; ovoid; acute.

WOOD. Heartwood pale brown tinged with red; sapwood white and thick; exceedingly durable; used locally for posts.

HABITAT. Upper Sonoran to Hudsonian zones; on exposed, dry, rocky, mountain slopes; usually in pure, open stands.

* * *

California Juniper

Juniperus californica Carr.

HABIT. A small tree, or shrubby as a result of large branches leaving the trunk near the ground.

LEAVES. 1/8 inch long; acute; yellow-green; conspicuously glandular on back; fringed on the margins.

FLOWERS. Dioecious.

FRUIT. Red-brown; glaucous; 3/8-3/4 inch long; maturing in 2 years. Seed: 1-2 in each cone; obtuse.

WOOD. Heartwood pale brown tinged with red; thin, white sapwood; durable; used locally for posts.

HABITAT. Upper Sonoran zone; on desert and mountain slopes; with pinyon and chaparral.

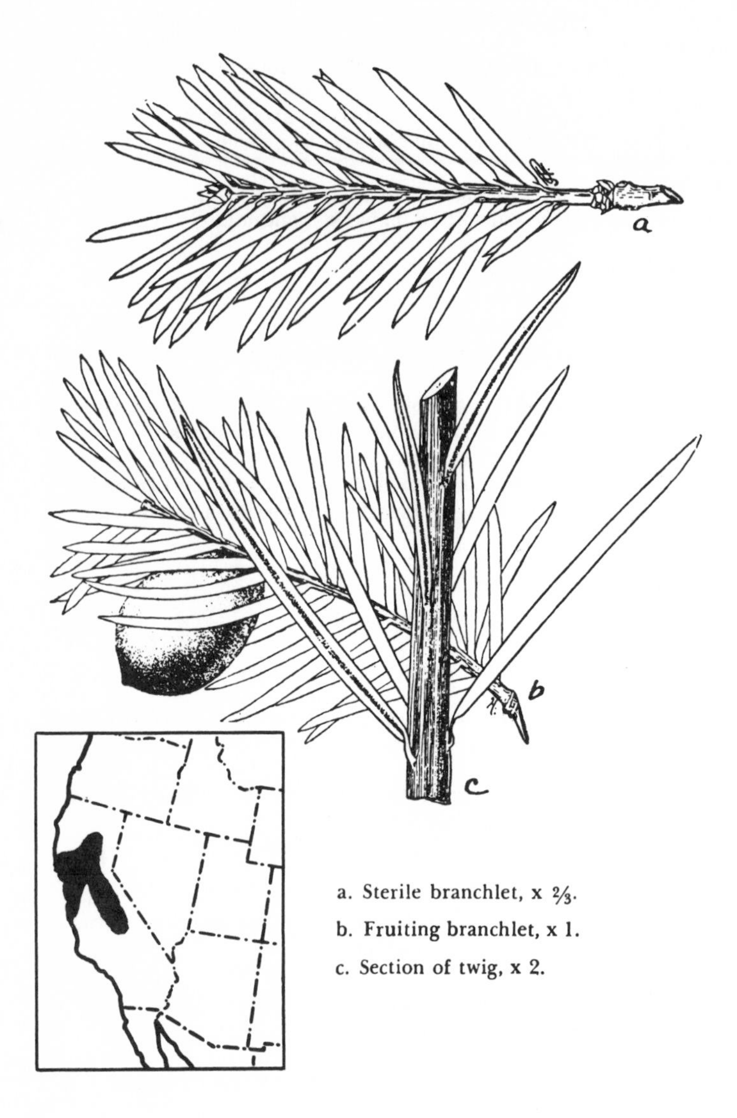

a. Sterile branchlet, x ⅔.
b. Fruiting branchlet, x 1.
c. Section of twig, x 2.

California Torreya. California Nutmeg

Torreya californica Torr. [*Tumion californicum* (Torr.) Greene]

HABIT. A rare, small tree 15-70 feet high and 1-2 feet in diameter; clear, often malformed bole; crown pyramidal, with slender, spreading branches.

LEAVES. Persistent many years; linear-lanceolate; spirally arranged, appearing 2-ranked; 1-3 1/2 inches long; acuminate and often bristle-tipped; flat; rigid; dark green and lustrous above; pale below, with 2 narrow, distinct bands of stomata; petioled; decurrent.

FLOWERS. Dioecious; male solitary, consisting of numerous stamens in whorls of 4; female in pairs, 1/4 inch long, the ovule surrounded by and finally enclosed in fleshy sac.

FRUIT. Drupelike; a single seed with a woody outer coat completely enclosed in a thin, fleshy, green to purple envelope; ellipsoidal; 1-1 1/2 inches long; maturing in 2 years, but requiring additional year before germinating.

TWIGS. Slender; drooping. Winter buds: small; ovoid; acute; of few overlapping, shining, opposite scales.

BARK. Thin (1/3-1/2 inch); gray-brown; in narrow, scaly ridges.

WOOD. Unimportant; soft; light clear yellow.

SILVICAL CHARACTERS. Tolerant; growth slow; long-lived; reproduction sparse, but stumps sprout vigorously; lateral roots.

HABITAT. Dependent on moist sites; in pure, dense thickets or mixed with hardwoods. In small, scattered, isolated areas within shaded portion on map.

* * *

Florida Torreya. Stinking-cedar

Torreya taxifolia Arn. [*Tumion taxifolium* (Arn.) Greene]

This is a rare, small, handsome tree found only along the bluffs of the Apalachicola River in Florida and southern Georgia. It differs from the western species in having needles slightly rounded on the back and about 1 1/2 inches long.

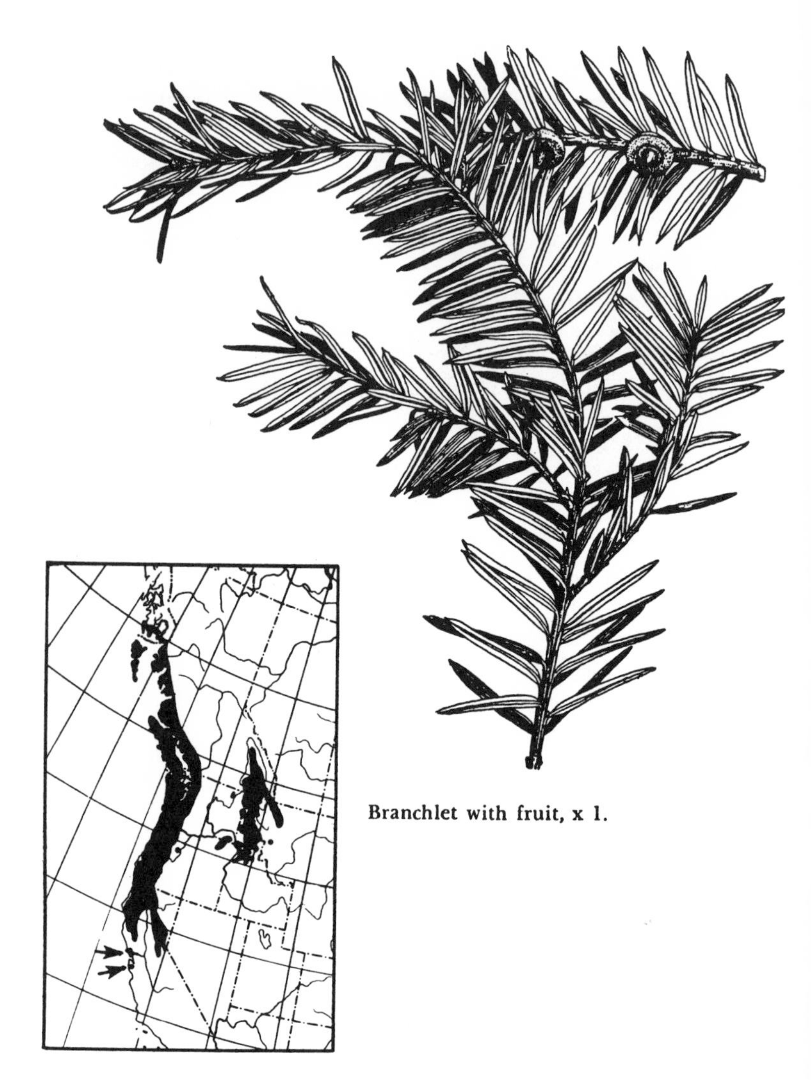

Branchlet with fruit, x 1.

Pacific Yew. Western Yew

Taxus brevifolia Nutt.

HABIT. A small tree or large shrub, 20-50 feet high and 1-2 feet in diameter; limby, often fluted or malformed trunks; large, open, conical crown, with long, slender, drooping branches; a sprawling shrub near timberline.

LEAVES. Persistent 5-12 years; linear-lanceolate; spirally arranged, appearing 2-ranked; 1/2-1 inch long; sharp-pointed; petiolate; green above, paler beneath.

FLOWERS. Dioecious; solitary; axillary; surrounded by scales of bud; male in globose heads of 6-14 stamens, yellowish; female single, greenish, the apical scale bearing a solitary erect ovule with a basal disk.

FRUIT. A single, erect, ovoid-oblong seed with a hard, bony shell; 1/3 inch long; exposed at apex, but partially or entirely surrounded by, but free from, the thickened, scarlet, fleshy aril-like disk of the flower; maturing in one season.

TWIGS. Slender; drooping. Winter buds: small; ovoid; obtuse; of numerous overlapping scales.

BARK. Very thin (1/4 inch); scaly; dark red-purple.

WOOD. Heavy; hard; strong; durable; heartwood bright orange to rose-red; sapwood thin, yellow; used for bows, canoe paddles, fence posts, turned articles.

SILVICAL CHARACTERS. Most tolerant forest tree of Northwest; growth slow; maturity reached in 250-350 years; reproduction scanty; deep, wide-spreading roots; stumps sprout.

HABITAT. Transition to Hudsonian zone; altitudinal range 2,000-8,000 feet; on deep, moist soils; in small groups or as an occasional understory tree in mixed forests.

* * *

Florida Yew

Taxus floridana Nutt.

This is a rare, small, bushy tree found only along the bluffs of the Apalachicola River in Florida. It differs from Pacific yew in having dark green, falcate needles, 3/4-1 inch long.

SALICACEAE

The Poplars and Cottonwoods

Characteristics of the Genus *Populus* L.

HABIT. Mostly large, fast-growing, deciduous trees.

LEAVES. Alternate; simple; turning yellow before falling in autumn; petioles mostly long and often laterally compressed; stipules present, falling as leaves unfold.

FLOWERS. Regular; dioecious; both sexes in drooping aments; appearing from separate buds, before the leaves; individual flowers solitary, apetalous, inserted on broad, cup-shaped disk, subtended by dilated, lobed, and often laciniate scale or bract; staminate with 4 to many stamens; pistillate, a single, 1-celled, usually sessile ovary with 2-4 placentas and 2-4 stigmas.

FRUIT. A 1-celled, 2-4-valved capsule containing numerous seeds. Seed: small; tufted with long silky hairs; extremely buoyant; vitality transient and must germinate within few days.

TWIGS. Pith homogeneous; angled or stellate in cross section. Winter buds: terminal present, covered by several scales.

BARK. Astringent; light-colored; deeply furrowed or smooth.

WOOD. Light; soft; weak; diffuse-porous; not durable; heartwood light brown; used for pulp, veneer, and lumber.

SILVICAL CHARACTERS. Intolerant; fast-growing; short-lived; reproduction widespread and abundant; stumps and roots sprout vigorously; extensive, wide-spreading root systems.

GENERAL. This genus contains about 35 species scattered over the Northern Hemisphere and in northern Africa; 10 species and several hybrids are native to North America; 2 European species have become naturalized and are included.

* * *

KEY TO THE SPECIES OF POPLARS AND COTTONWOODS

1. Twigs, buds, petioles, and lower surfaces of leaves white tomentose; widely planted and naturalized.
   2. Leaves irregularly sinuate-dentate or palmately 3-5-lobed, base usually rounded or flat . . . . *P. alba*, White Poplar, p. 123
   2. Leaves toothed, not lobed, base heart-shaped . . . . *P.* X *canescens*, Gray Poplar, p. 123
1. Twigs, buds, petioles, and leaves hairy or glabrous at maturity.
   3. Leaf petiole round in cross section, leaf base round or heart-shaped.
      4. Leaves broadly ovate, 4-7 inches long and nearly as broad, densely hairy when they unfold, but becoming glabrous; wet sites; Atlantic Coast, southeastern and central states . . . *P. heterophylla*, Swamp Cottonwood, p. 125
      4. Leaves ovate to lanceolate, 1/3 or more longer than broad.

5. Leaves lanceolate to ovate-lanceolate, 2-4 inches long; petioles short (less than 1/3 length of blade); buds 1/4-3/4 inch long, resinous; Rocky Mountains . . . . *P. angustifolia*, Narrowleaf Cottonwood, p. 129
5. Leaves ovate to ovate-lanceolate; petioles at least 1/2 as long as leaf blade.
    6. Leaves 3-6 inches long, margins finely toothed; buds 3/4-1 inch long, very resinous and aromatic.
        7. Capsule 3-valved, hairy; stigmas 3; 40-60 purple stamens; Pacific region . . . . . . *P. trichocarpa*, Black Cottonwood, p. 127
        7. Capsule 2-valved, glabrous; stigmas 2; 20-30 pink stamens; northern North America . . . . *P. balsamifera*, Balsam Poplar, p. 127
    6. Leaves 2-4 inches long, margins coarsely toothed; buds 1/2-3/4 inch long, slightly resinous, not aromatic; Rocky Mountains . . . . *P.* X *acuminata*, Lanceleaf Cottonwood, p. 129

3. Leaf petiole definitely flattened laterally.
    8. Leaves broadly ovate to suborbicular, 1-3 inches long, base rounded; buds almost nonresinous; capsules 2-valved, thin-walled; 6-12 stamens.
        9. Leaves finely toothed, glabrous; buds glabrous; Atlantic to Pacific . . . . *P. tremuloides*, Quaking Aspen, p. 123
        9. Leaves with few coarse teeth; buds hairy; eastern . . . . *P. grandidentata*, Bigtooth Aspen, p. 123
    8. Leaves deltoid to rhombic-ovate, 2-7 inches long; buds resinous; capsules 3- or 4-valved (2-valved in Black and Lombardy Poplars) thick-walled; 12-60 stamens.
        10. Leaves rhombic-ovate (rarely deltoid), long acuminate, base wedge-shaped or rounded, hybrids or naturalized trees widely planted through North America.
            11. Leaves longer than broad, 2-6 inches long, coarsely toothed. . . . . *P.* X *canadensis*, Carolina Poplar, p. 125
            11. Leaves about as long as broad, 2-4 inches long, finely toothed.
                12. Tree with wide-spreading, stout branches . . . . *P. nigra*, Black Poplar, p. 123
                12. Columnar tree; erect branches; narrow, spinelike crown . . . . *P. nigra italica*, Lombardy Poplar, p. 123
        10. Leaves deltoid in shape, 3-6 inches long; native trees.
            13. Leaves with glands at point where petiole and blade meet; margins ciliate, with numerous teeth.
                14. Leaves thick, very broad and truncate at base; twigs thick.
                    15. Buds glabrous; leaves finely toothed, eastern . . . . *P. deltoides*, Eastern Cottonwood, p. 125
                    15. Buds slightly hairy, leaves coarsely toothed; Great Plains to Rocky Mountains . . . . *P. sargentii*, Plains Cottonwood, p. 125
                14. Leaves thin, often rounded or wedge-shaped at base; twigs slender; Texas . . . . *P. palmeri*, Palmer Cottonwood, p. 125
            13. Leaves without glands at point where petiole and blade meet; margins not ciliate, with usually less than 15 teeth on a side.
                16. Pedicels longer than capsules; leaves with no more than 10 teeth on a side; Colorado, New Mexico, west Texas, Mexico . . . . *P. fremontii wislizenii*, Rio Grande Cottonwood
                16. Pedicels shorter than capsules; leaves, more than 10 teeth on a side; New Mexico west to California . . . . *P. fremontii*, Fremont Cottonwood, p. 125

QUAKING ASPEN

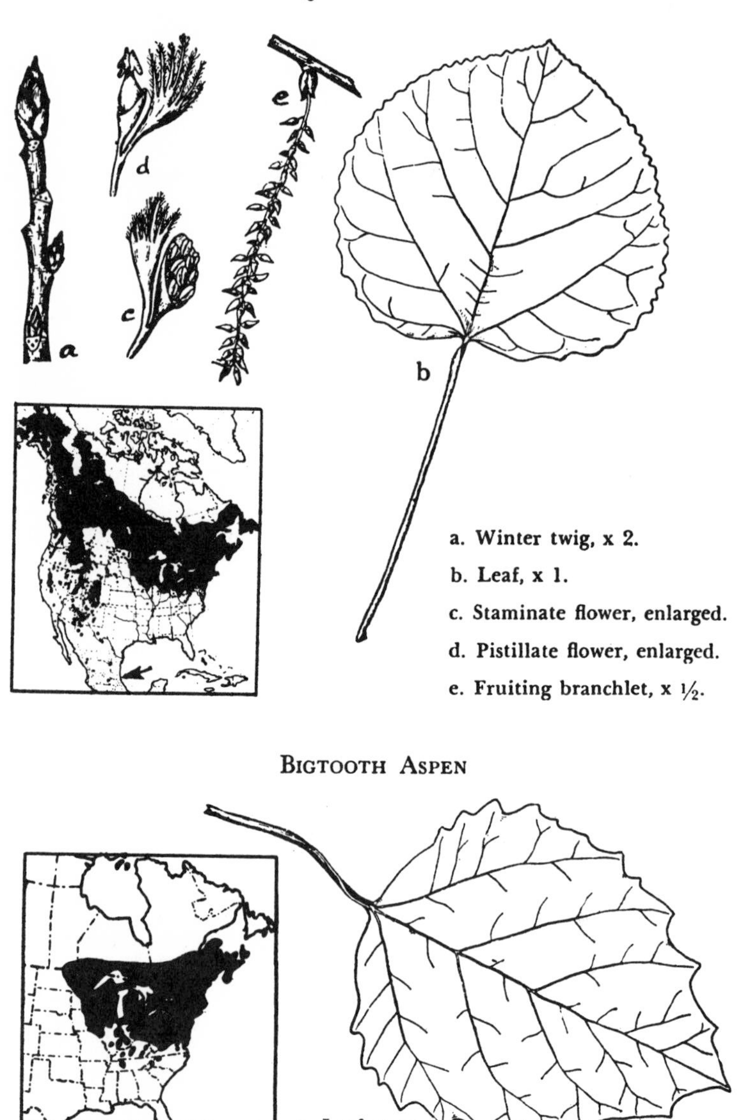

a. Winter twig, x 2.

b. Leaf, x 1.

c. Staminate flower, enlarged.

d. Pistillate flower, enlarged.

e. Fruiting branchlet, x ½.

BIGTOOTH ASPEN

a. Leaf, x ⅔.

SALICACEAE

Quaking Aspen. Poplar

Populus tremuloides Michx.

HABIT. A small tree 50-60 feet high and 1-2 feet in diameter; loose, rounded crown with slender branches.

LEAVES. Semiorbicular or broadly ovate; 1 1/2-3 inches in diameter; yellow-green; apex acute; base rounded; finely crenate-serrate with glandular teeth; thin and firm; glabrous; petioles flattened, 1 1/2-3 inches long.

FRUIT. Capsules narrowly conical; 1/4 inch long; curved; 2-valved; gray-hairy. Seed: light brown; 1/32 inch long.

TWIGS. Slender; round; bright red-brown and lustrous, becoming gray. Winter buds: terminal 1/4-1/2 inch long; conical; sharp-pointed; red-brown; sometimes slightly resinous.

BARK. Smooth; green-white to cream-colored.

SILVICAL CHARACTERS. Canadian and Hudsonian zones; reproducing vigorously on cutover or burned-over areas and forming a protective canopy for more tolerant species.

* * *

Bigtooth Aspen

Populus grandidentata Michx.

This medium-sized tree differs from quaking aspen in having coarsely sinuate-toothed leaves 2-5 inches long, stout twigs, and puberulous, dusty-gray buds 1/8 inch long.

* * *

The following European forms have become naturalized in parts of the United States. Their distinguishing features appear in the key:

1. White poplar, *P. alba* L., is a large, widely planted tree.
2. Gray poplar, *P.* X *canescens* (Ait.) Sm., is a hybrid of *alba* X *tremula.*
3. Black poplar, *P. nigra* L., is a wide-spreading tree.
4. Lombardy poplar, *P. nigra* var. *italica* Muenchh., is a hybrid clone of narrow columnar crown and ascending branches and is widely planted.

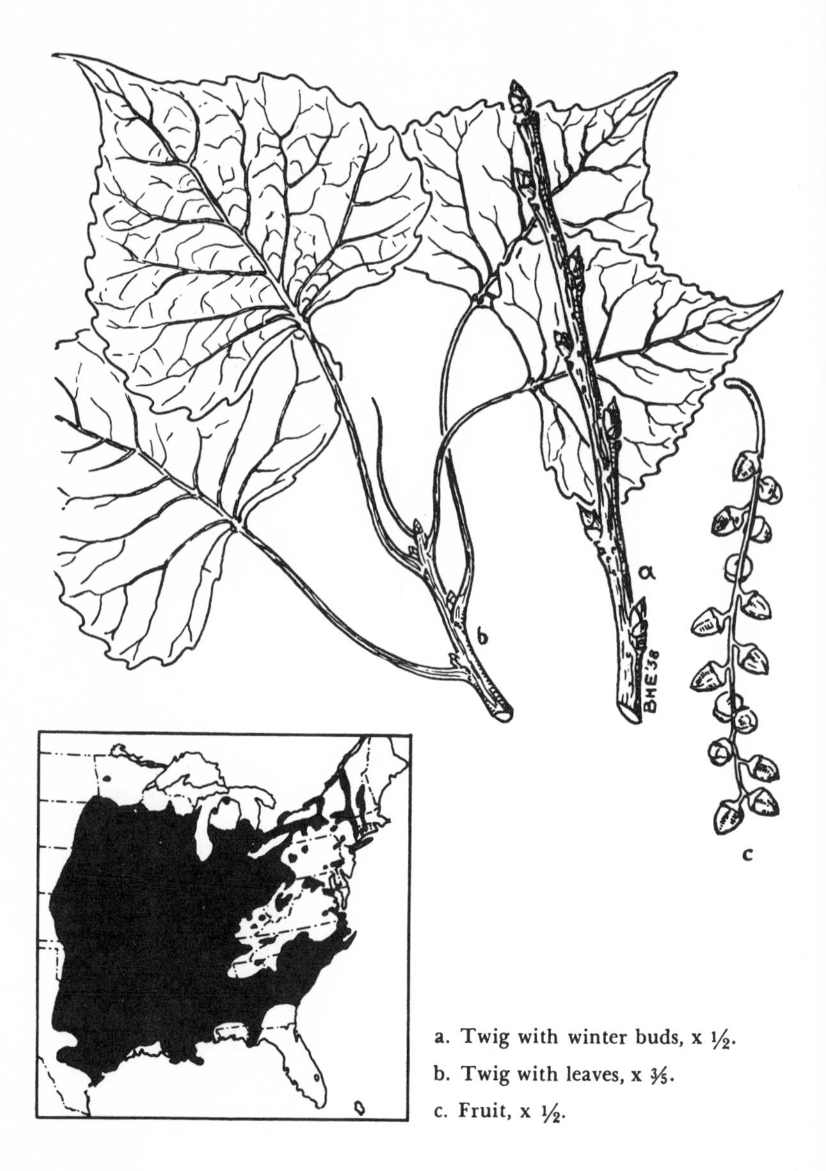

a. Twig with winter buds, x ½.

b. Twig with leaves, x ⅗.

c. Fruit, x ½.

Eastern Cottonwood

Populus deltoides Bartr.

HABIT. This important eastern species is a large tree 60-100 feet high and 3-5 feet in diameter (max. 175 by 11 feet); broad, open crown; large, spreading branches.

LEAVES. Broadly deltoid; 3-6 inches long; apex acuminate; coarsely crenate-serrate, teeth glandular; glabrous; petiole flattened, 1 1/2-3 inches long, with 2 glands at apex.

FRUIT. Capsules ovoid, 1/3 inch long; in catkins 8-12 inches long. Seed: light brown; hairy; about 1/16 inch long.

TWIGS. Stout; yellow-brown; angular; glabrous. Winter buds: 3/4 inch long; conical; acute; shiny-brown; resinous.

BARK. Yellow-green and smooth on young trunks, becoming thick ash-gray, and furrowed.

* * *

Five other cottonwoods and 2 hybrids having the deltoid leaves and other characters similar to *P. deltoides* are native and distinguished in the key:

1. Fremont cottonwood, *P. fremontii* S. Wats., from Colorado to California and south into Mexico.
2. Rio Grande cottonwood, *P. fremontii* var. *wislizenii* S. Wats. from Colorado and Utah south into New Mexico and Texas, also northern Mexico.
3. Swamp cottonwood, *P. heterophylla* L., in the East differs in having a rounded petiole and heart-shaped leaf base.
4. Palmer cottonwood, *P. palmeri* Sarg., in western Texas.
5. Plains cottonwood, *P. sargentii* Dode, in the Great Plains from Alberta to Texas.
   P. X *canadensis* Moench., the Carolina poplar, (*P. deltoides* X *nigra*) is extensively planted in cities.
   P. X *parryi* Sarg. (*P. fremontii* X *trichocarpa*) in California.

Black Cottonwood

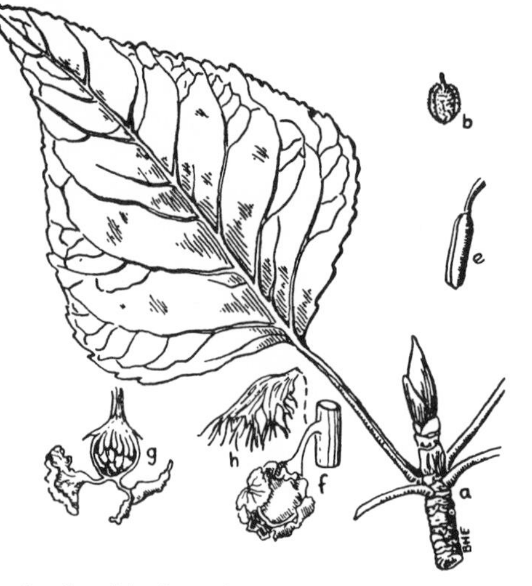

a. Leaf and bud, x ½.
b. Fruit (capsule), x 1.
c, d, e. Staminate flower.
c. Fringed scale, x 2.
d. Stamens on disc, x 2.
e. Single stamen, x 4.
f, g, h. Pistillate flower (after Longyear).
f. Single flower, x 2.
g. Cross section of ovary with stigmas attached.
h. Fringed scale, x 2.

Balsam Poplar

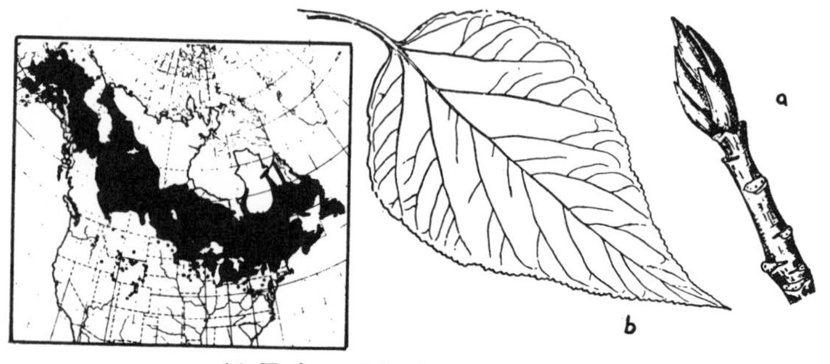

(a) Twig and bud, x ½; (b) Leaf, x ½.

Black Cottonwood
Populus trichocarpa Torr. & Gray

HABIT. The largest of American poplars and the largest broad-leaved tree in the Pacific Northwest, reaching 200 feet in height and 7-8 feet in diameter; long, clear, cylindrical trunk; crown open, rounded.

LEAVES. Ovate to ovate-lanceolate; 5-6 inches long and 3-4 inches wide; apex acute to long-acuminate; base rounded or slightly cordate; margin finely crenate; leathery; dark green above, silver-white to pale green below; petioles round, long.

FRUIT. Capsules subglobose; 1/3 inch long; 3-valved; hairy.

TWIGS. Moderately slender; round or slightly angled; red-brown at first, becoming glabrous and dark gray. Winter buds: terminal 3/4 inch long; ovoid; orange-brown; covered by fragrant, yellow-brown resin.

BARK. Smooth and greenish on young stems; becoming pale gray, 1-2 1/2 inches thick, deeply and sharply furrowed.

* * *

Balsam Poplar
Populus balsamifera L. (*Populus tacamahacca* Mill.)

HABIT. A medium-sized tree seldom over 60-80 feet high and 1-3 feet in diameter; open, narrow crown.

LEAVES. Broadly ovate to ovate-lanceolate; 3-6 inches long and 2-4 inches broad; apex acute to acuminate; base rounded or cordate; finely crenate-serrate; thin and firm; dark green above, much paler below; petioles round, slender, long.

FRUIT. Capsules ovoid; 1/4-1/3 inch long; 2-valved; glabrous; short-stalked. Seed: light brown; 1/12 inch long.

TWIGS. Moderately stout; round; red-brown. Winter buds: terminal 1 inch long; ovoid; chestnut-brown; saturated by fragrant, amber-colored resin.

BARK. On young trunks smooth and green-brown; on large trunks deeply furrowed and gray-black.

GENERAL. The variety *subcordata* Hylander differs from the species in having leaves that are usually broader with pubescent petioles. Balm-of-Gilead, a widely planted ornamental tree in the Northeast, is a clone from a single pistillate tree of this variety.

Narrowleaf Cottonwood

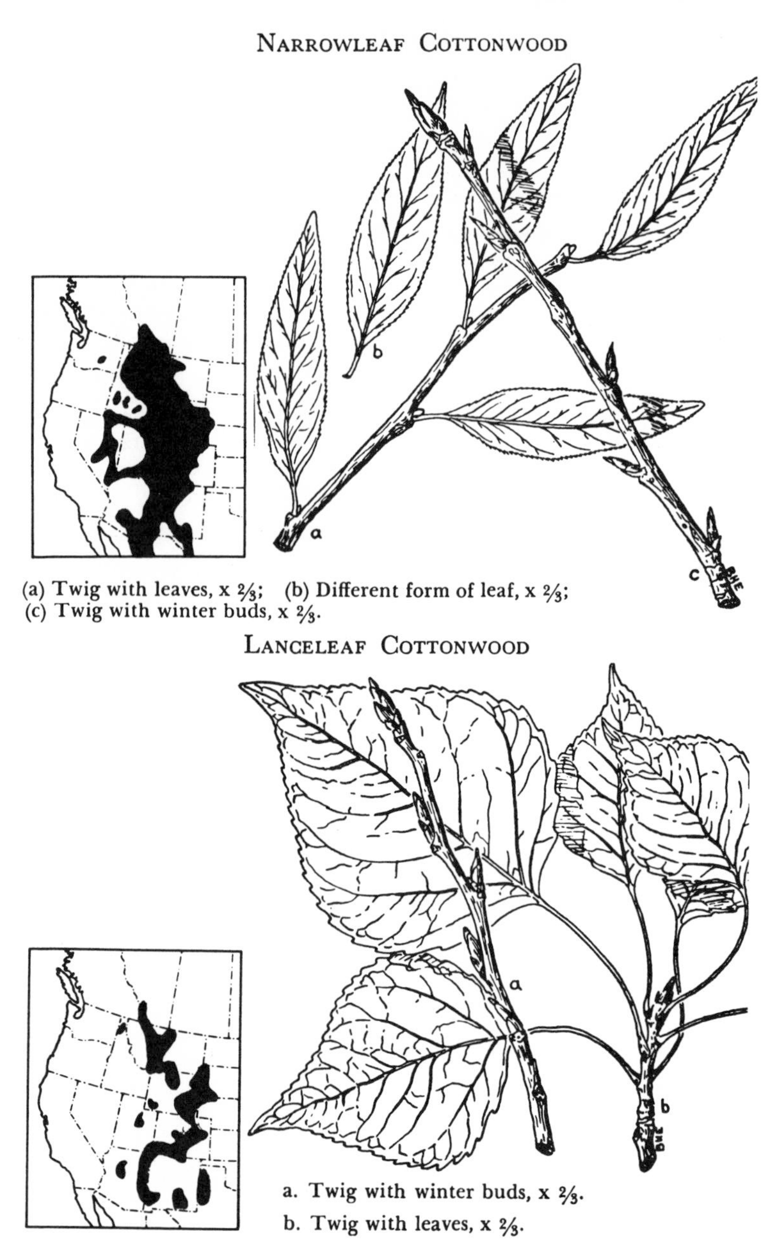

(a) Twig with leaves, x 2/3; (b) Different form of leaf, x 2/3; (c) Twig with winter buds, x 2/3.

Lanceleaf Cottonwood

a. Twig with winter buds, x 2/3.

b. Twig with leaves, x 2/3.

Narrowleaf Cottonwood

Populus angustifolia James

HABIT. A medium-sized tree 50-70 feet high and 12-20 inches in diameter; crown pyramidal with slender, erect branches.

LEAVES. Lanceolate to ovate-lanceolate; 2-4 inches long and 1/2-1 1/2 inches wide; apex long-tapering; finely to coarsely serrate; thin and firm; bright yellow-green above, paler below; petioles short (less than 1/3 length of blade), slender, somewhat flattened on upper side.

FRUIT. Capsules broadly ovoid; 1/4 inch long; 2-valved.

TWIGS. Slender; round; yellow-green, becoming ash-gray. Winter buds: terminal 1/4-3/4 inch long; slender; long-pointed; chestnut-brown; very resinous and somewhat aromatic.

BARK. Light yellow-green; 3/4-1 inch thick; smooth except near base of old trees where it is divided by shallow furrows.

* * *

Lanceleaf Cottonwood

Populus X *acuminata* Rydb.

HABIT. A medium-sized tree about 40 feet high and 1-2 feet in diameter; rounded crown with large branches.

LEAVES. Rhombic-lanceolate to ovate; 2-4 inches long and 3/4-2 inches wide; apex acuminate; base usually cuneate; coarsely crenate-serrate; rather thick; dark green above, dull green below; petioles nearly round, slender, 1-3 inches long.

FRUIT. Capsules oblong-ovoid; about 1/3 inch long; thin-walled; acute; 3- or rarely 2-valved; on short, slender stalks.

TWIGS. Slender; round or slightly 4-angled; pale yellow-brown. Winter buds: terminal about 1/3-3/4 inch long; narrow; acuminate; light chestnut-brown; rather resinous.

BARK. Nearly white and smooth on young trunks; on old trees gray-brown, 1/2 inch thick, deeply divided into flat ridges.

GENERAL. This tree is now generally regarded as a hybrid between *P. angustifolia* and *P. sargentii* or *P. fremontii.*

SALICACEAE

The Willows

Characteristics of the Genus *Salix* L.

HABIT. Shrubs or less frequently trees; latter often with several trunks from greatly extended rootstalk.

LEAVES. Alternate; simple; commonly lanceolate; margins entire or toothed; pinnately veined; sessile or short-petioled; stipules small and soon falling, except on vigorous shoots where they are leaflike and persistent.

FLOWERS. Regular; dioecious; in terminal and axillary aments, appearing with or before the leaves; individual flowers solitary, apetalous, on glandlike disk, subtended by pubescent, entire to dentate, deciduous, or persistent scale or bract; staminate with 1-2 or 3-12 stamens inserted on base of bract; pistillate a single, 1-celled, sessile or stalked ovary, containing 4-8 ovules on each of 2 placentas, style short, terminating in 2 short, 2-parted stigmas.

FRUIT. A 1-celled, 2-valved, acuminate capsule. Seed: small; dark brown; tufted with long, silky hairs; buoyant; of transient vitality.

TWIGS. Slender to stout; round; tough; often easily separated at junction with branch; marked by elevated leaf scars; pith homogeneous, terete. Winter buds: terminal absent; lateral covered by single caplike scale.

BARK. Astringent; scaly; variously colored.

WOOD. Light; soft; weak; usually brittle; durable in some species; heartwood pale brown, often tinged with red; used to small extent for athletic goods and charcoal.

SILVICAL CHARACTERS. Intolerant; fast-growing; remarkable vitality and sprouting vigorously from stumps or cuttings; usually swamp or moisture-loving plants.

GENERAL. This genus contains about 170 species scattered over the Northern Hemisphere, with about 100 species native to North America; identification of the various species is difficult; 38 native and 4 naturalized species have been listed as occasionally reaching tree size in North America, but only 15 of these native species are commonly trees and included in the key with the naturalized species.

KEY TO THE TYPICAL TREE SPECIES OF WILLOWS

1. Stamens 3 to 12; filaments free; flower bracts deciduous.
    2. Leaves green below; petioles without glands at base.
        3. Capsule hairy; twigs yellow-gray; southwestern . . . . *S. gooddingii* Ball, <u>Goodding Willow</u>
        3. Capsule glabrous; twigs red . . . . *S. nigra* Marsh., <u>Black Willow</u>, p. 133
    2. Leaves pale or bluish below; ovary and capsules glabrous.
        4. Petioles without glands at base of leaf.
            5. Twigs easily separable; flower bracts entire.
                6. Leaves 4-5 inches long, narrowly lanceolate to lanceolate; petioles under 1/2 inch; southeastern . . . . *S. caroliniana* Michx., <u>Coastal Plain Willow</u>
                6. Leaves 2 1/2-4 inches long, lanceolate to ovate-lanceolate; petioles 1/2-3/4 inch long; northern and central . . . . *S. amygdaloides* And., <u>Peachleaf Willow</u>, p. 133
            5. Twigs firmly attached; flower bracts toothed; southwestern.
                7. Leaves silver-white below; capsules short-stalked; stamens usually 3 . . . . *S. bonplandiana* H. B. K., <u>Bonpland Willow</u>
                7. Leaves glaucous or yellow-hairy below; capsules long-stalked; stamens 4-6; southwestern . . . . . . *S. laevigata* Bebb, <u>Red Willow</u>
        4. Petioles glandular; twigs separable; flower bracts toothed; western . . . . *S. lasiandra* Benth., <u>Pacific Willow</u>
1. Stamens 2.
    8. Widely planted and naturalized ornamentals; capsules glabrous.
        9. Tree with distinctive "weeping" habit; leaves glabrous; capsules sessile . . . . *S. babylonica* L., <u>Weeping Willow</u>
        9. Branches not drooping; capsule short-stalked.
            10. Leaves with silky hairs on both sides; twigs green to bright yellow . . . . *S. alba* L., <u>White Willow</u>
            10. Leaves nearly glabrous; twigs brownish. . . . . *S. fragilis* L., <u>Crack Willow</u>
    8. Native species often shrubby and seldom planted.
        11. Bracts deciduous, yellowish; filaments free, more or less hairy; stipe of ovary much shorter than bract.
            12. Leaves 1/3-1 1/3 inches long, linear; stigmas linear; southwestern. . . . . *S. taxifolia* H. B. K., <u>Yewleaf Willow</u>
            12. Leaves over 1 1/3 inches long, mostly linear-lanceolate; stigmas short.
                13. Leaves white-hairy below; stipules minute or lacking; Western. . . . . *S. exigua* Nutt., <u>Coyote Willow</u>
                13. Leaves nearly glabrous below; stipules large, deciduous; North America . . . . *S. interior* Rowlee, <u>Sandbar Willow</u>
        11. Bracts persistent, darker; filaments glabrous.
            14. Ovary and capsule glabrous; bracts dark brown.
                15. Leaves acute or rounded, thin; filaments united at base; western . . . . *S. lasiolepis* Benth., <u>Arroyo Willow</u>
                15. Leaves acuminate, thick and leathery; filaments free.
                    16. Leaves 1 1/2-2 inches long, oblanceolate; twigs glabrous; western . . . . . . *S. mackenzieana* Barr., <u>Mackenzie Willow</u>
                    16. Leaves 4-6 inches long, narrow-lanceolate; twigs hairy; central United States . . . . *S. eriocephaia* Michx., <u>Missouri River Willow</u>
            14. Ovary and capsule hairy.
                17. Stipe of ovary longer than light, hairy, flower bracts; North America . . . . *S. bebbiana* Sarg., <u>Bebb Willow</u>
                17. Stipe shorter than brown, hairy, flower bracts; western . . . . *S. scouleriana* Barr., <u>Scouler Willow</u>

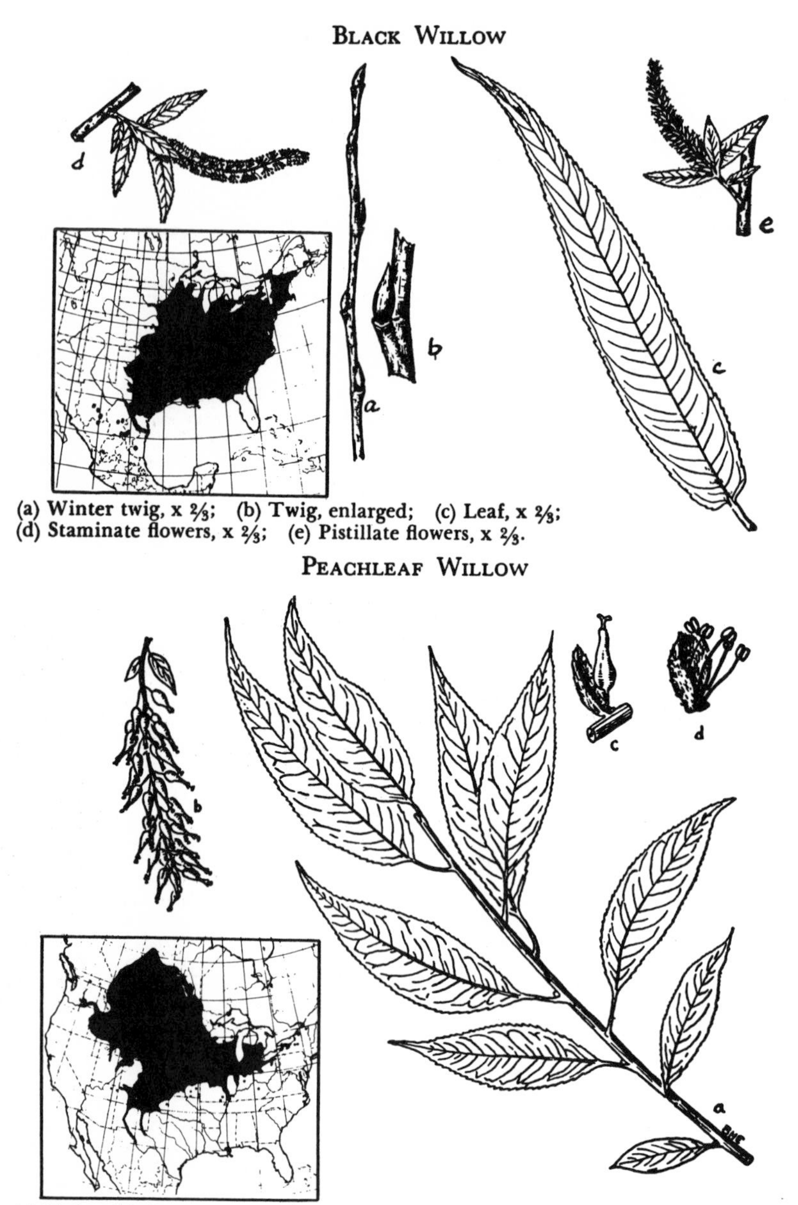

(a) Winter twig, x ⅔; (b) Twig, enlarged; (c) Leaf, x ⅔; (d) Staminate flowers, x ⅔; (e) Pistillate flowers, x ⅔.

(a) Twig with leaves, x ⅔; (b) Fruit, x ⅔; (c) Pistillate flower, x 4; (d) Staminate flower, x 4.

SALICACEAE

Black Willow

Salix nigra Marsh.

HABIT. A small to large tree (max. 120 by 4 feet); trunk often divided; crown broad and open with stout branches.

LEAVES. Lanceolate; 3-6 inches long; acuminate; finely serrate; thin; bright green above, paler beneath; petiole short.

FRUIT. Capsules ovoid-conic; 1/4 inch long; short-stalked; glabrous. Seed: minute; hairy-tufted.

BARK. Nearly black; thick; deeply divided into furrows separating thick, scaly ridges.

SILVICAL CHARACTERS. Intolerant; fast-growing; stream banks and lake shores; with mixed hardwoods.

GENERAL. This is the largest American willow and the only one important for its wood products. It is used for lumber, veneer, pulp, charcoal, and artificial limbs.

* * *

Peachleaf Willow

Salix amygdaloides Anderss.

HABIT. A tree rarely 60-70 feet high and 2 feet in diameter; trunk single, columnar; rather narrow, rounded crown.

LEAVES. Lanceolate to ovate-lanceolate; 2-5 inches long; acuminate; finely serrate; thin and firm; light green above, pale and glaucous below; petioles slender, 1/2-3/4 inch long.

FRUIT. Capsules globose-conic; 1/4 inch long; long-stalked; glabrous; light yellow-red. Seed: minute; hairy-tufted.

BARK. Brown, often tinged with red; 1/2-3/4 inch thick; divided by irregular furrows into broad, flat, connecting ridges.

SILVICAL CHARACTERS. Intolerant; moist sites along banks of streams.

GENERAL. This willow is the only species native to the Rocky Mountain region that is typically a tree.

JUGLANDACEAE

The Walnuts

Characteristics of the Genus *Juglans* L.

HABIT. Small to large handsome trees; crowns broad and open; stout, ascending branches.

LEAVES. Alternate; pinnately compound; deciduous; native species with 9-23 nearly sessile, oblong-lanceolate, finely serrate, acute to acuminate leaflets; stipules absent.

FLOWERS. Monoecious; apetalous; appearing with or after the leaves; staminate in preformed unbranched aments, with 8-40 stamens; pistillate in spikes of 2-8 flowers, each with a 2-celled ovary and 2 divergent, plumose stigmas.

FRUIT. A drupelike nut encased in a semifleshy, indehiscent husk; maturing in one year; nut thick-shelled, containing solitary, 2-4-lobed, sweet, oily, edible seed.

TWIGS. Stout; round; with characteristic chambered pith. Winter buds: terminal present; few-scaled. Leaf scars: conspicuous; three groups of U-shaped bundle scars.

WOOD. Dark-colored; durable; diffuse-porous; valuable.

SILVICAL CHARACTERS. Intolerant; deep taproot; nuts disseminated largely by rodents or streams.

GENERAL. Of the six native species only two eastern ones are of commercial importance and treated separately. The four southwestern species are distinguished in the key. They are:

1. California Walnut, *Juglans californica* S. Wats.
2. Hinds Walnut, *Juglans hindsii* Jeps.
3. Arizona Walnut, *Juglans major* (Torr.) Heller.
4. Little Walnut, *Juglans microcarpa* Berl. (*J. rupestris* Engelm.)

The Persian or English walnut, *Juglans regia* L., is widely planted in warmer parts of the country and is included in the key. The variety with thin-shelled nuts is commonly marketed and the wood is widely used under the name of Circassian walnut.

1. Leaves with 9-29 toothed leaflets; nut with thick, bony partitions; native species.
   2. Fruit ovoid-oblong in clusters of 3-5; band of pale hair separating leaf scar from bud; leaflets 11-17, oblong-lanceolate; heartwood light brown, soft; eastern . . . . *J. cinerea*, Butternut, p. 137
   2. Fruit globose, solitary or in pairs; buds not separated by hairy band from leaf scar; heartwood dark brown, hard.
      3. Fruit 1 1/2-2 inches in diameter.
         4. Nut deeply ridged; 15-23 leaflets; eastern . . . . *J. nigra*, Black Walnut, p. 137
         4. Nut smooth or faintly grooved; 15-19 leaflets; California . . . . *J. hindsii*, Hinds Walnut, p. 134
      3. Fruit 1-1 1/2 inches long; nut grooved; leaflets 9-13 (rarely to 19), oblong-lanceolate; Colorado, New Mexico, and Arizona . . . . *J. major*, Arizona Walnut, p. 134
      3. Fruit 1/2-3/4 inch long; nut grooved.
         5. Leaflets 17-23, narrow-lanceolate; Oklahoma, Texas, and New Mexico . . . . *J. microcarpa*, Little Walnut, p. 134
         5. Leaflets 11-15 (rarely to 19), oblong-lanceolate; California . . . . *J. californica*, California Walnut, p. 134
1. Leaves with 5-9 (rarely to 13), entire or nearly entire leaflets; fruit globose, 1 1/2-2 inches in diameter; nut wrinkled, with thin partitions; widely planted in warm areas . . . . *J. regia*, English Walnut, p. 134

* * *

MYRICACEAE

Waxmyrtle. Bayberry
Myrica L.

Three species of this usually shrubby genus may form small trees. This genus is easily distinguishable by its simple, alternate, aromatic, persistent leaves, which bear minute, resinous dots; dioecious or monoecious flowers in aments; small, globose, drupaceous fruit, covered with wax and containing a thick-walled nut; slender twigs without terminal buds and with scaly lateral buds; thin, smooth, whitish to gray bark. The fruit wax is used in making candles.

1. Southeastern species; flowers dioecious.
   2. Leaves oblanceolate, coarsely toothed, yellow-green with conspicuous orange glands; fruit 1/8 inch in diameter, coated with thick, blue wax; coastal plain from New Jersey to Florida, west to Arkansas and Texas . . . . *Myrica cerifera* L., Southern Waxmyrtle
   2. Leaves oblong-obovate, dark green.
      3. Leaves entire, glabrous; fruit 1/3-1/2 inch in diameter thinly coated with white wax; Florida to Mississippi . . . . *Myrica inodora* Bartr., Odorless Bayberry
      3. Leaves toothed, glabrous or pubescent below, fruit 1/8 inch in diameter thickly coated with whitish wax; coastal plain New Jersey to Louisiana . . . . *Myrica heterophylla* Raf., Evergreen Bayberry
1. Pacific Coast species; flowers monoecious; leaves 2-4 inches long, oblanceolate to oblong-lanceolate, remotely toothed, with minute black glands below; fruit 1/4 inch in diameter, purple, papillose, thinly coated with gray wax; coastal areas in Pacific states . . . . *Myrica californica* Cham., Pacific Waxmyrtle

Black Walnut

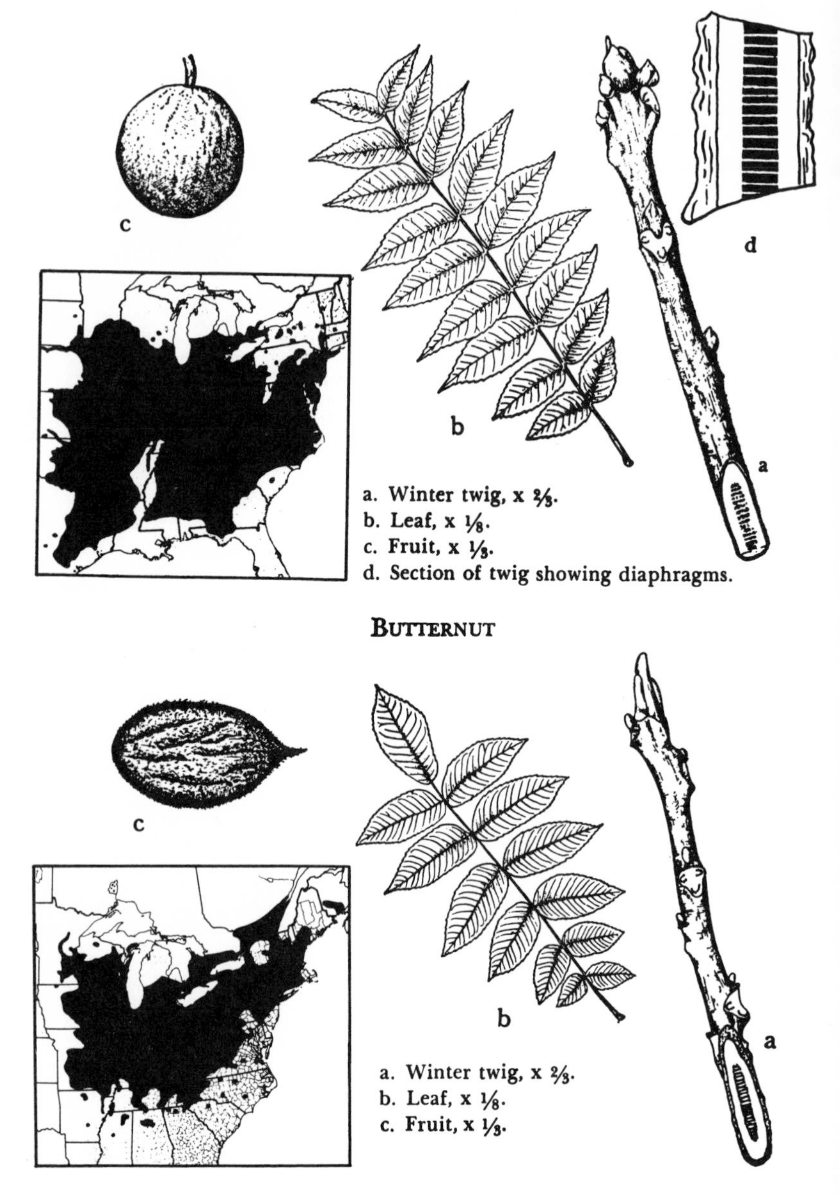

a. Winter twig, x ⅔.
b. Leaf, x ⅛.
c. Fruit, x ⅓.
d. Section of twig showing diaphragms.

Butternut

a. Winter twig, x ⅔.
b. Leaf, x ⅛.
c. Fruit, x ⅓.

JUGLANDACEAE

Black Walnut
Juglans nigra L.

HABIT. A tree 60-90 feet high and 2-3 feet in diameter (max. 150 by 6 feet); trunk straight and clear; crown broad and open.

LEAVES. 1-2 feet long, 15-23 leaflets (terminal leaflet often suppressed); leaflets 3-4 inches long, ovate-lanceolate, finely serrate, yellow-green, glabrous above and hairy below.

FRUIT. Globose; 1 1/2-2 inches in diameter; solitary or in clusters of 2-3; husk thick, yellow-green, pubescent; nut shell corrugated.

TWIGS. Stout; light brown; leaf scar obcordate, without downy pad at top; chambered pith, buff-colored with thin diaphragms. Winter buds: terminal 1/3 inch long, ovoid, blunt, hairy; lateral buds smaller, often superposed.

BARK. Thick; dark brown to nearly black; deep, narrow, interlacing furrows.

WOOD. Heavy; hard; strong; very valuable and important.

SILVICAL CHARACTERS. Fast-growing; long-lived; on fertile, moist soils; in mixed stands; reproduction depends largely on rodents.

* * *

Butternut
Juglans cinerea L.

HABIT. A tree 30-60 feet high and 1-2 feet in diameter (max. 110 by 3 feet); trunk short; crown broad and open.

LEAVES. 1-2 1/2 feet long with 11-17 leaflets; leaflets 2-4 inches long, oblong-lanceolate, serrate, yellow-green, rugose above and hairy below.

FRUIT. Oblong-ovoid; 1 1/2-2 1/2 inches long; solitary or in clusters of 2-5; husk green-brown, sticky, glandular-hairy; nut shell deeply corrugated.

TWIG. Stout; green to red-brown; leaf scar straight across top with dense hairy pad at top; chambered pith, dark chocolate-brown with thick diaphragms. Winter buds: terminal 1/2-3/4 inch long, conical, flattened, hairy; lateral buds smaller, rusty-tomentose.

BARK. Rather thick; light gray; smooth on young trunks, becoming shallowly furrowed with broad ridges.

WOOD. Rather light and soft; light brown; substitute for walnut.

SILVICAL CHARACTERS. Fast-growing; short-lived; varied sites; mixed stands; becoming rare in South.

The Hickories

Characteristics of the Genus *Carya* Nutt.

HABIT. Medium-sized trees with clear straight trunks and open spreading crowns.

LEAVES. Alternate; pinnately compound; deciduous; leaflets 3-17, short-stalked to nearly sessile, ovate to obovate or lanceolate, finely serrate, tips acute to acuminate, bases inequilateral, stipules absent.

FLOWERS. Monoecious; apetalous; appearing after the leaves; staminate in 3-branched aments, with 3-10 stamens; pistillate in spikes of 2-10 flowers, each with a 1-celled ovary and 2 sessile stigmas.

FRUIT. Nut enclosed in a semiwoody, partly dehiscent, 4-valved husk; maturing in one year; nut shell thin to thick, hard, ribbed, smooth to rugose; sweet to bitter.

TWIGS. Mostly rather stout; round; leaf scars large, heart-shaped, or 3-lobed, with numerous bundle scars; pith solid, angular to star-shaped. Winter buds: terminal present, much larger than laterals, with imbricated or valvate scales.

BARK. Smooth and gray on young trunks, becoming scaly or rough; very hard and compact.

WOOD. Highly important; very heavy; hard; strong; light colored; not durable; semiring-porous; used for handles, tools, etc.

SILVICAL CHARACTERS. Moderately intolerant; slow-growing; long-lived; taproot system; varied moist sites.

GENERAL. Twelve species are recognized in the United States (some authors recognize 19 species); several, together with numerous varieties and hybrids, are difficult to identify. Eight species come under the group known as "true hickories," characterized by numerous, imbricated bud scales, 3-9 leaflets, and essentially unwinged fruit husks; and four species are classified as "pecan hickories," characterized by 4-6 valvate bud scales, 5-17 (mostly more than 7) often falcate leaflets, and fruit husks that are typically broadly winged at the sutures.

KEY TO THE SPECIES OF HICKORIES

1\. Bud naked or with 4-6 valvate scales; leaflets 7-17, usually lanceolate, often falcate, about of same size; fruit husk winged. . . . . Pecans

2\. Nuts about twice as long as broad, sweet, thin-shelled; buds brown, covered with yellow hairs; leaflets 9-17; bark thick, furrowed; Illinois to Alabama, Texas . . . . *C. illinoensis*, Pecan, p. 145

2\. Nuts about as long as broad; leaflets 7-13; bark scaly.

3\. Leaflets 7-9 (rarely 5-11), ovate-lanceolate to obovate; buds brown, hairy; nut sweet, with thick, hard shell; South Carolina to Arkansas and south . . . . *C. myristicaeformis*, Nutmeg Hickory, p. 145

3\. Leaflets lanceolate to ovate-lanceolate; nut bitter.

4\. Leaflets 7-13 (usually 9-11); buds red-brown; Virginia to Illinois and south . . . . *C. aquatica*, Water Hickory, p. 145

4\. Leaflets 7-9 (usually 7); buds bright yellow; Quebec to Minnesota and south . . . . *C. cordiformis*, Bitternut Hickory, p. 145

1\. Bud with more than 6 overlapping scales; leaflets 3-9, the uppermost largest; husk usually without wings; nut thick-shelled, sweet. . . . . True Hickories

5\. Terminal buds 1/2-1 inch long; twigs thick; fruit 1-2 1/2 inches long.

6\. Bark on old trunks shaggy with loose plates; fruit husk 1/4-1/2 inch thick; upper 3 leaflets conspicuously larger than others; twigs and petioles glabrous to hairy.

7\. Leaflets usually 7 (rarely 5-9), hairy below; twigs pale orange, hairy; shell very thick; New York to Kansas and south . . . . *C. laciniosa*, Shellbark Hickory, p. 141

7\. Leaflets usually 5 (rarely 7), glabrous or slightly hairy below; twigs red-brown, usually glabrous; Quebec to Minnesota and south . . . . *C. ovata*, Shagbark Hickory, p. 141

6\. Bark on old trunks close; fruit husk 1/8-1/4 inch thick; leaflets 5-9, of about same size; twigs and petioles often woolly; Massachusetts to Nebraska and south . . . . *C. tomentosa*, Mockernut Hickory, p. 143

5\. Terminal buds 1/4-1/2 inch long; twigs moderately thick to slender; fruit 3/4-1 1/2 inches long; bark on old trunks close; fruit husk 1/12-1/4 inch thick.

8\. Winter buds and leaves glabrous or pale hairy.

9\. Buds, leaves, and petioles essentially glabrous; husk of fruit very thin (1/12-1/8 inch); leaflets 5 (rarely 3-7); fruit obovoid, pyriform or oval; Maine to Minnesota and south . . . . *C. glabra*, Pignut Hickory, p. 143

9\. Winter buds and leaves with silvery or pale hairs; husk of fruit 1/8-1/4 inch thick; leaflets 7 (rarely 5-9).

10\. Leaves covered with silvery scales in spring; petiole and rachis hairy; fruit ellipsoidal to globose; Virginia to Tennessee and south . . . . *C. pallida*, Sand Hickory, p. 143

10\. Leaves woolly in spring; petioles glabrous; fruit obovoid; Arkansas to Texas, Mississippi . . . . *C. leiodermis*, Swamp Hickory, p. 143

8\. Buds, twigs, and leaves red-hairy; leaflets 5-7.

11\. Fruit obovoid; Florida . . . . *C. floridana*, Scrub Hickory, p. 143

11\. Fruit subglobose, obovoid, or pyriform; Indiana to Kansas, Texas, Louisiana . . . . *C. texana*, Black Hickory, p. 143

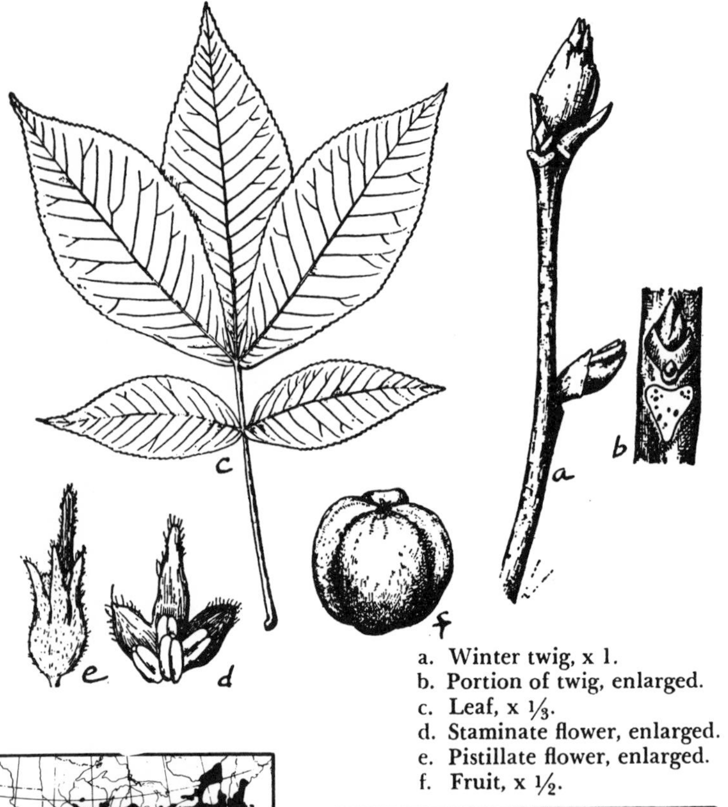

a. Winter twig, x 1.
b. Portion of twig, enlarged.
c. Leaf, x 1/3.
d. Staminate flower, enlarged.
e. Pistillate flower, enlarged.
f. Fruit, x 1/2.

SHAGBARK HICKORY

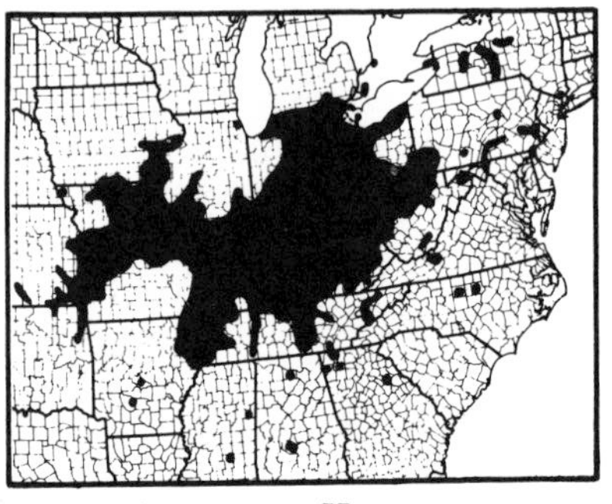

SHELLBARK HICKORY

Shagbark Hickory

Carya ovata (Mill.) K. Koch [*Hicoria ovata* (Mill.) Britt.] [*Carya carolinae-septentrionalis* (Ashe) E. & G.]

HABIT. A tree 60-80 feet high and 1-2 feet in diameter (max. 120 by 4 feet); narrow, open crown and large branches.

LEAVES. 8-14 inches long with 5 (rarely 7) leaflets; leaflets obovate to ovate-lanceolate, the terminal 5-7 inches long and usually much longer than the lateral; finely serrate margins; thick, dark green and glabrous above, yellow-green and usually glabrous below.

FRUIT. Globose to ellipsoidal; 1-2 1/2 inches long (mostly about 1 1/2 inches); husk 1/4-1/2 inch thick, completely dehiscent; nut 4-ribbed with thick shell and sweet seed.

TWIGS. Stout, gray to red-brown, more or less hairy. Winter buds: terminal 1/2-3/4 inch long; broadly ovoid; 3-4 visible, overlapping, brown, pubescent, loosely fitting scales.

BARK. Very characteristic on mature trunks; 3/4-1 inch thick; gray; breaking into thin plates 1-3 feet long free at one or both ends, giving tree typical shaggy appearance.

SILVICAL CHARACTERS. This important tree has a varied habitat, being typically mixed with oak on upland slopes in the north, while in the south it is commonly on moist alluvial soils with several hardwoods.

* * *

Shellbark Hickory

Carya laciniosa (Michx.) Loud. [*Hicoria laciniosa* (Michx.) Sarg.]

This species has similar shaggy bark and differs from shagbark hickory in having larger leaves 15-22 inches long, with usually 7 leaflets that are velvety below, orange twigs, and somewhat larger fruit (1 3/4-2 1/2 inches long) with a 4-6-ribbed nut.

MOCKERNUT HICKORY

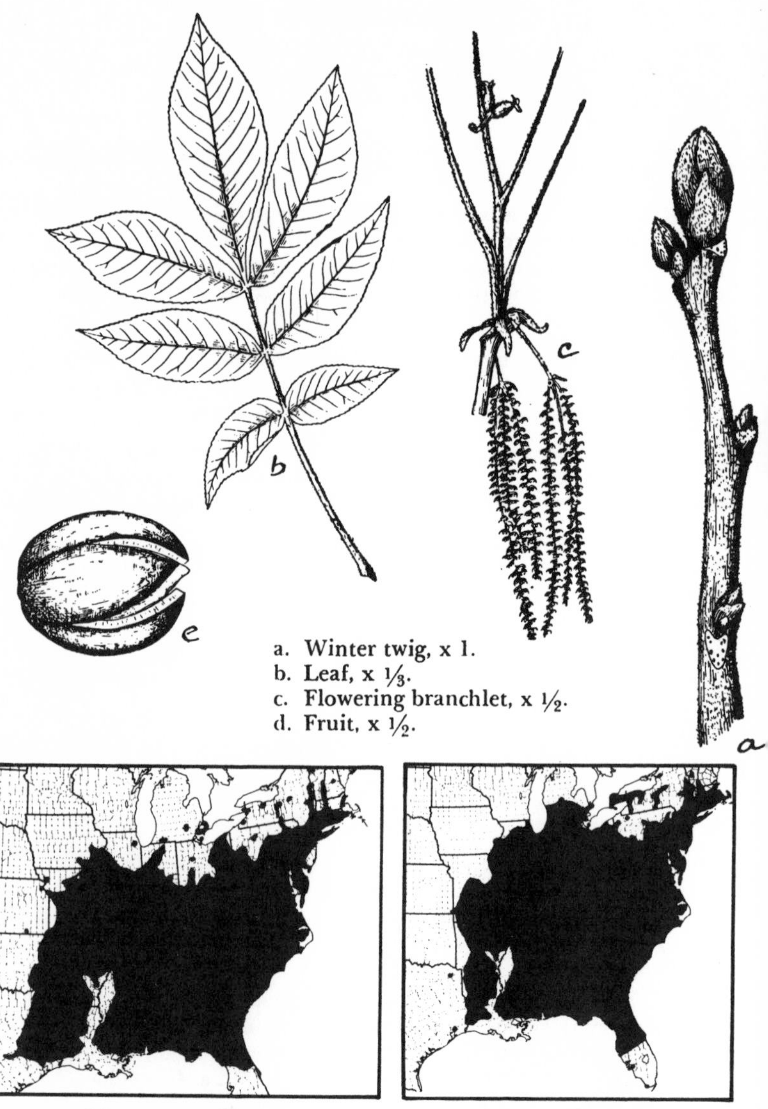

a. Winter twig, x 1.
b. Leaf, x ⅓.
c. Flowering branchlet, x ½.
d. Fruit, x ½.

MOCKERNUT HICKORY

PIGNUT HICKORY

JUGLANDACEAE

Mockernut Hickory

Carya tomentosa Nutt. [*Hicoria alba* (L.) Britt.]
[*Carya alba* (L.) Nutt.]

HABIT. A tree 50-70 feet high and 1-2 1/2 feet in diameter (max. 100 by 4 feet); broad, open crown with large branches.

LEAVES. 8-12 inches long with 7-9 (rarely 5) leaflets; leaflets lanceolate to obovate-oblanceolate, the terminal 4-7 inches long and somewhat larger than the lateral; finely to coarsely serrate margins; thick, dark yellow-green and glabrous above, paler and hairy below; glandular, resinous, and fragrant.

FRUIT. Globose to obovoid; 1 1/2-2 inches long; husk 1/8-1/4 inch thick, splitting nearly to the base; thick shell; sweet seed.

TWIGS. Stout; gray to red-brown; hairy. Winter buds: terminal 1/2-3/4 inch long; subglobose; red-brown; hairy.

BARK. Thick; firm; dark gray; shallow furrows and low, interlacing ridges on mature trunks.

SILVICAL CHARACTERS. Typical of well-drained upland slopes; associated with oak and other hardwoods.

* * *

Pignut Hickory

Carya glabra (Mill.) Sweet [*Carya ovalis* (Wang.) Sarg.]

Pignut hickory differs from mockernut in having 5 (rarely 7) essentially glabrous leaflets; glabrous terminal buds 1/4-1/2 inch long; fruit usually about 1 inch long, oval, pyriform, or obovoid, with a thin husk (under 1/8 inch) often splitting only part way to the base.

* * *

Four other unimportant species of "true hickories" have furrowed and firm to scaly bark and are distinguished in the key. They are:

1. Scrub Hickory, *Carya floridana* Sarg., on dry Florida hills.
2. Swamp Hickory, *Carya leiodermis* Sarg., on wet sites in Arkansas, Texas, Louisiana, and Mississippi.
3. Sand Hickory, *Carya pallida* (Ashe) Engl. & Graebn., on dry sites from Virginia and Tennessee to Florida and Alabama.
4. Black Hickory, *Carya texana* Buckl., on dry uplands from Indiana to Kansas to Louisiana and Texas.

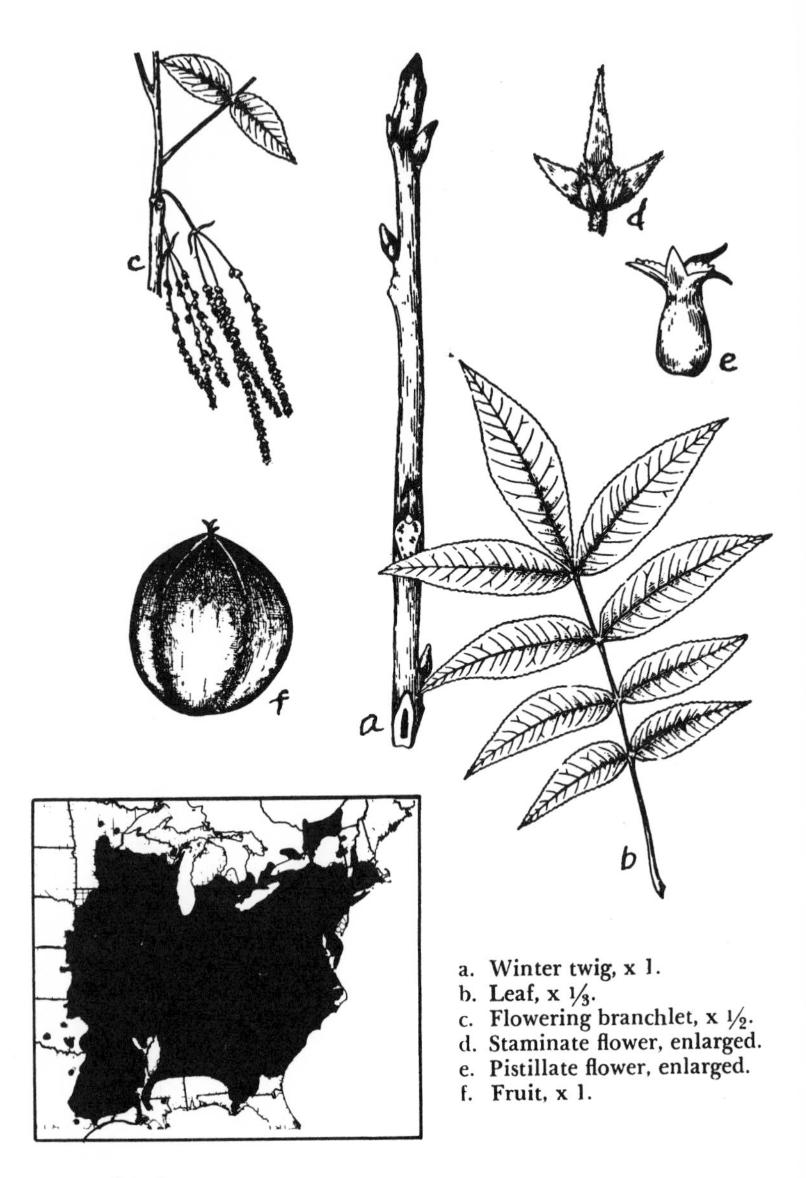

a. Winter twig, x 1.
b. Leaf, x 1/3.
c. Flowering branchlet, x 1/2.
d. Staminate flower, enlarged.
e. Pistillate flower, enlarged.
f. Fruit, x 1.

Bitternut Hickory

Carya cordiformis (Wang.) K. Koch [*Hicoria cordiformis* (Wang.) Britt.]

HABIT. A tree 50-75 feet high and 1-2 feet in diameter (max. 100 by 3 feet); broad, open crown with slender ascending branches.

LEAVES. 6-10 inches long with 7-11 leaflets; leaflets lanceolate to oblong-lanceolate, the terminal 3-6 inches, slightly larger than lateral, finely to coarsely serrate margins, bright green and glabrous above, paler below.

FRUIT. Subglobose; about 1 inch long; husk thin (less than 1/8 inch) with yellow-green, scurfy pubescence, splitting halfway to base; 4-winged above middle; nut smooth, thin-shelled, bitter.

TWIGS. Rather stout; green to gray-brown. Winter buds: terminal 1/3-3/4 inch long, flattish, long-pointed, sulfur-yellow, scurfy-pubescent, valvate scales.

BARK. Thick; firm; gray; shallow furrows and ridges.

SILVICAL CHARACTERS. Adapted to variety of sites; most abundant and uniformly distributed of the hickories; with mixed hardwoods.

* * *

There are three other native species of "pecan hickories":

1. Pecan, *Carya illinoensis* (Wang.) K. Koch (*Carya pecan* Engl. & Graebn.), famed for its nuts, is the largest of the hickories (max. 180 by 6 feet). This species is characterized by having 9-17 lanceolate, usually falcate leaflets; fruit ellipsoidal, 1-2 1/2 inches long, with thin dark brown husk, splitting from apex to base; nut nearly smooth with sweet seed; buds 1/4-1/2 inch long, yellow-brown, and scurfy. Mississippi River Valley from southern Wisconsin to Texas, Nebraska, and Alabama and widely planted through the South.
2. Water Hickory, *Carya aquatica* (Michx.) Nutt., is similar to pecan but differs in having a smaller (1 1/2 inches long) fruit, with a thin nearly black. husk that splits halfway to the base and an obovoid, 4-ribbed nut with a bitter seed. In swamps from Virginia to Texas and north to Illinois.
3. Nutmeg Hickory, *Carya myristicaeformis* (Michx.) Nutt., is a rather rare tree with 5-9 ovate-lanceolate leaflets and an ellipsoidal, 4-winged fruit 1 1/2 inches long, with a very thin husk that splits to the base. In swamps or moist sites from South Carolina to Oklahoma and south to Alabama and Texas.

The Birches

Characteristics of the Genus *Betula* L.

HABIT. Small to medium-sized graceful trees and shrubs; crown on young trees narrow, pyramidal, symmetrical; branches short and slender, more or less erect on young trees and becoming horizontal or pendulous on older trees.

LEAVES. Alternate; simple; deciduous; mostly ovate to triangular; acute to acuminate; serrate, dentate, or lobulate; deciduous; petioled; stipules fugacious; scarious.

FLOWERS. Regular; monoecious; apelatous; appearing before or with leaves; staminate in 1-3-clustered, long, pendulous aments produced early the previous season, every bract with 3 individual flowers, each of 4 stamens adnate to a 4-parted calyx; pistillate in solitary, small, slender aments on ends of spurlike lateral branches below staminate flowers, individual flowers naked, clusters of 3, subtended by 3-lobed bract.

FRUIT. Small, compressed, laterally winged nutlet; erect or pendent strobiles; scales shaped like fleur-de-lis, deciduous from persistent cone axis at maturity, releasing nutlets; maturing in fall of first year (one species in spring).

TWIGS. Slender; round; horizontal lenticels and small leaf scars having 3-bundle scars; spur shoots with paired leaves commonly present on old growth; pith small, round, homogeneous. Winter buds: terminal absent; lateral, acute, 3 (rarely 4) visible scales; twig lengthening by one of upper lateral buds.

BARK. Smooth, papery (or in sweet birch furrowed); resinous; marked by horizontally elongated lenticels; often peeling off in thin, papery layers.

WOOD. Strong; heavy; hard; diffuse-porous; light-colored; some species highly valued for timber.

SILVICAL CHARACTERS. Mostly fast-growing and short-lived; adapted to planting on poor, sandy, or boggy soil; many used for ornamental planting because of handsome foliage and showy bark; lateral root systems.

GENERAL. This genus contains about 40 species of trees and shrubs scattered through the Northern Hemisphere; the European white birth (*Betula pendula* Roth.) and especially its cut-leaf, weeping variety (*dalecarlica* Schn.) are often planted in this country; eleven species are native to North America, 7 of which form trees.

KEY TO THE SPECIES OF BIRCHES

1. Leaves with 9-12 pairs of veins; fruit erect, sessile, or nearly so, oblong-ovoid; wing about as broad as nutlet; twigs with wintergreen flavor; important eastern timber species.
   2. Bark dark, furrowed, not separating into papery layers; bracts of fruit glabrous . . . . *B. lenta*, Sweet Birch, p. 153
   2. Bark dirty-yellow, scaly, separating into thin, papery layers; bracts of fruit hairy . . . . *B. alleghaniensis*, Yellow Birch, p. 151
1. Leaves, 8 or less pairs of veins (7-9 in River Birch); fruit peduncled, oblong or cylindric; twigs, no wintergreen flavor.
   3. Bark red-brown to chestnut-brown.
      4. Bark separating into papery layers; fruit erect; wing narrower than nutlet; eastern . . . . *B. nigra*, River Birch, p. 151
      4. Bark not separating into papery layers; fruit pendent; wing broader than nutlet; western.
         5. Fruit 1/2-3/4 inch; leaves wedge-shaped at base; west Canada, Alaska . . . . *B.* X *eastwoodiae*, Yukon Birch, p. 155
         5. Fruit 1-1 1/4 inches long; leaves usually rounded at base; western United States and southwestern Canada . . . . *B. occidentalis*, Water Birch, p. 155
   3. Bark distinctly white in color.
      6. Bark chalky-white, not separating freely into papery layers; leaves triangular, bright green; Delaware, Pennsylvania to eastern Canada . . . . *B. populifolia*, Gray Birch, p. 153
      6. Bark lustrous, pink-white, separating freely into papery layers; leaves usually ovate to oval.
         7. Young twigs hairy, nearly nonglandular; bracts of fruit with middle lobe much longer than lateral; northern transcontinental . . . . *B. papyrifera*, Paper Birch, p. 149
         7. Young twigs glabrous and resinous-glandular; bracts of fruit with middle lobe not longer than lateral lobes.
            8. Leaves hairy along veins on lower surface; small tree from Vermont north to east Canada . . . . *B. caerulea-grandis*, Blueleaf Birch, p. 149
            8. Leaves strictly glabrous; a handsome, widely cultivated, Eurasian tree; often with pendulous branches and lobed leaves . . . . *B. pendula*, European Birch, p. 146

* * *

LEITNERIACEAE

Corkwood

Leitneria floridana Chapm.

This shrub or small tree with extremely soft wood is the only species in the family. It is rare and local in the coastal plain of southeastern Georgia, western Florida, and southeastern Texas; also in eastern Arkansas and southeastern Missouri. Simple, alternate, deciduous leaves are 4-6 inches long, lanceolate and entire. Fruit is an oval, pointed, dry drupe 3/4 inch long with conspicuously netted veins. Twigs are stout with small, scaly buds; marked by conspicuous circular lenticels and elevated leaf scars.

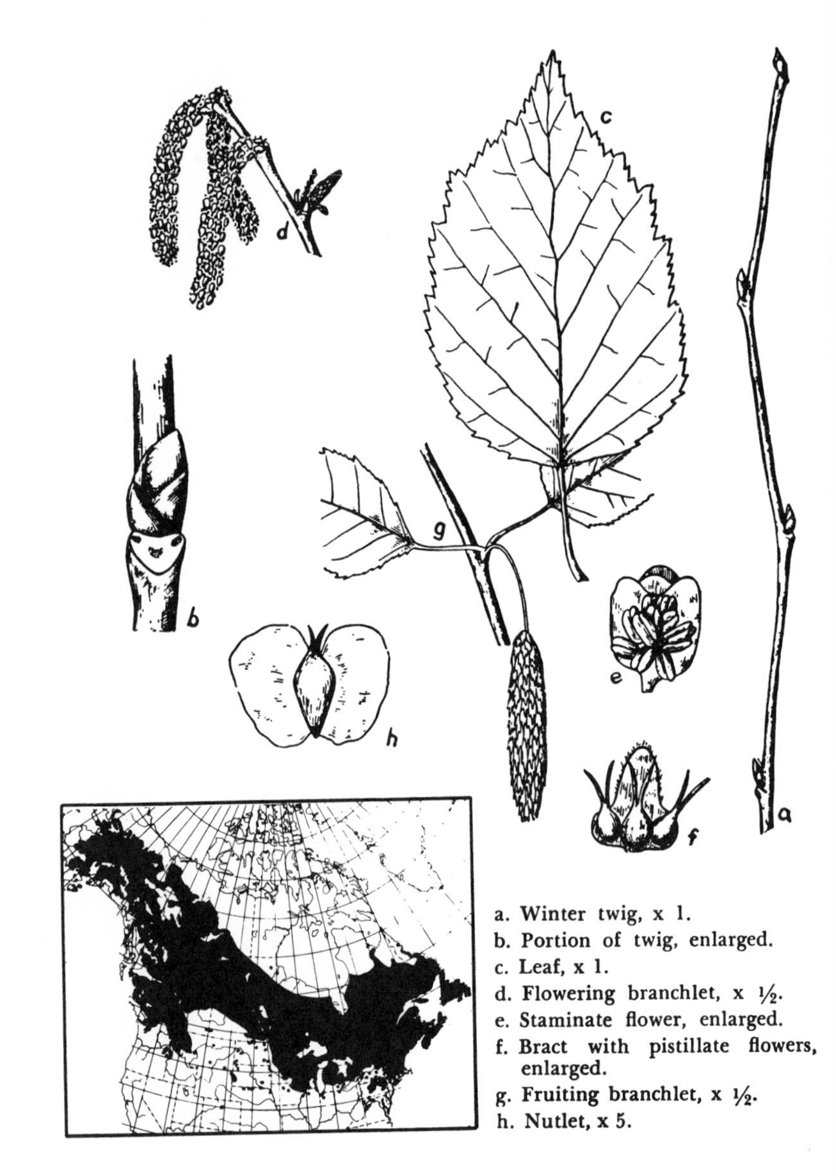

a. Winter twig, x 1.
b. Portion of twig, enlarged.
c. Leaf, x 1.
d. Flowering branchlet, x ½.
e. Staminate flower, enlarged.
f. Bract with pistillate flowers, enlarged.
g. Fruiting branchlet, x ½.
h. Nutlet, x 5.

BETULACEAE

Paper Birch

Betula papyrifera Marsh.

HABIT. A tree 60-70 feet high and 2-3 feet in diameter (max. 120 by 4 feet); old trees with open crowns and short, pendulous branches.

LEAVES. Ovate; 2-5 inches long; coarsely, irregularly, and usually doubly serrate; dull dark green and glabrous above; light yellow-green, black-glandular, glabrous or puberulous below.

FRUIT. Strobiles cylindrical, pendent, on slender peduncles, 1-1 1/2 inches long; scales about as long as broad, puberulous; nutlet narrower than wings.

TWIGS. At first green, hairy, and marked by scattered, orange-colored, oblong lenticels; becoming dark orange-brown and lustrous; not aromatic. Winter buds: 1/4 inch long; obovoid; acute; dark chestnut-brown; glabrous.

BARK. Cream-white; separating into thin, papery layers; marked by long, narrow, raised lenticels; inner bark orange.

WOOD. Moderately important; used for pulp, lumber, turned articles, fuel.

SILVICAL CHARACTERS. Canadian zone; intolerant; short-lived; spreading roots; reproduction vigorous, taking over extensive areas following fire; rich or sandy soils.

GENERAL. Six varieties have been differentiated and are included in the distribution map:

1. *commutata* (Reg.) Fern. from Idaho to Washington and British Columbia.
2. *cordifolia* (Reg.) Fern. from Labrador to Minnesota and New York.
3. *humilis* (Reg.) F. & R. in western Canada and Alaska.
4. *kenaica* (Evans) Henry in Alaska.
5. *montanensis* (But.) Sarg. in Montana and Idaho.
6. *subcordata* (Rydb.) Sarg. from Idaho to Oregon and western Canada.

* * *

Blueleaf Birch, *Betula caerulea-grandis* Blanchard, is a small tree found in Nova Scotia, Maine, and Vermont. It is characterized by ovate, blue-green leaves and pink-white bark that is not papery, or is exfoliating but slightly.

Yellow Birch

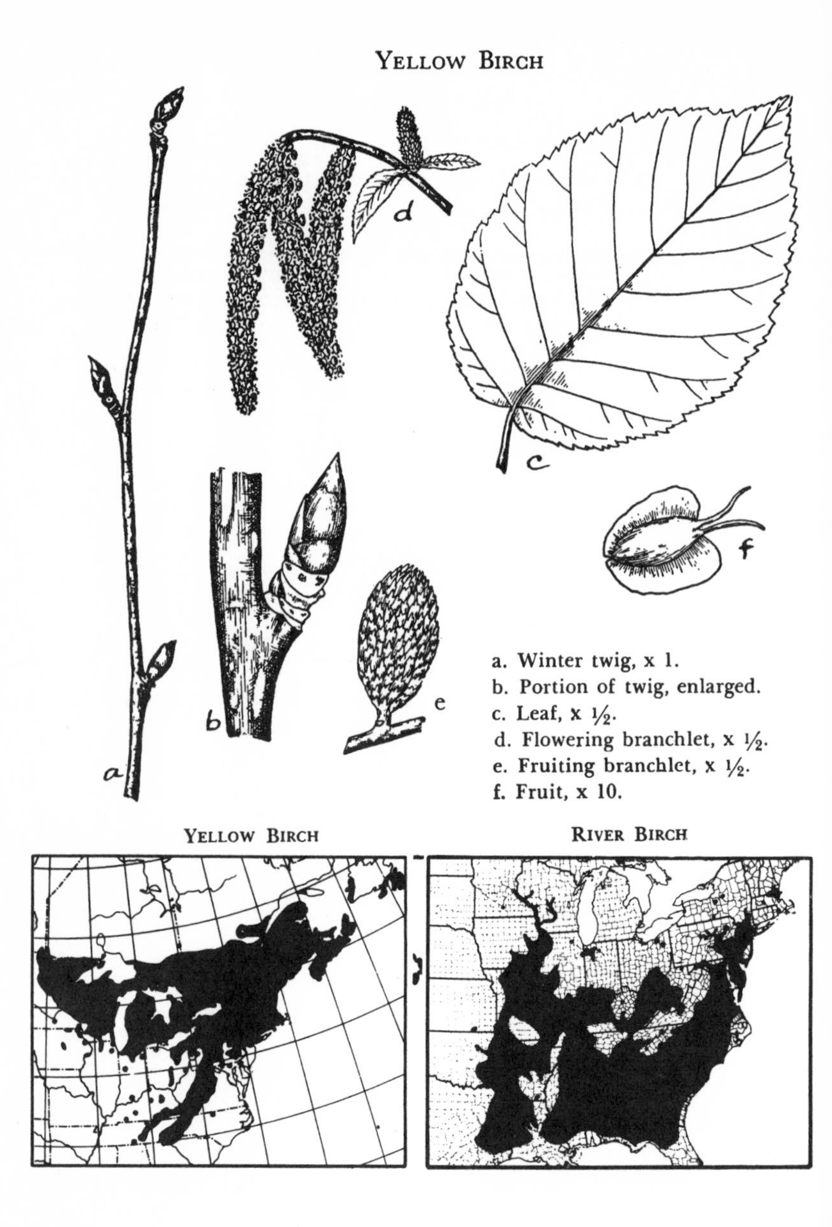

a. Winter twig, x 1.
b. Portion of twig, enlarged.
c. Leaf, x ½.
d. Flowering branchlet, x ½.
e. Fruiting branchlet, x ½.
f. Fruit, x 10.

Yellow Birch

Betula alleghaniensis Britt. (*Betula lutea* Michx.)

HABIT. A tree 60-80 feet high and 1-2 feet in diameter (max. 100 by 4 feet); rounded, open crown and pendulous branches.

LEAVES. Ovate to oblong-ovate; 3-5 inches long; sharply doubly serrate; rounded, inequilateral base; dull, dark green and glabrous above, paler below; slightly aromatic; 9-11 pairs of veins.

FRUIT. Strobiles ovoid, erect, nearly sessile, 1-1 1/2 inches long; scales hairy, longer than broad, wide lateral lobes, tardily deciduous; nutlet about as broad as wing.

TWIGS. Smooth; lustrous; yellow-brown to dark brown; slight wintergreen taste (source of oil of wintergreen). Winter buds: lateral 1/4 inch long; ovate; acute; chestnut-brown; ciliate on scale margins.

BARK. Thin; dirty yellow to bronze-colored; separating horizontally into thin, papery, curled strips.

WOOD. Very important, supplying about 75 percent of birch lumber; used for furniture, flooring, veneer, etc.

SILVICAL CHARACTERS. Intermediate in tolerance; maturity reached in about 150 years; growth rather rapid; reproduction vigorous; shallow roots; restricted to cool, moist sites; in mixtures with conifers and hardwoods.

* * *

River Birch

Betula nigra L.

This unimportant species is a medium-sized tree typical of stream banks. It is quite similar to yellow birch in appearance and silvical characters and differs from it in the following ways:

LEAVES. Rhombic-ovate, often lobed; base wedge-shaped; more or less hairy below; 5-9 pairs of veins.

FRUIT. Unique in that it matures in May or June; strobiles oblong, erect, on stout peduncles; scales with narrow, hairy, erect lobes.

TWIGS. Red-brown; hairy at first; not aromatic. Winter buds: 1/4 inch long; woolly through summer.

BARK. Thin; salmon-pink to red-brown; separating into thin, papery, curled strips; furrowed and scaly at base of old trees.

Sweet Birch

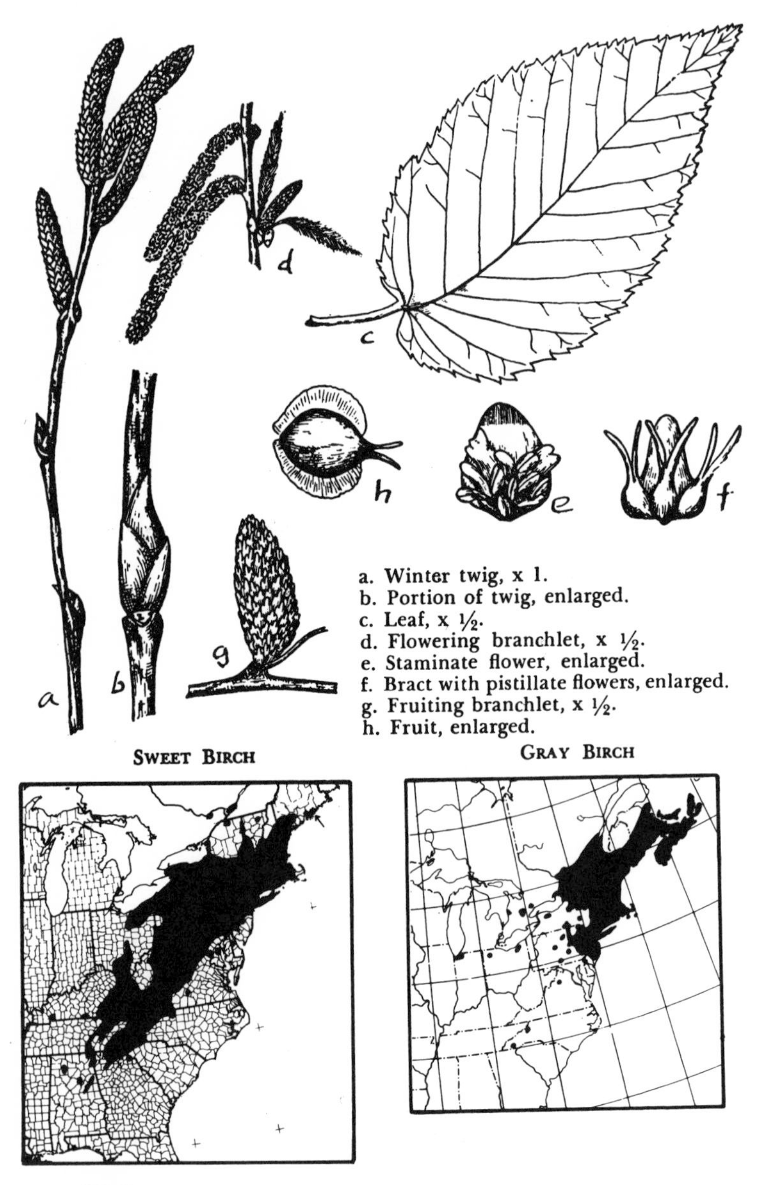

a. Winter twig, x 1.
b. Portion of twig, enlarged.
c. Leaf, x ½.
d. Flowering branchlet, x ½.
e. Staminate flower, enlarged.
f. Bract with pistillate flowers, enlarged.
g. Fruiting branchlet, x ½.
h. Fruit, enlarged.

BETULACEAE

Sweet Birch

Betula lenta L.

HABIT. A medium-sized tree 50-75 feet high and 1-2 feet in diameter (max. 80 by 4 1/2 feet); rounded, open crown with pendulous branches.

LEAVES. Ovate to oblong-ovate; 2 1/2-4 inches long; usually singly serrate; heart-shaped or unequally rounded at base; dull, dark green, and glabrous above, paler below; strongly aromatic; 9-11 pairs of veins.

FRUIT. Strobiles oblong-ovoid, erect, nearly sessile, 1-1 1/2 inches long; scales glabrous, longer than broad, wide lateral lobes; nutlet about as broad as wing.

TWIGS. Smooth; lustrous; light red-brown; strong wintergreen taste (source of oil of wintergreen). Winter buds: 1/4 inch long; conical; acute; red-brown; mostly glabrous.

BARK. 1/2-3/4 inch thick; nearly black; not papery; breaking into irregular, thin, scaly plates; smooth on young trees.

WOOD. Important and similar to yellow birch.

SILVICAL CHARACTERS. Intermediate in tolerance; maturity reached in about 150 years; growth rather slow; reproduction not abundant; typical on moist sites, but hardy on poor, dry soils; a scattered tree with mixed hardwoods.

* * *

Gray Birch

Betula populifolia Marsh.

This unimportant, small tree is typical on poor sites in the Northeast. In its silvical characters it is similar to paper birch.

LEAVES. Triangular to rhombic; long-pointed; doubly serrate; 5-9 pairs of veins; petioles slender, elongated.

FRUIT. Strobiles cylindrical, pendent, on slender peduncles; scales about as long as broad with recurved lateral lobes; nutlet slightly narrower than wing.

TWIGS. Resinous-glandular; red-yellow; not aromatic. Winter buds: 1/4 inch long; ovoid; chestnut-brown; gummy.

BARK. Thin; gray-white; close and firm; not papery, or exfoliating very slightly; black, triangular patches below branches.

WATER BIRCH

a. Twig with leaves and fruit, x 1.

b. Staminate aments, x ½.

c. Scale of fruit, x 4.

Water Birch. Red Birch

Betula occidentalis Hook. (*Betula fontinalis* Sarg.)

HABIT. A shrub or small tree 20-25 feet high and 12-14 inches in diameter; broad, open crown with ascending branches; frequently in crowded, dense thickets.

LEAVES. Ovate; 1-2 inches long and 3/4-1 inch wide; acute or acuminate; sharply and often doubly serrate; sometimes slightly lobed; thin and firm; glabrous; dark dull green above, pale yellow-green and minutely glandular below; turning dull yellow in autumn; petioles stout, 1/3-1/2 inch long, light yellow, glandular-dotted; stipules bright green, slightly ciliate.

FRUIT. Strobiles cylindrical, pendent (rarely erect) long-stalked, 1-1 1/4 inches long; scales ciliate; nutlet slightly narrower than wing.

TWIGS. At first light green and glandular, becoming dark red-brown; marked by horizontal lenticels. Winter buds: 1/4 inch long; ovoid; acute; very resinous; chestnut-brown.

BARK. Thin (1/4 inch); smooth; lustrous dark bronze; marked by pale horizontal lenticels, becoming 6-8 inches long and 1/4 inch wide on old trunks.

WOOD. Rather light and soft; strong; heartwood light brown; sapwood thick, light-colored; not important; used locally for fencing and fuel.

SILVICAL CHARACTERS. Upper Sonoran and transition zones; intolerant; shallow root system; reproduction abundant in moist, mineral soil; generally along borders of streams in moist mountain valleys and canyons.

* * *

Yukon birch, *Betula* X *eastwoodiae* Sarg., a hybrid of *B. glandulosa* and *B. papyrifera* is a small tree or shrub found in swampy sites in northern Alberta and the Yukon. It is characterized by broadly ovate to elliptic leaves 1-1 1/2 inches long; pendulous, cylindric strobiles 3/4 inch long, with glabrous scales longer than broad; and close, chestnut-brown bark not readily separating into papery scales. The bog or resin birch, *Betula glandulosa* Michx., is a shrubby circumpolar form of cold sites.

BETULACEAE

The Alders

Characteristics of the Genus *Alnus* B. Ehrh.

HABIT. Deciduous shrubs or small to medium-sized trees.

LEAVES. Alternate; simple; usually serrate or dentate; pinnately veined; falling without change of color; petioled; stipules fugacious, ovate, acute, scarious.

FLOWERS. Regular; monoecious; apetalous; mostly appearing before or with leaves (rarely opening in autumn); in 1-3-flowered cymes; formed during previous season; staminate in long, pendulous aments, every scale bearing 3-6 flowers, each flower subtended by 3-5 bractlets and composed of 4-parted calyx and 4 (rarely 1-3) stamens; pistillate in erect, stalked, ovoid or oblong aments, appearing below staminate flowers, individual flowers in pairs, composed of a naked ovary surmounted by 2 stigmas and subtended by 2-4 bractlets.

FRUIT. Small, flat, chestnut-brown, wingless or laterally winged nutlet, bearing remnants of style at apex; in persistent, semiwoody strobiles, each scale bearing 2-4 nutlets.

TWIGS. Slender to moderately stout; round; reddish or tinged with red; marked by raised leaf scars with 3-bundle scars and lenticels; pith homogeneous, triangular in cross section. Winter buds: terminal absent; lateral typically stalked and 2-3-scaled; usually red; twig lenthening by one of upper lateral buds.

BARK. Astringent; mostly gray; smooth, except at the base of trunks of large trees.

WOOD. Light; soft; straight-grained; diffuse-porous; durable in water; heartwood red-brown; sapwood very thick and whitish.

SILVICAL CHARACTERS. Tolerant to intolerant; rather short-lived; shallow, spreading roots; on moist or wet sites, commonly along streams or on mountain slopes.

GENERAL. This genus contains about 30 species scattered through the cooler portions of the Northern Hemisphere and extending into the mountains of South America; 8 species attaining tree size are native to North America, although 7 of these are typically shrubs and only *Alnus rubra* is of commercial importance. The European alder, *Alnus glutinosa* (L.) Gaertn., has become naturalized in the eastern United States.

KEY TO THE SPECIES OF ALDERS

1. Buds sessile, 3-6-scaled, dark purple; pistillate ament enclosed in bud during winter; leaves ovate, lobulate, usually irregularly singly toothed; Alaska to California and east to Montana . . . . *Alnus sinuata* (Reg.) Rydb., Sitka Alder, p. 161

1. Buds stalked, 2-3 scaled; pistillate ament appearing in fall (sometimes enclosed during winter in red alder).

    2. Leaves with 8-15 pairs of veins.

        3. Leaves lobulate, doubly toothed; buds reddish.

            4. Wing of nut broad; leaves ovate to elliptic, often red-hairy below, slightly revolute; important tree; Alaska to California and east to Idaho . . . . *Alnus rubra* Bong., Red Alder, p. 159

            4. Wing of nut reduced to narrow border; leaves neither red-hairy nor revolute; shrubs or rarely small trees to 30 feet.

                5. Twigs marked with orange lenticels; Rocky Mountains and Pacific Coast from Yukon to Mexico . . . . *Alnus tenuifolia* Nutt., Thinleaf Alder, p. 161

                5. Twigs marked with white lenticels; Newfoundland to Saskatchewan, south to Pennsylvania and Nebraska, also in Europe and Asia . . . . *Alnus rugosa* (D. R.) Spr., Speckled Alder, p. 161

        3. Leaves finely to coarsely toothed.

            6. Western shrubs or small trees; leaves coarsely toothed.

                7. Leaves oval or ovate; Idaho to Washington and south to lower California . . . . *Alnus rhombifolia* Nutt., White Alder

                7. Leaves oblong-lanceolate; New Mexico, Arizona, south into Mexico . . . . *Alnus oblongifolia* Torr., Arizona Alder

            6. Eastern shrubs, rarely small trees; leaves finely toothed.

                8. Leaves with straight veins; several pistillate catkins per bud; Maine to Minnesota and south to Florida and Texas . . . . *Alnus serrulata* (Alt) Willd., Hazel Alder

                8. Leaves with arching veins; 1-3 pistillate catkins per bud; Delaware, Maryland, Oklahoma . . . . *Alnus maritima* (Marsh.) Muhl., Seaside Alder

    2. Leaves with 5-6 pairs of veins, oval to orbicular, coarsely and doubly toothed; native to Eurasia but naturalized in the northeastern United States . . . . *Alnus glutinosa* (L.) Gaertn., European Alder

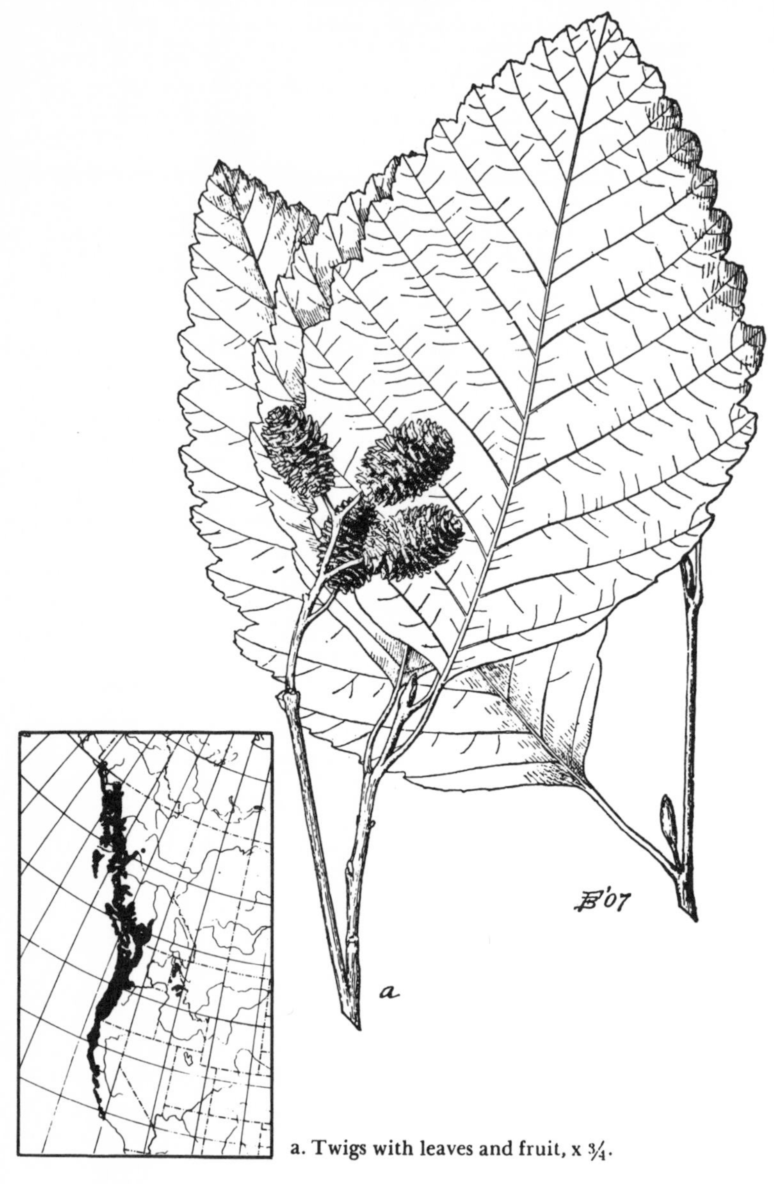

a. Twigs with leaves and fruit, x ¾.

BETULACEAE

Red Alder

Alnus rubra Bong.

HABIT. A medium-sized tree 80-100 feet high and 1-3 feet in diameter (max. 130 by 5 feet); clear trunk; crown narrow and rounded with pendulous branches.

LEAVES. Ovate to elliptic; 3-6 inches long and 1 1/2-3 inches wide; obtuse or rounded at base; apex acute; margins slightly crenately lobed with doubly dentate gland-tipped teeth; thick and firm; dark green and glabrous above, paler and rusty-pubescent below; petiole orange, round, grooved, 1/4-3/4 inch long.

FLOWERS. Preformed; the staminate aments 1 1/4 inches long, becoming 4-6 inches long, orange; pistillate opening in spring, 1/3-1/2 inch long with dark red scales and bright red style.

FRUIT. Strobiles 1/2-1 1/4 inches long; woody; persistent; ovoid to oblong; on orange peduncles or rarely sessile; truncate scales with much thickened, rugose tips; nutlet nearly circular with encircling membranous wing or 2 lateral wings.

TWIGS. Slender to rather stout; light green and tomentose at first, becoming bright red and lustrous during second year. Winter buds: lateral 1/3-2/3 inch long; stalked; dark red; pale, scurfy pubescence.

BARK. Thin; smooth; nearly white to blue-gray outer bark and bright red-brown inner bark; sometimes roughened by small, warty excrescences; breaking into large, flat plates on old trees.

WOOD. Important (probably the most important hard wood on the Pacific Coast); light; soft; prominent rays; used for furniture, veneer, and novelties.

SILVICAL CHARACTERS. Intermediate in tolerance; rapid growth; short-lived, maturity being reached in 60-90 years; lateral roots; aggressive reproduction, often taking over burned or logged areas; on varied sites, but typical of stream banks; in pure or mixed stands.

Thinleaf Alder

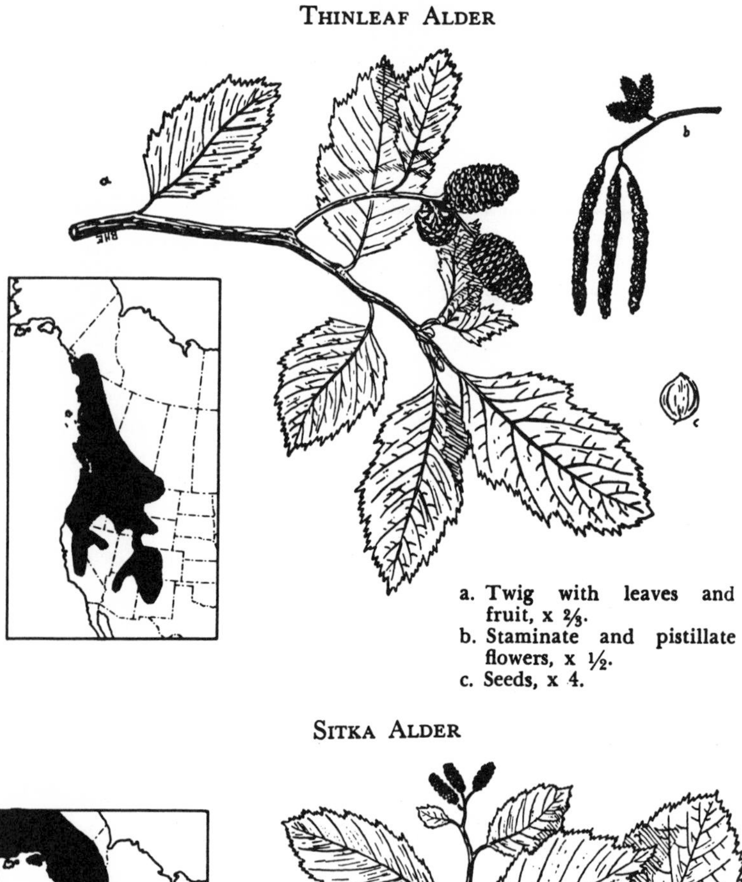

a. Twig with leaves and fruit, x ⅔.
b. Staminate and pistillate flowers, x ½.
c. Seeds, x 4.

Sitka Alder

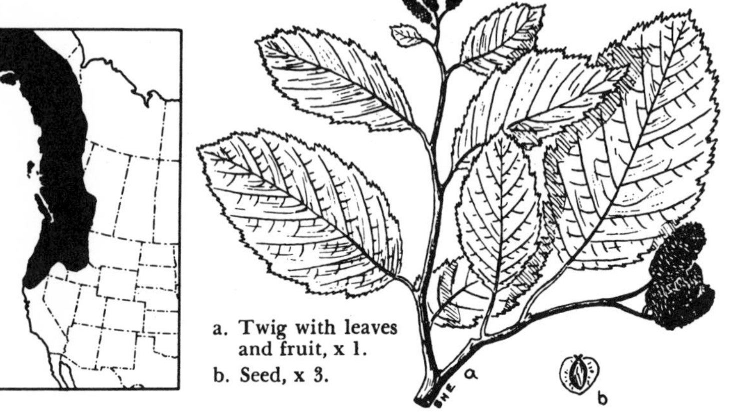

a. Twig with leaves and fruit, x 1.
b. Seed, x 3.

Thinleaf Alder. Mountain Alder

Alnus tenuifolia Nutt.

HABIT. A shrub or small tree occasionally 30 feet high and 6-8 inches in diameter; crown narrow and round-topped.

LEAVES. Ovate-oblong; 2-4 inches long; slightly, acutely and laciniately lobed; doubly serrate; thin and firm; dark green and glabrous above, pale yellow-green below.

FRUIT. Nut nearly circular; wing reduced to thin, membranaceous border; strobiles 1/3-1/2 inch long, obovoid-oblong; scales truncate, much thickened, 3-lobed at apex.

TWIGS. Slender; marked by few, large orange-colored lenticels. Winter buds: stalked; 1/4-1/3 inch long; bright red; puberulous.

BARK. Thin; smooth; red-brown.

SILVICAL CHARACTERS. Tolerant when young, becoming intolerant with age; on banks of mountain streams; the common alder of the Rocky Mountain region; closely related and considered by some authors to be a synonym of the eastern speckled alder, *Alnus rugosa.*

* * *

Sitka Alder

Alnus sinuata (Reg.) Rydb. (*Alnus sitchensis* Sarg.)

HABIT. A shrub or small tree rarely 40 feet high and 7-8 inches in diameter; crown narrow and open.

LEAVES. Ovate; 3-6 inches long; acute; usually divided into numerous, short, lateral lobes; sharply and doubly serrate; membranaceous; yellow-green above, pale below.

FRUIT. Nut oval, about as wide as its wings; strobiles 1/2-3/4 inch long and about 1/3 inch wide; truncate scales thickened at apex; on slender peduncles; in elongated, leafy panicles.

TWIGS. Slender; large, pale lenticels. Winter buds: sessile, 1/2 inch long; acuminate; dark purple; finely pubescent.

BARK. Thin; blue-gray; bright red inner bark.

SILVICAL CHARACTERS. Tolerant when young, becoming intolerant with age; in moist flats and along stream borders.

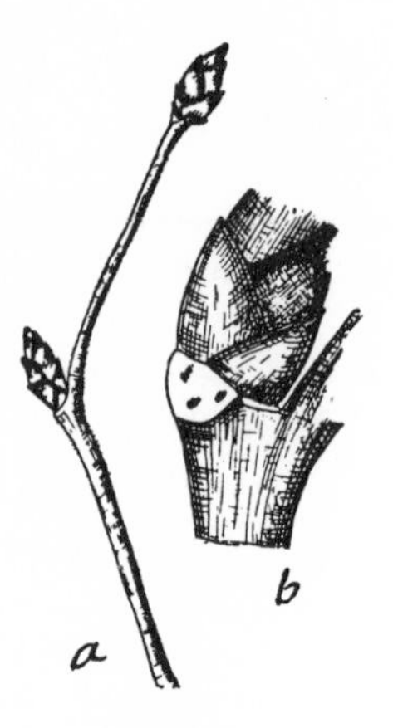

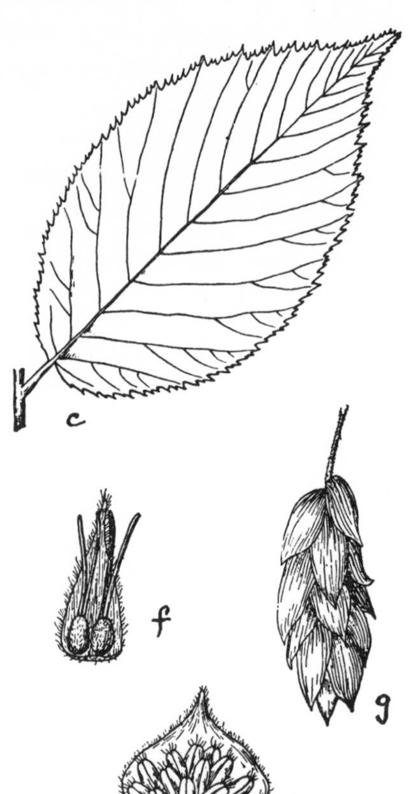

a. Winter twig, x 1.

b. Portion of twig, enlarged.

c. Leaf, x ½.

d. Flowering branchlet, x ½.

e. Staminate flower, enlarged.

f. Pistillate flower, enlarged.

g. Fruit, x ½.

Eastern Hophornbeam

Ostrya virginiana (Mill.) K. Koch

HABIT. A small tree 20-40 feet high and 8-18 inches in diameter (max. 65 by 3 feet); rounded crown and slender, pendulous, often contorted branches.

LEAVES. Alternate; simple; oblong-ovate; 3-5 inches long with 11-15 pairs of veins; deciduous; acuminate; sharply doubly serrate margins; thin and tough; dull dark green above, paler and somewhat hairy below; stipules fugacious.

FLOWERS. Regular; monoecious; perianth absent; appearing with leaves; staminate in long, drooping, cylindrical aments in clusters of 3, each flower containing 3-14 stamens crowded on a hairy receptacle; pistillate in erect, loose, paired aments, each flower enclosed in a hairy, saclike involucre.

FRUIT. Small (1/4 inch), 1-celled, 1-seeded, ovoid, flat, unwinged nut; enclosed in enlarged (1 inch), pale, membranaceous, involucre of flower; in loose, suspended strobiles 1-1 1/2 inches long, resembling clusters of hops.

TWIGS. Slender; round; red-brown becoming dark brown; leaf scars with 3 bundle scars. Winter buds: terminal absent; lateral 1/8-1/4 inch long; ovoid; acute; bud scales with greenish base and brown tips and characteristic striations.

BARK. Thin; gray-brown; broken into small, shaggy plates that give it characteristic shreddy appearance.

WOOD. Very heavy; hard; strong; unimportant; resembling hickory and known as ironwood; little used.

SILVICAL CHARACTERS. Tolerant; slow-growing; lateral roots; varied sites, but typical on dry slopes.

* * *

Knowlton Hophornbeam

Ostrya knowltonii Cov.

This rare, local tree is found in mountains and canyons of western Texas, southeastern New Mexico, northern Arizona, and southeastern Utah. It differs from *O. virginiana* in having leaves 1-2 inches long with 5-8 pairs of veins.

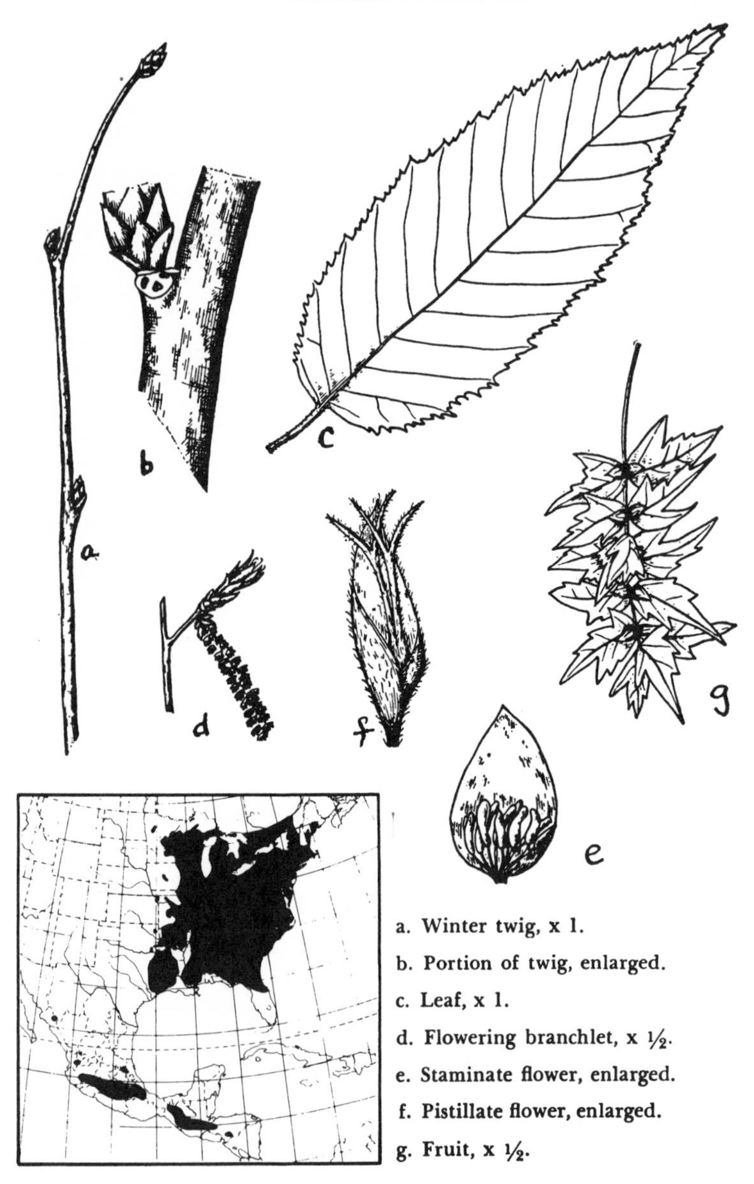

a. Winter twig, x 1.
b. Portion of twig, enlarged.
c. Leaf, x 1.
d. Flowering branchlet, x ½.
e. Staminate flower, enlarged.
f. Pistillate flower, enlarged.
g. Fruit, x ½.

BETULACEAE

American Hornbeam. Bluebeech

Carpinus caroliniana Walt.

HABIT. A large shrub or small bushy tree rarely 40 feet high and 1 foot in diameter; trunk characteristically fluted; crown close and flat-topped with slender, zigzag branches.

LEAVES. Alternate; simple; ovate to oval; 2-4 inches long; two distinct sizes; deciduous; acuminate; sharply doubly serrate; thin and tough; dull dark green above, paler below.

FLOWERS. Regular; monoecious; perianth absent; appearing with leaves; staminate in aments 1 1/2 inches long, with green, hairy scales, each flower with 3-20 stamens; pistillate in loose aments 1/2-3/4 inch long, each bearing 2 pistils with scarlet styles.

FRUIT. Small (1/3 inch), 1-seeded, ovoid, flat, brown, unwinged nut; attached to base of halberd-shaped, leafy involucre; in loose strobiles.

TWIGS. Slender; round; pale green at first, becoming lustrous and red-brown; leaf scars with 3 bundle scars. Winter buds: terminal absent; lateral 1/8 inch long; rusty brown; acute; scaly; usually 4-sided in cross section.

BARK. Thin; smooth; blue-gray; often mottled with light or dark patches; fluted trunk an unmistakable character.

WOOD. Heavy; hard; strong; unimportant.

SILVICAL CHARACTERS. Tolerant; slow-growing; lateral roots; on varied sites but prefers moist soil.

* * *

Hazelnut

Corylus L.

Three shrubby species are native to North America and one of these, *Corylus cornuta* var. *californica* (A. OC.) Sharp, may occasionally reach tree size in California. The other two species are small eastern shrubs known for their edible nuts, American hazelnut (*C. americana* Walt.) and beaked hazelnut (*C. cornuta* Marsh. var. *cornuta*). The giant filbert, *Corylus maxima* Mill., of southern Europe is planted in warm parts of the United States. The hazelnuts are characterized by deciduous, alternate, simple, oval to ovate, doubly serrate leaves; scaly lateral buds; and an ovoid nut surrounded by a leafy, toothed involucre.

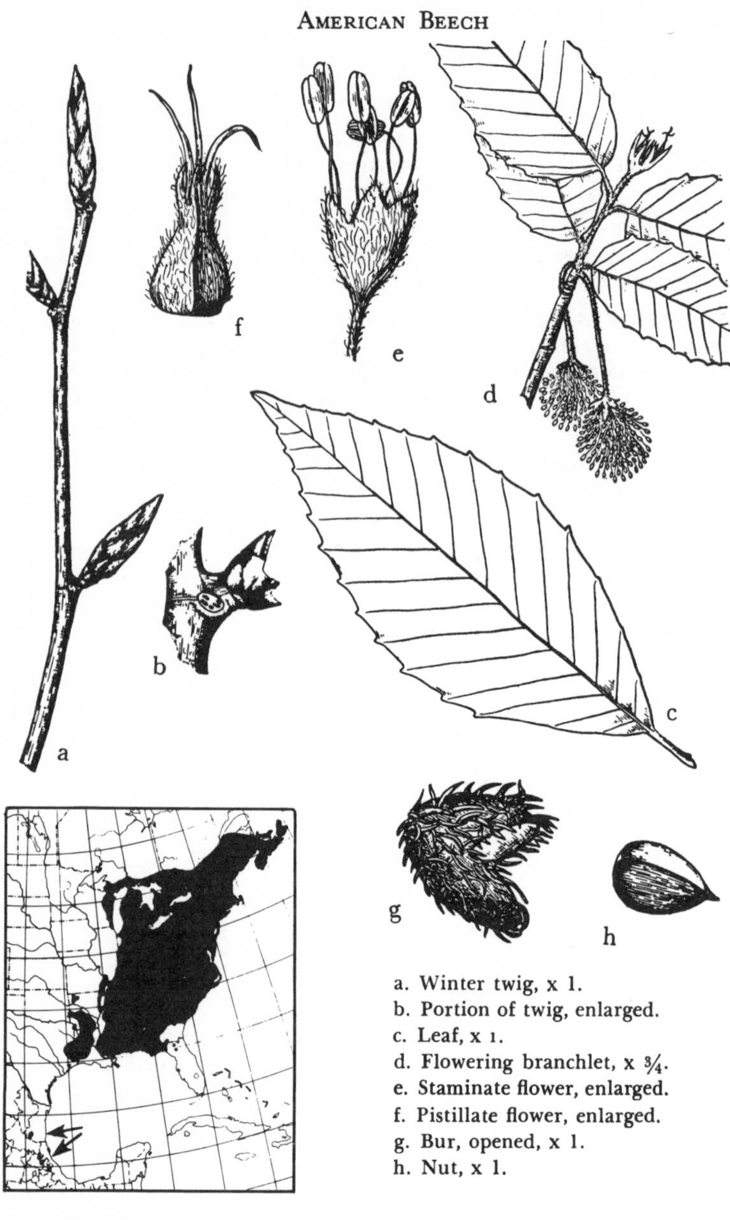

a. Winter twig, x 1.
b. Portion of twig, enlarged.
c. Leaf, x 1.
d. Flowering branchlet, x ¾.
e. Staminate flower, enlarged.
f. Pistillate flower, enlarged.
g. Bur, opened, x 1.
h. Nut, x 1.

FAGACEAE

American Beech

Fagus grandifolia Ehrh.

HABIT. A beautiful, medium-sized tree 60-80 feet high and 2-3 feet in diameter (max. 120 by 5 1/2 feet); clear straight trunk; compact, rounded crown with slender, spreading branches.

LEAVES. Alternate; simple; deciduous; elliptical to oblong-ovate; 2 1/2-6 inches long; penniveined; coarsely serrate with sharp teeth; acuminate; thin; dark blue-green above and yellow-green and lustrous below; short-petioled; stipules small.

FLOWERS. Monoecious; apetalous; appearing after the leaves open; staminate in globose heads 1 inch in diameter, each with a 4-8-lobed calyx and 8-16 stamens; pistillate in 2-4-flowered spikes, each with a 4-5-lobed calyx, a 3-celled ovary, and 3 styles.

FRUIT. A triangular, edible nut 1/2-3/4 inch long; 2-3 completely enclosed in a woody, 4-parted bur or involucre covered with weak, unbranched spines; maturing in one year.

TWIGS. Slender; round; lustrous; olive green, becoming ash-gray; occasionally zigzag; pith round; leaf scars small with many bundle scars. Winter buds: terminal present; 3/4-1 inch long; slender; lance-shaped; numerous scales.

BARK. Thin; close; smooth; light blue-gray; often mottled; similar in appearance on old and young trunks.

WOOD. Important; strong; heavy; hard; diffuse-porous; light color; not durable; used for flooring, handles, furniture.

SILVICAL CHARACTERS. Very tolerant; rather fast-growing; long-lived, attaining an age of about 400 years; shallow, lateral roots; reproduction good; roots sprout vigorously; on moist sites; in mixture with numerous hardwoods.

GENERAL. This genus contains about 10 species scattered through the Northern Hemisphere, with only one native to North America. The important, commonly planted, European beech, *Fagus sylvatica* L. has darker bark and smaller leaves 2-4 inches long, with smaller rounded teeth and hairy margins and veins on the lower surface; three common varieties are: *atropunicea* West., with dark bronze-purple leaves; *pendula* Loud., with the weeping habit of drooping branches; and *laciniata* Vig., with deeply cut leaves.

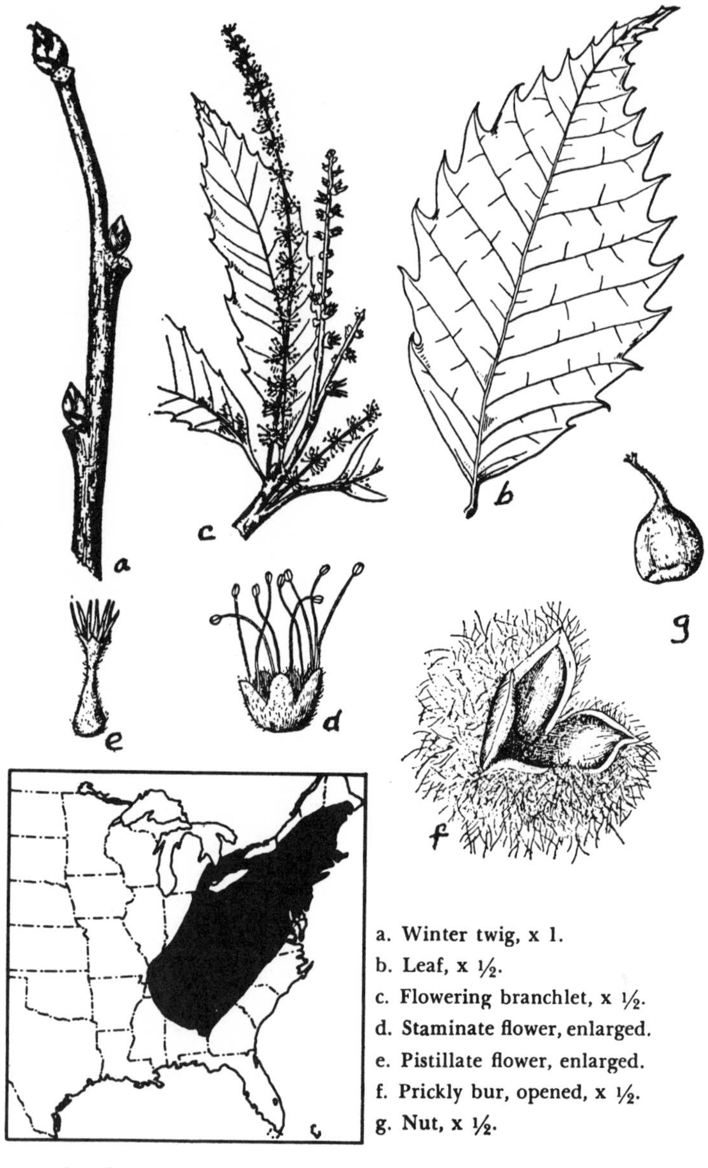

a. Winter twig, x 1.
b. Leaf, x ½.
c. Flowering branchlet, x ½.
d. Staminate flower, enlarged.
e. Pistillate flower, enlarged.
f. Prickly bur, opened, x ½.
g. Nut, x ½.

FAGACEAE

American Chestnut

Castanea dentata (Marsh.) Borkh.

HABIT. A handsome tree 60-90 feet high and 2-4 feet in diameter (max. 120 by 10 feet); frequently divided trunk; broad, open crown with large, horizontal branches.

LEAVES. Alternate; simple; deciduous; oblong-lanceolate; 5 1/2-8 inches long; penniveined; coarsely serrate with sharp, glandular teeth; acuminate; thin; yellow-green; glabrous.

FLOWERS. Monoecious; apetalous; appearing after leaves; staminate and bisexual aments; staminate in 3-7-flowered cymes; pistillate flowers solitary or clusters of 2-3.

FRUIT. Rounded, flattened, chestnut-brown, edible nut 1/2-1 inch long; 2-3 nuts completely enclosed in a 2-4-valved, globose bur, 2-2 1/2 inches in diameter and covered with prickly, sharp, branched spines; maturing in one year.

TWIGS. Rather stout; round; lustrous; chestnut-brown; glabrous; oval leaf scars with many bundle scars; pith stellate. Winter buds: terminal absent; lateral ovoid, acute, brown, 1/4 inch long, 2-3 visible scales.

BARK. Thick; gray-brown; furrowed, with broad, flat ridges.

WOOD. Formerly important; soft; very durable; ring-porous; used for furniture, poles, posts, ties, tannin.

SILVICAL CHARACTERS. Intermediate in tolerance; long-lived; taproot; stump sprouts vigorously; threatened with extinction by chestnut blight, a fungal bark disease; varied sites; with mixed hardwoods.

* * *

The Chinkapins

Four other native species of *Castanea* form small trees or shrubs in the southern states. These are known as Chinkapins and differ from chestnut in having hairy or woolly leaves, woolly twigs and buds, and smaller, single-seeded burs.

1. Leaves silky-hairy to nearly glabrous below, 3-4 inches long; spines on bur scattered and short; North Carolina to Arkansas and south . . . . *C. alnifolia* var. *floridana* Sarg., Florida Chinkapin
1. Leaves woolly below.
   2. Spines on bur crowded and slender.
      3. Leaves 5-9 inches long, bristle-tipped; Missouri, Arkansas, Oklahoma . . . . *C. ozarkensis* Ashe, Ozark Chinkapin
      3. Leaves 3-6 inches, short, rigid teeth; New Jersey to Missouri, south . . . . *C. pumila* (L.) Mill., Allegheny Chinkapin
   2. Bur spines scattered, short, stout; leaves 2-3 inches; coastal plain, Virginia to Texas . . . . *C. ashei* Sudw., Ashe Chinkapin

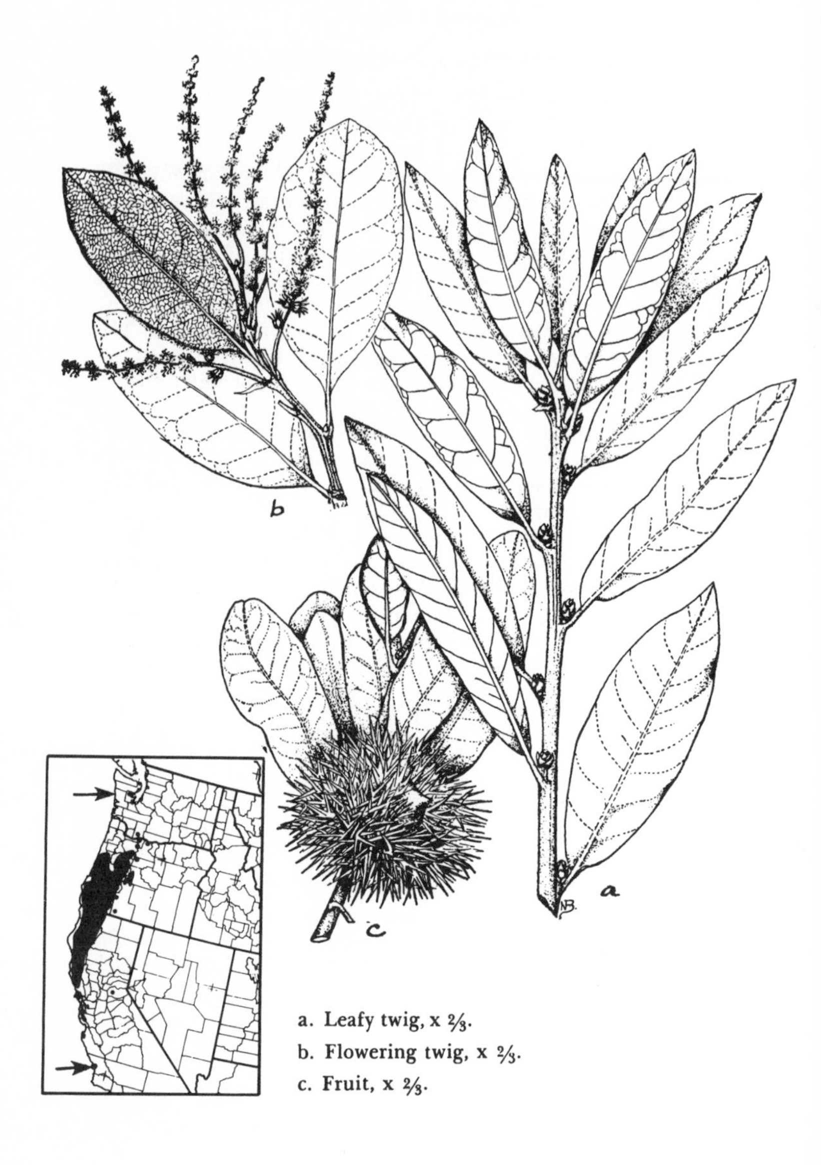

a. Leafy twig, x 2/3.
b. Flowering twig, x 2/3.
c. Fruit, x 2/3.

FAGACEAE

Golden Chinkapin

Castanopsis chrysophylla (Dougl.) A. DC.

HABIT. A small to medium-sized tree 60-80 feet high and 12 feet in diameter (max. 125 by 3 1/2 feet), a large shrub in Washington; trunk clear and straight; dense, rounded crown with stout, spreading branches.

LEAVES. Alternate; simple; persistent 2-3 years; lanceolate to oblong-ovate; 2-6 inches long; penniveined; entire and often revolute; usually acuminate; thick and leathery; dark green and lustrous above; covered with small golden-yellow scales below; stipules fugacious.

FLOWERS. Monoecious; apetalous; appearing after the leaves; in staminate and bisexual aments; the staminate aments in 3-flowered cymes, each flower with 10-12 stamens; the pistillate flowers solitary or in clusters of 2-3, with 3-celled ovaries.

FRUIT. A rounded, broadly ovoid, yellow-brown nut 1/2 inch long; 1-2 nuts enclosed in a 4-valved, globose bur 1-1 1/2 inches in diameter and covered with prickly, dense, unbranched spines; maturing in 2 years.

TWIGS. Rather slender; rigid; round; covered at first with golden-yellow scales, becoming scurfy and red-brown; pith stellate. Winter buds: terminal present; ovoid; 1/4 inch long; light brown; numerous overlapping scales.

BARK. Thick; dark red-brown; deeply furrowed.

WOOD. Unimportant; intermediate in strength and hardness; ring-porous; occasionally used for lumber.

SILVICAL CHARACTERS. Tolerant; growth rather rapid; long-lived, reaching 500 years; taproot in early life, giving way to deep lateral system; often in pure stands on poor, dry sites and as an understory tree in redwood and Douglas-fir stands.

GENERAL. There are about 30 Asiatic species of this genus, with only 2 species native to the West Coast of the United States. *Castanopsis sempervirens* Dudl. is a small alpine shrub in California and southern Oregon.

Tanoak

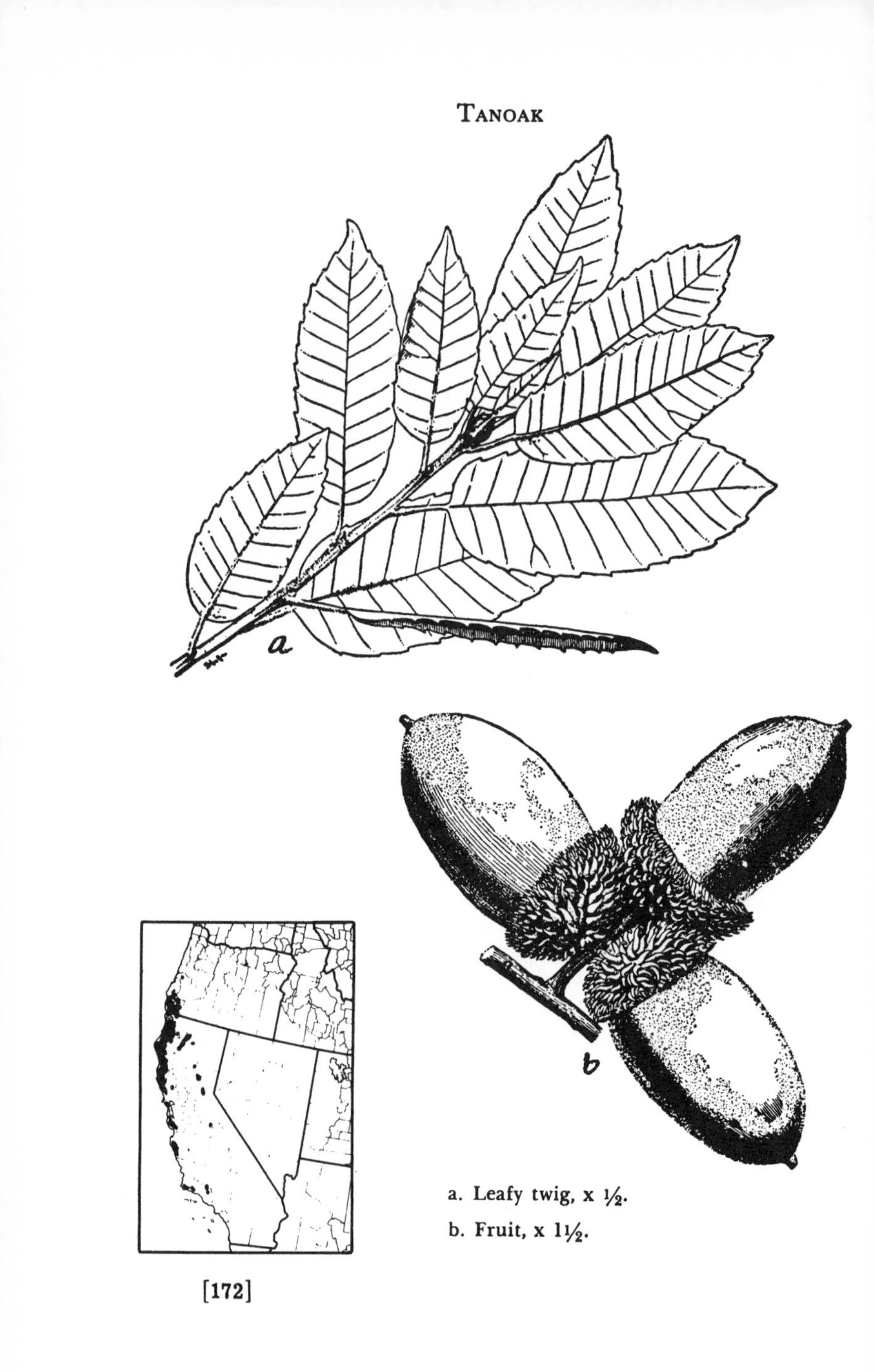

a. Leafy twig, x ½.

b. Fruit, x 1½.

Tanoak. Tanbark-oak

Lithocarpus densiflorus (Hook. & Arn.) Rehd.

HABIT. A medium-sized tree 70-90 feet high and 2-3 feet in diameter (max. 150 by 8 feet); trunk clear; crown rounded with ascending branches.

LEAVES. Alternate; simple; persistent 3-4 years; oblong to oblong-lanceolate; 3-5 inches long; penniveined; repand or dentate to entire-revolute; acute or rounded; thick; pale green and usually glabrous above; brown, woolly pubescent below in spring, becoming blue-white and nearly glabrous; stipules persistent on upper leaves.

FLOWERS. Monoecious; apetalous; appearing after the leaves; in staminate and bisexual aments; staminate flowers crowded, each with 10 stamens; the pistillate flowers solitary with 3-celled ovary.

FRUIT. An oval, bitter, yellow-brown nut (acorn) 3/4-1 inch long; enclosed at base in a shallow, tomentose, spiny, cuplike involucre, lined with a lustrous red pubescence; maturing in 2 years.

TWIGS. Stout; round; covered for 1-3 years with a thick tomentum of fascicled hairs, becoming red-brown, often with a glaucous bloom. Winter buds: terminal present; ovoid; obtuse; 1/4 inch long; covered by a few tomentose scales.

BARK. Thick; red-brown; deeply furrowed with broad, rounded, scaly ridges; high in tannin content.

WOOD. Unimportant; hard and strong; ring-porous; little used.

SILVICAL CHARACTERS. Tolerant; growth moderate; maturity reached in 200-300 years; deep taproot system; typical on moist, well-drained sites; in mixture with conifers and hardwoods; coppices vigorously and reproduces well by seed.

GENERAL. There are about 100 Asiatic species of this genus found in warm climates. Only the one species is native to North America.

FAGACEAE

The Oaks

Characteristics of the Genus *Quercus* L.

HABIT. Deciduous or evergreen trees or shrubs; with astringent properties; pubescence of fascicled hairs.

LEAVES. Alternate; simple; deciduous or persistent; shape and size often variable on same tree; stipulate.

FLOWERS. Regular; monoecious; vernal; staminate in clustered aments, individual flowers with a 4-7-lobed calyx enclosing 6 (rarely 2-12) stamens; pistillate solitary or in 2 to many-flowered spikes, individual flowers with a 6-lobed calyx surrounding a 3-celled (rarely 4-5-celled) ovary, 1-2 ovules in each cell, the whole partly enclosed in involucre.

FRUIT. An acorn; 1-seeded by abortion; maturing in 1-2 years; partially enclosed by scaly cup (modified involucre).

TWIGS. Slender to stout; pith homogeneous, stellate; marked by pale lenticels, semicircular leaf scars, many bundle scars. Winter buds: clustered at end of twig; terminal present; many chestnut-brown scales imbricated in 5 ranks.

BARK. Scaly or dark and furrowed.

WOOD. Heavy; hard; strong; prominent rays.

SILVICAL CHARACTERS. From intermediate in tolerance (many white oaks) to intolerant (most red and live oaks); long-lived; generally with deep taproot systems; hybridizes frequently.

GENERAL. This is a variable genus, and it is often difficult to distinguish between the species. Fifty-eight native species are recognized as reaching tree size; in addition there are numerous hybrids and varieties as well as many shrubby species. Five shrubby southwestern forms reported to become small trees in Texas, New Mexico, or Arizona are not keyed out. These are *Q. graciliformis* Mull., *Q. gravesii* Sudw., *Q. grisea* Liebm., *Q. harvardii* Rydb., and *Q. laceyi* Small. The key distinguishes between the tree species, and all oaks of importance are described and pictured in more detail. The native oaks logically break themselves down into 2 subgenera, the white oaks and the red oaks, each of which is further subdivided into 2 or 3 groups. A recent study indicates that all the live oaks should be placed under the red oak group on the basis of wood structure.

The English oak, *Quercus robur* L. is commonly planted and has leaves that resemble white oak but are smaller (2 1/2-5 inches long). It is further characterized by fruit borne on peduncles 1-3 inches long.

KEY TO THE SPECIES OF OAKS

I. Leaves turning brown in the fall of the first year, neither thick and leathery nor spiny-toothed.
  II. Leaves or leaf lobes not bristle-tipped; acorn maturing in 1 year, usually sweet; nut shell smooth inside . . . . White Oaks
    III. Leaves deeply lobed or rarely entire . . . . True White Oaks, p. 175
    III. Leaves coarsely toothed or shallowly lobed. . . . . Chestnut Oaks, p. 176
  II. Leaves or leaf lobes commonly bristle-tipped; acorn maturing in 2 years, usually bitter; nut shell woolly inside . . . . Red Oaks
    IV. Leaves lobed, rarely broadly ovate and entire . . True Red Oaks, p. 176
    IV. Leaves entire and mostly narrow . . . . Willow Oaks, p. 178
I. Leaves remaining green on the tree at least until the new leaves appear the following spring, usually thick and leathery, margins entire or spiny-toothed . . . . Live Oaks, p. 178

True White Oaks

1. Eastern species (east of the Rocky Mountains).
  2. Leaves entire or slightly lobed.
    3. Acorns enclosed only at base by thin cup.
      4. Acorn 3/4-1 inch long; leaves mostly blue-green; shrubby; western Texas, Oklahoma . . . . *Q. durandii* var. *brevilobata* (Torr.) Palmer, Bigelow Oak
      4. Acorn 1/2-2/3 inch long; leaves yellow-green; large tree; South Carolina to Florida, west to Texas, Arkansas . . . . *Q. durandii* Buckl., Durand Oak
    3. Acorns enclosed 1/4-2/3 length in thick cup.
      5. Acorn 1/3-1/2 inch long; leaves blue-green; western Texas, Oklahoma . . . . *Q. mohriana* Buckl., Mohrs Oak
      5. Acorn 5/8 inch long; leaves yellow-green; South Carolina to Florida . . . . *Q. chapmanii* Sarg., Chapman Oak
  2. Leaves more or less deeply lobed.
    6. Acorns enclosed in cup, or cup fringed.
      7. Acorn cup conspicuously fringed on margin; leaves nearly divided in two by deep sinuses . . . . *Q. macrocarpa*, Bur Oak, p. 181
      7. Acorn cup not fringed and nearly covering acorn; leaves irregularly lobed . . . . *Q. lyrata*, Overcup Oak, p. 183
    6. Acorns enclosed 1/4-1/2 length, in unfringed cup.
      8. Leaves regularly, usually deeply, 7-9-lobed, glabrous below; buds nearly glabrous . . . . *Q. alba*, White Oak, p. 181
      8. Leaves typically cross-shaped, 5-lobed, woolly below; buds hairy . . . . *Q. stellata*, Post Oak, p. 183
1. Western species (Rocky Mountains and Pacific Coast).
  9. Leaves undulately lobed with acute lobes or entire, blue-green, 1-1 1/2 inches long; acorn enclosed only at base in cup.
    10. Leaves not spiny-toothed; Colorado to Nevada and south . . . . *Q. undulata* Torr., Wavyleaf Oak
    10. Leaves spiny-toothed; Texas to Arizona . . . . *Q. pungens* Liebm., Sandpaper Oak
  9. Leaves pinnately and usually deeply lobed, rounded lobes.
    11. Acorn under 3/4 inch long, enclosed in cup for 1/2 or more of length; leaves yellow-green, 3-7 inches long, usually shrubby; Rocky Mountains north to Wyoming, Nevada . . . . *Q. gambelii*, Gambel Oak, p. 185
    11. Acorn 3/4 inch long or longer, enclosed in cup for less than 1/2 length; Pacific Coast.

12. Acorns 1-2 1/4 inches long, slender; deep, cuplike bowl; leaves 2 1/2-3 inches long; California . . . . *Q. lobata*, California White Oak, p. 187

12. Acorns 3/4-1 1/2 inches long, ovoid; very shallow cup.

13. Leaves dark yellow-green, 3-6 inches long; petioles hairy, 1/2-1 inch long; British Columbia to California . . . . *Q. garryana*, Oregon White Oak, p. 187

13. Leaves blue-green, 2-4 inches long; shallowly lobed to sinuately toothed to entire; petioles woolly, 1/4-1/2 inch long; California . . . . *Q. douglasii*, Blue Oak, p. 185

Chestnut Oaks (Eastern Trees)

1. Fruit long-stalked (1-4 inches); leaves coarsely sinuate-toothed to shallowly lobed, usually with 6-8 pairs of veins . . . *Q. bicolor*, Swamp White Oak, p. 191

1. Fruit short-stalked or sessile; leaves coarsely toothed.

2. Leaves 4-9 inches long with 9-17 pairs of veins; large trees.

3. Leaves obovate, white-hairy below and dark green above; acorn 1-1 1/2 inches long in thick, subsessile cup with wedge-shaped scales . . . . *Q. michauxii*, Swamp Chestnut Oak, p. 189

3. Leaves mostly elliptical, slightly hairy below and yellow-green above; acorn in thin cup with partially fused scales.

4. Leaves with rounded teeth, acorn 1-1 1/2 inches long; cup on short stalk (1/4-1/2 inch long) buds 1/4 inch long . . . . *Q. prinus*, Chestnut Oak, p. 189

4. Leaves with sharp teeth; acorn 1/2-3/4 inch long; cup subsessile; buds 1/8 inch long . . . . *Q. muehlenbergii*, Chinkapin Oak, p. 191

2. Leaves 2 1/2-5 inches long with 3-8 pairs of veins and usually sharp teeth; acorn 2/5-3/5 inch long, thick cup; shrubby or rarely a small tree; Massachusetts to North Carolina, west to Nebraska, Texas . . . . *Q. prinoides* Willd., Dwarf Chinkapin Oak

True Red Oaks

1. California and Oregon species; leaves usually 7-lobed; scales of fruit cup loosely imbricated . . . . *Q. kelloggii*, California Black Oak, p. 201

1. Eastern species (not reaching the Rocky Mountains).

2. Leaves deeply lobed.

3. Leaves white- to red-woolly below.

4. Leaves usually 5-lobed, 2-5 inches long; shrub to small tree; Maine to North Carolina, Kentucky . . . . *Q. ilicifolia* Wang., Bear Oak

4. Leaves of 2 forms, either 3-lobed at apex or deeply 5-11-lobed, 6-8 inches long . . . . *Q. falcata*, Southern Red Oak, p. 195

3. Leaves green below, often with hairy tufts in axils.

5. Leaves dull, 7-11-lobed, nearly glabrous, sinuses extending about 1/2 way to midrib; acorn enclosed at base or to 1/3 length in thick cup . . . . *Q. rubra*, Northern Red Oak, p. 193

5. Leaves lustrous, mostly 5-7-lobed, sinuses extending more than 1/2 way to midrib.

  6. Acorn cup shallow, covering basal 1/3 or less of acorn; acorn cup scales tightly imbricated.

    7. Leaves 6-8 inches long, many toothed; acorns 3/4-1 1/4 inches, oblong-ovoid . . . . *Q. shumardii*, Shumard Oak, p. 195

    7. Leaves 2 1/2-6 inches long; acorn not over 1/2 inch.

      8. Leaves 5-7-lobed, 4-6 inches long; acorn hemispheric, 1/2 inch. . . . . *Q. palustris*, Pin Oak, p. 199

      8. Leaves 3-5-lobed, 2 1/2 inches long; acorn ellipsoidal to globose, 1/3-1/2 inch long; only in Georgia . . . . *Q. georgiana* Curtis, Georgia Oak

  6. Acorn cup covering at least the basal 1/3 of acorn.

    9. Acorn cup with loosely imbricated scales, forming a free margin on rim; winter buds woolly or red-hairy.

      10. Leaves usually 7-lobed; petioles 1-2 inches long; winter buds gray, tomentose . . . . . *Q. velutina*, Black Oak, p. 193

      10. Leaves 3-5-lobed; petioles 1/4-3/4 inches long; winter buds rusty-pubescent . . . . *Q. laevis*, Turkey Oak, p. 201

    9. Acorn cup with tightly imbricated scales; winter buds not woolly or red-hairy.

      11. Buds white-hairy; acorn, rings near apex, 1/2-1 inch long . . . . *Q. coccinea*, Scarlet Oak, p. 197

      11. Buds not white-hairy; acorn without rings.

        12. Acorn ellipsoidal, 1/2-3/4 inch; Lake states . . . . *Q. ellipsoidalis*, Northern Pin Oak, p. 199

        12. Acorn oblong, 3/4-1 1/4 inches long; lower Mississippi Valley . . . . *Q. nuttallii*, Nuttall Oak, p. 199

2. Leaves 3-lobed at apex or entire.

  13. Mature leaves distinctly hairy below.

    14. Leaves gray to rusty-woolly below, of two types; acorn 1/2 inch long, enclosed in thin cup for 1/3 or less of length . . . . *Q. falcata*, Southern Red Oak, p. 195

    14. Leaves tawny-hairy below and scurfy; acorn 3/4 inch long, enclosed in thick cup for 1/3-1/2 its length . . . . *Q. marilandica*, Blackjack Oak, p. 197

  13. Mature leaves nearly glabrous and green below; acorn enclosed in thin cup for 1/3 or less of length.

    15. Leaves rounded at base; acorn 1/4-1/3 inch long; Georgia, Florida, Alabama, Arkansas . . . . *Q. arkansana* Sarg., Arkansas Oak

    15. Leaves wedge-shaped at base, lobing variable; acorn 1/3-2/3 inch long . . . . *Q. nigra*, Water Oak, p. 203

Willow Oaks (Eastern Species)

1. Leaves mostly linear-lanceolate, glabrous to woolly below; acorn 1/3-1/2 inch long, green-brown . . . . *Q. phellos*, Willow Oak, p. 203

1. Leaves mostly oblong-ianceolate to oblong-obovate.

  2. Leaves white-woolly below; North Carolina to Florida, west to Texas . . . . *Q. incana*, Bluejack Oak, p. 205

  2. Leaves glabrous to hairy below.

    3. Leaves hairy below; acorn dark chestnut-brown, 1/2-3/4 inch long . . . . *Q. imbricaria*, Shingle Oak, p. 205

    3. Leaves essentially glabrous below; acorn nearly black, 1/2 inch long.

      4. Leaves elliptical to oblong-ovate, nearly evergreen . . . . *Q.* X *laurifolia*, Laurel Oak, p. 205

      4. Leaves variable, spatulate to obovate to linear, sometimes lobed, deciduous . . . . *Q. nigra*, Water Oak, p. 203

Live Oaks

1. Southeastern Live Oaks.

  2. Shell of nut glabrous inside, acorn maturing in 1 year, on long stalks; leaves 2-5 inches, elliptical to ovate . . . . *Q. virginiana*, Live Oak, p. 207

  2. Shell of nut woolly inside, acorn maturing in 2 years, short-stalked; leaves obovate, revolute, 1/2-2 inches long; South Carolina to Florida, west to Mississippi . . . . *Q. myrtifolia* Willd., Myrtle Oak

1. Western Live Oaks.

  3. California Live Oaks.

    4. Shell of nut glabrous on inside, acorn maturing in 1 year.

      5. Leaves about 3/4 inch long; usually a shrub; through California . . . . *Q. dumosa* Nutt., California Scrub Oak

      5. Leaves 1-3 inches long; southwestern California . . . . *Q. engelmannii* Greene, Engelmann Oak

    4. Shell of nut woolly inside, acorn maturing in 2 years (except California Live Oak).

      6. Acorn 3/4-1 1/2 inches long, narrow; more than 1/2 covered by thin cup . . . . *Q. wislizenii*, Interior Live Oak, p. 211

      6. Acorn less than 1/2 covered by cup.

        7. Acorn 3/4-1 1/2 inches long, narrow; leaves glabrous to hairy below . . . . *Q. agrifolia*, California Live Oak, p. 211

        7. Acorn 1/2-2 inches long, oblong to ovoid; leaves woolly below.

          8. Leaves oblong-lanceolate; restricted to islands off coast of southern California . . . *Q. tomentella* Engelm., Island Live Oak

          8. Leaves oblong-ovate; Washington to California, Arizona . . . . *Q. chrysolepis*, Canyon Live Oak, p. 213

  3. Southwestern Live Oaks (Mexican border region with only *Q. turbinella* extending north to Nevada, Utah, Colorado).

    9. Acorn shell glabrous inside, light brown outside, nut maturing in 1 year.

      10. Fruit in long-stalked clusters of 2 or more; leaves broadly obovate, yellow-hairy below . . . . *Q. reticulata* H. & B., Netleaf Oak

      10. Fruit sessile or nearly so, single or paired; leaves oblong to ovate.

11. Leaves with many spiny teeth, 1/2-1 1/4 inches long . . . . Q. *turbinella* Greene, Shrub Live Oak

11. Leaves entire or with few short teeth.

 12. Leaves glabrous below, 1-2 inches long; acorn enclosed about 1/3 length in cup slightly fringed at rim . . . . Q. *oblongifolia*, Mexican Blue Oak, p. 207

 12. Leaves hairy below; acorn enclosed in unfringed cup about 1/2 its length.

  13. Leaves 1/2-1 inch long; acorn 1/2-2/3 inch long; Arizona . . . . Q. *toumeyi* Sarg., Toumey Oak

  13. Leaves 1-4 inches long; acorn 3/4-1 inch long; Arizona, New Mexico . . . Q. *arizonica*, Arizona White Oak, p. 209

9. Acorn shell woolly on inner surface.

 14. Acorn maturing in 1 year, dark brown to nearly black; leaves nearly glabrous below . . . . Q. *emoryi*, Emory Oak, p. 209

 14. Acorn maturing in 2 years, light brown; leaves woolly below.

  15. Leaves lanceolate to elliptic, persistent until second spring . . . . Q. *hypoleucoides*, Silverleaf Oak, p. 213

  15. Leaves oblong-ovate to elliptic, persistent 3-4 years . . . . Q. *chrysolepis*, Canyon Live Oak, p. 213

White Oak

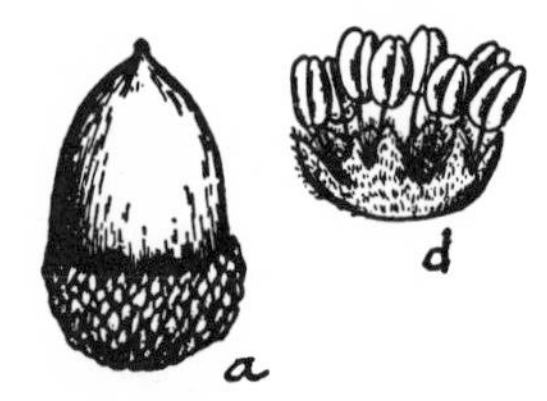

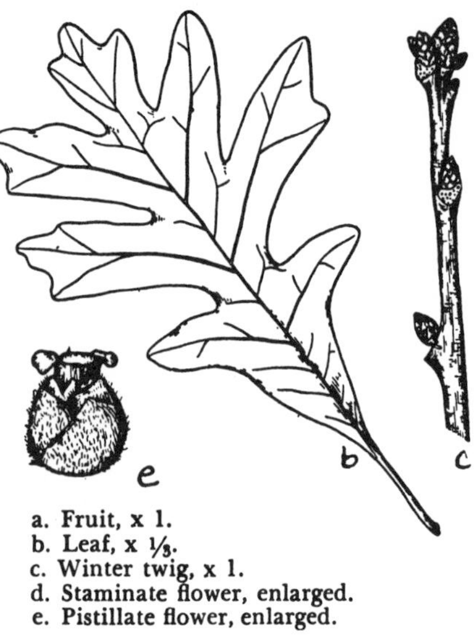

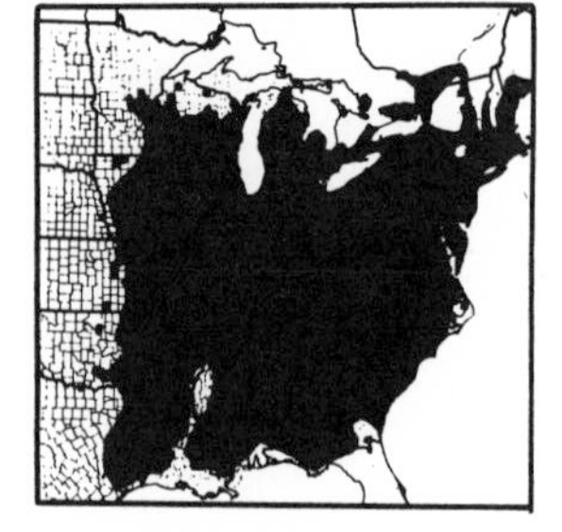

a. Fruit, x 1.
b. Leaf, x 1/3.
c. Winter twig, x 1.
d. Staminate flower, enlarged.
e. Pistillate flower, enlarged.

Bur Oak

a. Fruit, x 1.
b. Leaf, x 1/3.
c. Flowering branchlet, x 1/2.

White Oak

Quercus alba L.

HABIT. A medium-sized tree 80-100 feet high and 3-4 feet in diameter (max. 150 by 9 feet); crown rounded, heavy branches.

LEAVES. Deciduous; oblong to obovate; 5-9 inches long; 7-9-lobed (rarely 5-lobed), lobes either broad with shallow sinuses or narrow with deep sinuses extending nearly to the midrib; apex of lobes rounded; thin and firm; bright green and glabrous above, paler below.

FRUIT. Maturing in first year; sessile or short-stalked; acorn 1/2-3/4 inch long, light brown, oblong, enclosed for about 1/4 its length in bowllike cup with thickened warty scales; kernel sweet.

TWIGS. Rather stout; reddish. Winter buds: terminal 1/8-3/16 inch long; ovoid to globose; obtuse; red-brown; nearly glabrous.

BARK. Light ashy gray; typically irregularly plated with loose plates, but sometimes furrowed with narrow ridges.

WOOD. The most important species of oak; heavy; hard; strong; heartwood light brown, quite durable; used for furniture, cooperage, finish, ties, etc.

SILVICAL CHARACTERS. Intermediate in tolerance; rather slow-growing; dry to moist sites; long-lived, attaining ages of 600 years.

* * *

Bur Oak

Quercus macrocarpa Michx.

HABIT. A handsome medium-sized tree 60-80 feet high and 2-3 feet in diameter (max. 170 by 7 feet); crown rounded with large, heavy branches.

LEAVES. Deciduous; oblong to obovate; 6-10 inches long; characteristically 5-9-lobed, rounded lobes, the 2 center sinuses nearly dividing the leaf into 2 halves; thick; dark green and lustrous above, paler and hairy below.

FRUIT. Maturing in one year; sessile or long-stalked; acorn 3/5-2 inches long, ellipsoidal, brown, enclosed for 1/3 to all of length in characteristic fringe-margined cup; kernel sweet.

TWIGS. Stout; yellow-brown to gray; often with characteristic corky wings. Winter buds: 1/8-1/4 inch long; hairy.

BARK. Thick; gray-brown; deeply furrowed and ridged.

WOOD. Important; similar to *Q. alba.*

SILVICAL CHARACTERS. Similar to *Q. alba;* moist bottomlands to dry hills in northwest; extending farthest west of eastern oaks.

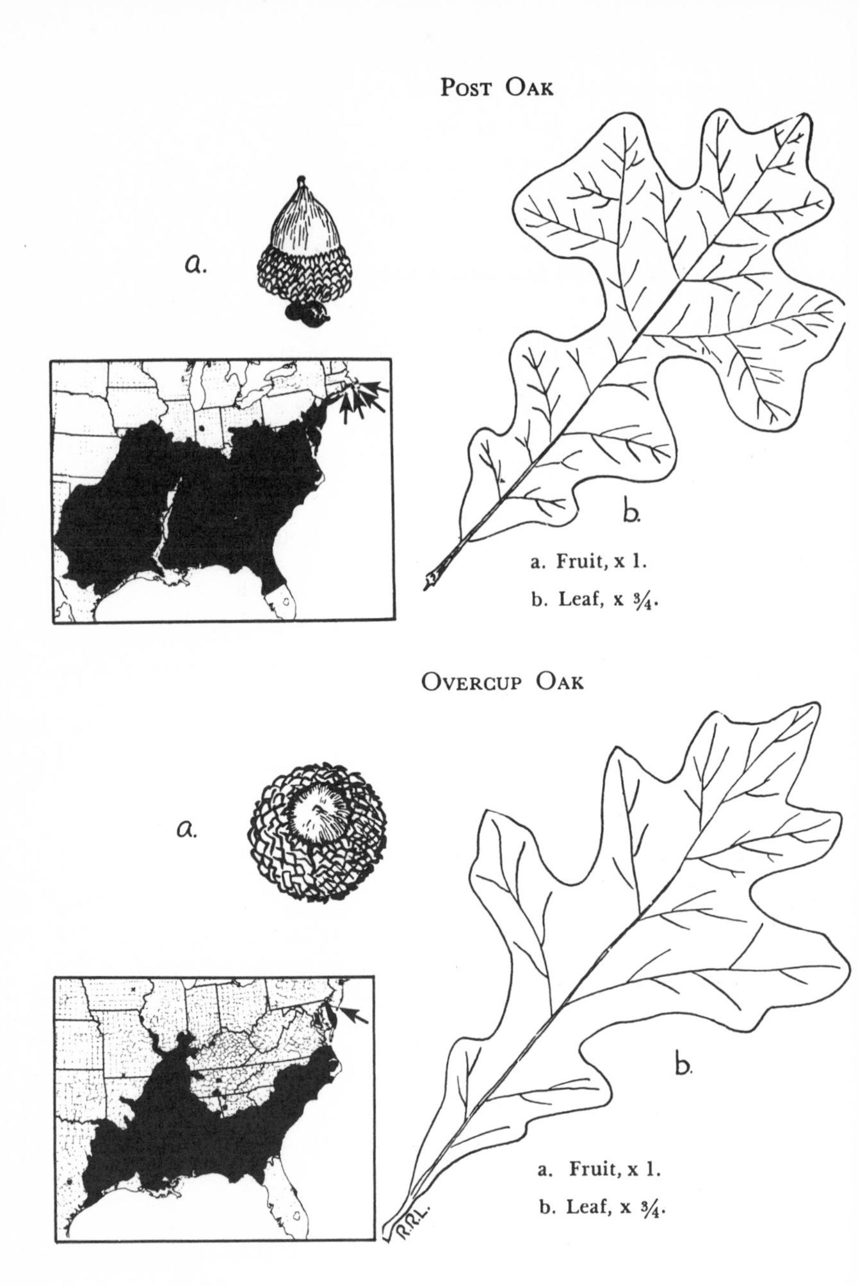

POST OAK

a. Fruit, x 1.
b. Leaf, x ¾.

OVERCUP OAK

a. Fruit, x 1.
b. Leaf, x ¾.

Post Oak

Quercus stellata Wang.

HABIT. A small to medium-sized tree 40-50 feet high and 1-2 feet in diameter (max. 100 by 4 1/2 feet), on poor sites a shrub; crown rounded, few large branches.

LEAVES. Deciduous; oblong to obovate; 4-6 inches long; variable but typically deeply 5-lobed, rounded middle lobes opposite giving a crosslike appearance; thick; dark green above, tawny-tomentose below.

FRUIT. Maturing in one year; sessile or nearly so; acorn 1/2-2/3 inch long, red-brown, oval, enclosed for 1/3-1/2 length in usually thin-scaled, bowl-shaped cup; kernel sweet.

TWIGS. Stout; more or less tomentose. Winter buds: 1/8-1/4 inch long; hairy; chestnut-brown.

BARK. Gray-brown; furrowed; without loose plates of *Q. alba.*

WOOD. Important; similar to *Q. alba;* often poor quality.

SILVICAL CHARACTERS. Similar to *Q. alba;* site varies from dry sandy plains to moist river bottoms; a variable species with several varieties and hybrids.

* * *

Overcup Oak

Quercus lyrata Walt.

HABIT. A tree 40-60 feet high and 2-3 feet in diameter (max. 100 by 5 feet); crown open and rounded with large, crooked branches.

LEAVES. Deciduous; oblong to obovate; 6-10 inches long; deeply 5-9-lobed, broad irregular sinuses and rounded lobes, giving the leaf an irregular, deeply cut appearance; dark green above, paler below; glabrous or white-hairy.

FRUIT. Maturing in one year; sessile or nearly so; acorn 1/2-1 inch long, chestnut-brown, subglobose; nearly enclosed in deep, thin cup with unfringed margin; kernel sweet.

TWIGS. Rather stout; gray-brown; usually glabrous. Winter buds: 1/8 inch long; mostly woolly; rounded; red-brown.

BARK. Similar to *Q. alba.*

WOOD. Important; similar to *Q. alba;* poor quality.

SILVICAL CHARACTERS. Similar to *Q. alba;* typical of wet, poorly drained clay soils.

Gambel Oak

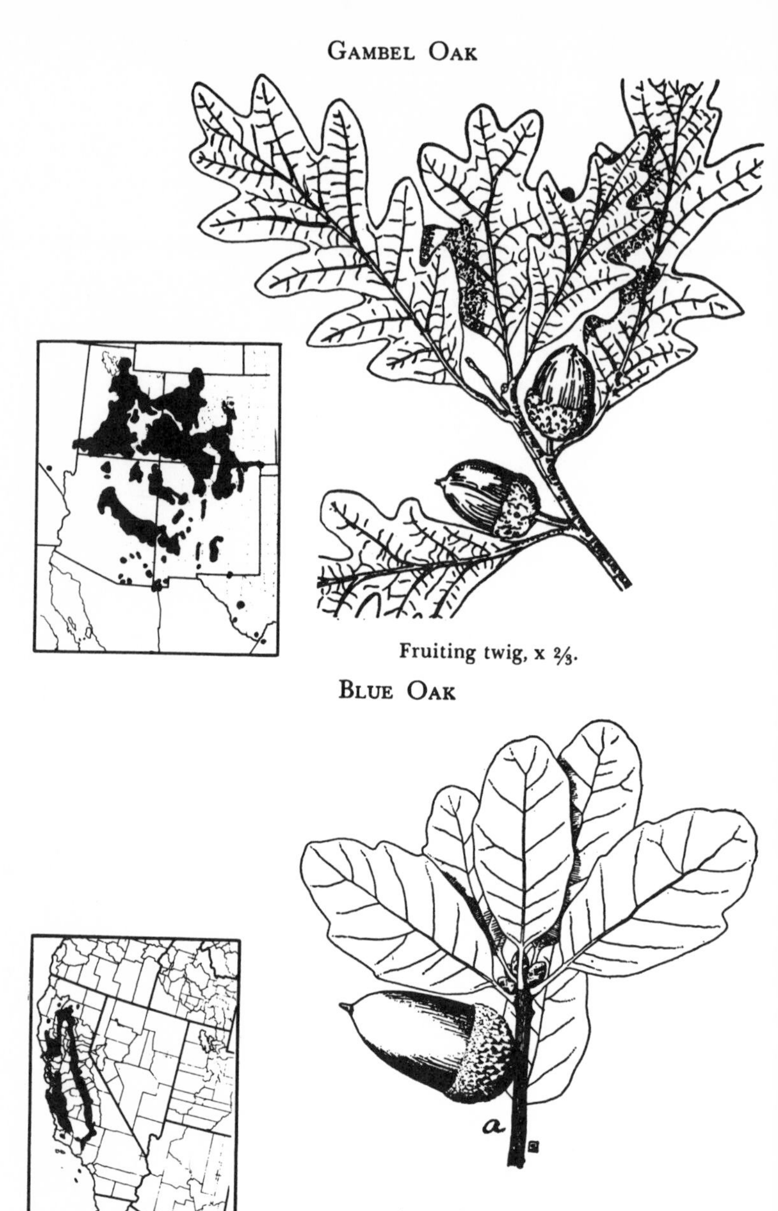

Fruiting twig, x ⅔.

Blue Oak

(a) Fruiting twig, x ⅔.

FAGACEAE

Gambel Oak

Quercus gambelii Nutt. (*Quercus utahensis* Rydb.) (*Quercus leptophylla* Rydb.) (*Quercus novomexicana* Rydb.)

HABIT. A shrub or small tree 25-35 feet high and 1 foot in diameter (max. 60 by 5 feet); crown open with thick, erect branches.

LEAVES. Deciduous; oblong to obovate; 2 1/2-7 inches long; deeply 7-9-lobed, rounded lobes; thick; dark green above, paler and hairy below.

FRUIT. Maturing in one year; sessile or nearly so; acorn 3/5-3/4 inch long; brown; ovoid; enclosed for about 1/4-1/2 its length in thick-scaled, hairy cup.

TWIGS. Stout; orange-brown to red-brown; hairy. Winter buds: 1/8-1/4 inch long; brown; hairy.

BARK. Similar to *Q. alba.*

WOOD. Unimportant; similar to *Q. alba;* used for fuel.

SILVICAL CHARACTERS. Similar to *Q. alba;* on dry foothills and canyon walls, the only abundant deciduous oak in the low Rocky Mountain forests; has been subdivided into several species by some authors.

* * *

Blue Oak

Quercus douglasii Hook. & Arn.

HABIT. A shrub to medium-sized tree 50-80 feet high and 2-3 feet in diameter; dense rounded crown.

LEAVES. Deciduous; oblong to obovate; 2-5 inches long; entire to sinuate-toothed to shallowly and irregularly 4-5-lobed; lobes and teeth rounded but often mucronate; dark blue-green above, paler and hairy below.

FRUIT. Maturing in one year; sessile or nearly so; acorn 3/4-1 1/4 inches long, chestnut-brown, ellipsoidal; enclosed at base in thin, shallow, hairy cup; kernel sweet.

TWIGS. Stout; red-brown. Winter buds: 1/8-1/4 inch long; bright red.

BARK. Thin; gray; scaly; similar to *Q. alba.*

WOOD. Unimportant; similar to the red oaks and placed in that group by some authors.

SILVICAL CHARACTERS. Intolerant; on dry foothill sites.

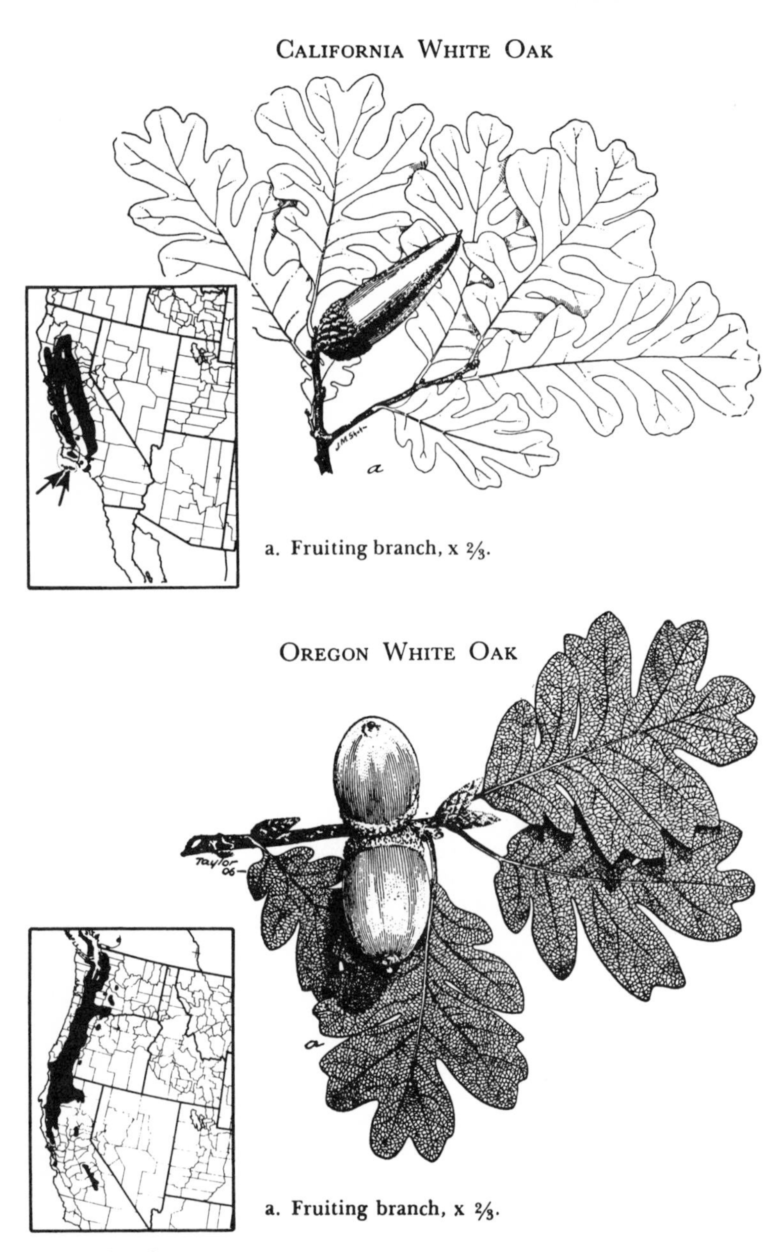

California White Oak

a. Fruiting branch, x 2/3.

Oregon White Oak

a. Fruiting branch, x 2/3.

California White Oak

Quercus lobata Née

HABIT. The largest western oak, commonly 60-80 feet high and 3-5 feet in diameter (max. 125 by 9 feet); massive trunk dividing near the ground into large, spreading limbs; wide-spreading, drooping crown.

LEAVES. Deciduous; oblong to obovate; 2 1/2-3 1/2 inches long; deeply 7-11-lobed, rounded lobes; dark green and hairy above, pale and hairy below.

FRUIT. Maturing in one year; sessile or nearly so; acorn characteristic, 1 1/4-2 1/4 inches long, elongate-conic, green becoming brown; enclosed for 1/3 length in hairy cup with scales free at tip; kernel sweet.

TWIGS. Slender; gray to red-brown; hairy. Winter buds: 1/4 inch long; orange-brown; ovoid; hairy.

BARK. Thick; light gray; scaly or broken into square plates.

WOOD. Unimportant; similar to *Q. alba.*

SILVICAL CHARACTERS. Similar to *Q. alba;* fast-growing; fertile loams to poor dry soils.

* * *

Oregon White Oak

Quercus garryana Dougl.

HABIT. A tree 50-70 feet high and 2-3 feet in diameter (max. 120 by 8 feet); broad, compact crown with large branches.

LEAVES. Deciduous; oblong to obovate; 3-6 inches long; deeply 5-9-lobed with rounded lobes; thick and leathery; dark green above, paler and somewhat hairy below.

FRUIT. Maturing in one year; sessile or nearly so; acorn 1-1 1/4 inches long, ovoid; enclosed at base or to 1/3 of length in shallow cup with hairy, thickened scales; glabrous; kernel sweet.

TWIGS. Stout; hairy and orange at first, becoming red-brown and glabrous. Winter buds: 1/3-1/2 inch long; densely woolly.

BARK. Similar to *Q. alba.*

WOOD. Unimportant; similar to *Q. alba.*

SILVICAL CHARACTERS. Similar to *Q. alba;* moist to dry sites; a low shrub in north and at higher altitudes.

Chestnut Oak

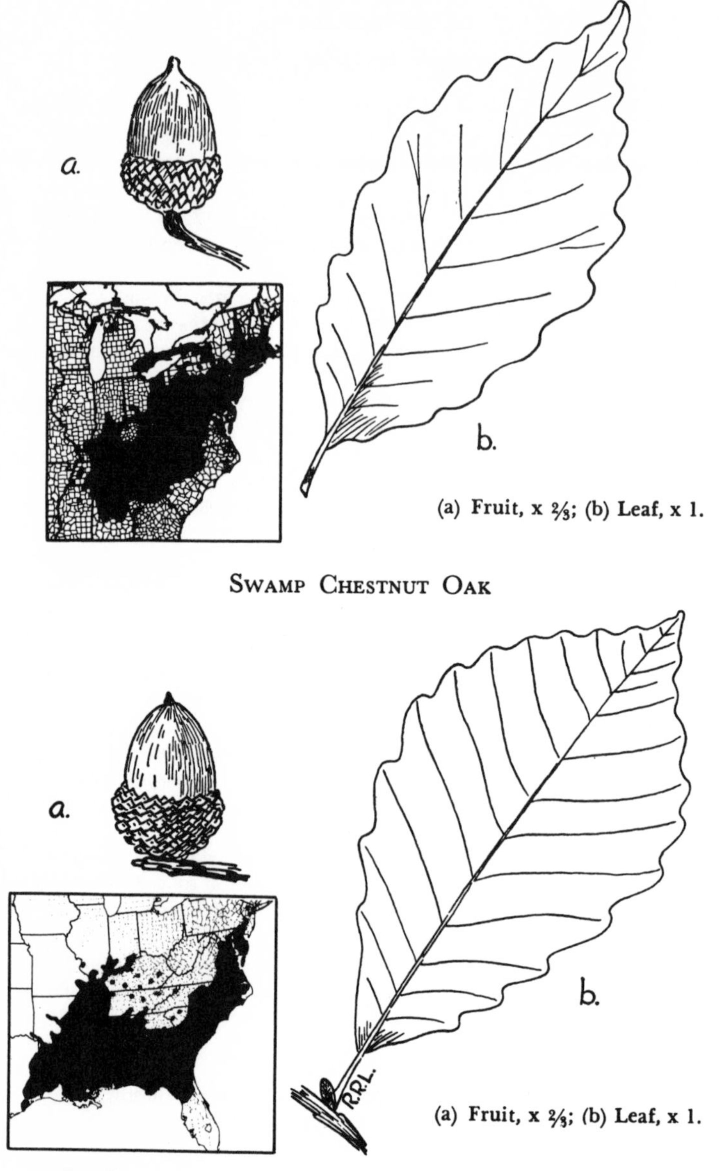

(a) Fruit, x ⅔; (b) Leaf, x 1.

Swamp Chestnut Oak

(a) Fruit, x ⅔; (b) Leaf, x 1.

Chestnut Oak

Quercus prinus L. (*Quercus montana* Willd.)

HABIT. Medium-sized tree 50-70 feet high, 2-3 feet in diameter (max. 100 by 6 1/2 feet); trunk often forking shortly above ground into large ascending limbs; crown broad, open.

LEAVES. Deciduous; elliptical to obovate; 4-9 inches long; 1 1/2-3 inches wide; coarsely serrate or crenate; yellow-green and lustrous above, paler, finely hairy below; petioles 1 inch.

FRUIT. Maturing in one year; short stalk; acorn 1-1 1/2 inches, oval to ovoid, chestnut-brown, lustrous; enclosed for 1/2 or less of length (sometimes only at base) in thin cup with scales more or less fused; kernel sweet.

TWIGS. Stout; orange to red-brown. Winter buds: 1/4-1/2 inch; conical; acute; bright chestnut-brown; soft-hairy; ciliate.

BARK. Thick; dark gray-brown to black; deeply furrowed with broad, rounded, scaly ridges; important source of tannin.

WOOD. Important; similar to *Q. alba.*

SILVICAL CHARACTERS. Similar to *Q. alba;* on poor, dry sites; in pure or mixed stands.

* * *

Swamp Chestnut Oak

Quercus michauxii Nutt. (*Quercus prinus* L.)

HABIT. A medium-sized tree 60-80 feet high and 2-3 feet in diameter (max. 120 by 10 feet); straight, clear trunk; crown narrow, rounded, compact.

LEAVES. Deciduous; obovate; 5-8 inches long; 3-4 1/2 inches wide; coarsely dentate or crenate, often with gland-tipped teeth; dark green and lustrous above, pale green and hairy below, petioles 3/4 inch long.

FRUIT. Maturing in one year; sessile or nearly so; acorn 1-1 1/2 inches, ovoid, bright brown; enclosed 1/3 or less of length in thick cup with distinct wedge-shaped scales; kernel sweet.

TWIGS. Stout; green to red-brown. Winter buds: 1/4 inch long; ovoid; acute; red; thin, hairy scales with pale margins.

BARK. Rather thick; ash-gray; scaly or irregularly furrowed with scaly, narrow ridges.

WOOD. Important; similar to *Q. alba.*

SILVICAL CHARACTERS. Similar to *Q. alba;* on moist to wet, often inundated sites; in mixed stands.

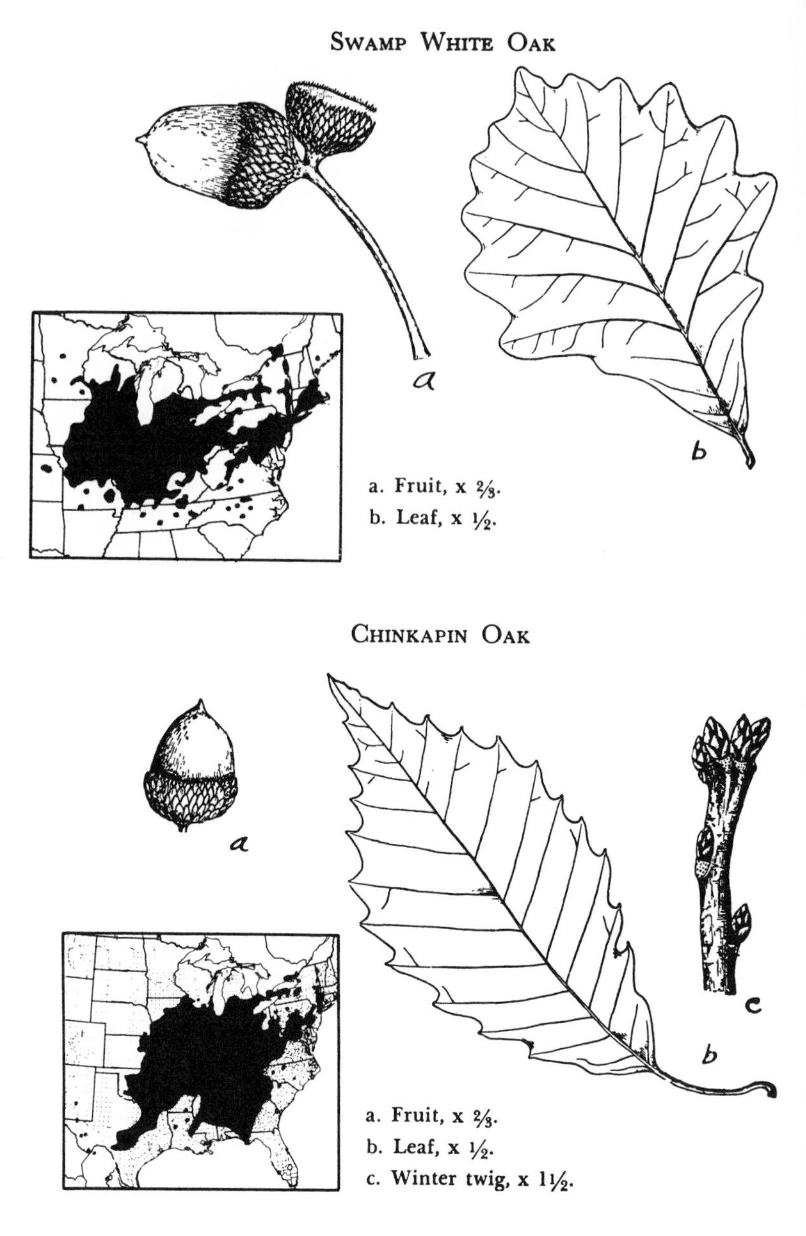

Swamp White Oak

a. Fruit, x ⅔.
b. Leaf, x ½.

Chinkapin Oak

a. Fruit, x ⅔.
b. Leaf, x ½.
c. Winter twig, x 1½.

Swamp White Oak

Quercus bicolor Willd.

HABIT. A medium-sized tree 50-70 feet high and 2-3 feet in diameter (max. 100 by 5 1/2 feet); trunk often limby; crown irregular, open.

LEAVES. Deciduous; obovate to oblong-obovate; 5-7 inches long; 2-4 inches wide; coarsely sinuate-crenate or dentate with glandular teeth or often shallowly, irregularly lobed; dark green and lustrous above, paler and slightly hairy to woolly below.

FRUIT. Maturing in one year; characteristic stalks 1-4 inches long; acorn 3/4-1 1/4 inches long, ovoid, light brown; enclosed for about 1/3 length in thick, light brown, slightly fringed, hairy cup; kernel sweet.

TWIGS. Slender to rather stout; dark brown. Winter buds: 1/16-1/8 inch long; orange-brown; globose; nearly glabrous.

BARK. Thick; dark brown; deeply furrowed into blocky or long, scaly ridges; on younger parts separated into curly, papery scales.

WOOD. Moderately important; similar to *Q. alba.*

SILVICAL CHARACTERS. Similar to *Q. alba;* moist to wet sites.

* * *

Chinkapin Oak

Quercus muehlenbergii Engelm.

HABIT. A medium-sized tree 50-70 feet high and 1-3 feet in diameter (max. 160 by 4 1/2 feet); narrow, rounded crown with rather short, ascending branches.

LEAVES. Deciduous; obovate to oblong-lanceolate; 4-7 inches long; coarsely serrate with sharp, gland-tipped teeth; thick; lustrous yellow-green above, paler and hairy below.

FRUIT. Maturing in one year; sessile or nearly so; acorn 1/2-3/4 inch, chestnut-brown to black, ovoid; enclosed about 1/2 length in thin, hairy cup with small scales; kernel sweet.

TWIGS. Slender; orange to gray-brown; glabrous. Winter buds: 1/8 inch long; orange-brown; acute; conical.

BARK. Thin; ash-gray; roughly shallowly furrowed or scaly.

WOOD. Moderately important; similar to *Q. alba.*

SILVICAL CHARACTERS. Similar to *Q. alba;* often on dry sites.

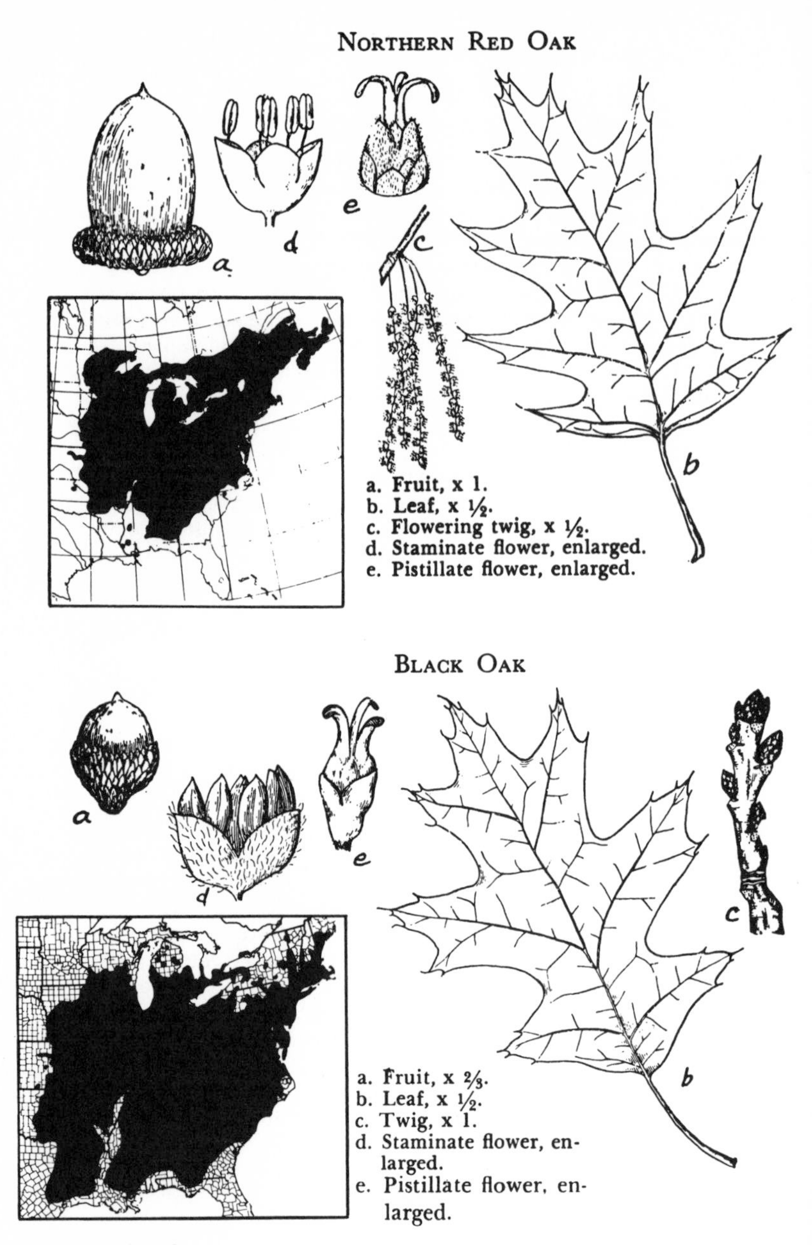

a. Fruit, x 1.
b. Leaf, x ½.
c. Flowering twig, x ½.
d. Staminate flower, enlarged.
e. Pistillate flower, enlarged.

a. Fruit, x ⅔.
b. Leaf, x ½.
c. Twig, x 1.
d. Staminate flower, enlarged.
e. Pistillate flower, enlarged.

Northern Red Oak

Quercus rubra L. (*Quercus borealis* Michx.)

HABIT. A medium-sized tree 60-80 feet high and 2-4 feet in diameter (max. 150 by 6 feet); broad, rounded crown.

LEAVES. Deciduous; oblong to obovate; 5-9 inches long; 7-11-lobed with narrow sinuses extending halfway to midrib; lobes pointing forward, toothed, and bristle-tipped; dull green above, paler and nearly glabrous below except for axillary tufts.

FRUIT. Maturing in 2 years; sessile or nearly so; acorn 5/8-1 inch long, pale brown, oblong to ovoid; enclosed at base or to 1/3 length in thick cup with closely appressed scales (shallow cup form known as var. *maxima*); inner surface of nut shell woolly; kernel bitter, white.

TWIGS. Stout; red-brown; lustrous. Winter buds: 1/4-3/8 inch long; not strongly angled; lustrous; red; hairy toward tip of scales.

BARK. Thick; nearly black; shallow furrows with wide, smooth, light, flat ridges; inner bark light red.

WOOD. Important; less strong and durable than *Q. alba.*

SILVICAL CHARACTERS. Moderately intolerant; sprouts vigorously; on moist sites; rather fast-growing; in mixed stands.

* * *

Black Oak

Quercus velutina Lam.

HABIT. A medium-sized tree 50-70 feet high and 2-3 feet in diameter (150 by 8 feet); irregular, rounded crown.

LEAVES. Deciduous; obovate to oblong; 5-9 inches long; 5-7-lobed with broad sinuses extending 2/3-7/8 of way to midrib; lobes toothed and bristle-tipped; lustrous and dark green above, paler or coppery below with more or less scurfy pubescence and prominent axillary tufts.

FRUIT. Maturing in 2 years; sessile or nearly so; acorn 1/2-3/4 inch long, red-brown, often striate, ovoid; enclosed for 1/3 to 1/2 length in bowl-shaped cup of thin, loose, dull, woolly scales with free tips; kernel bitter, yellow.

TWIGS. Stout; red-brown; glabrous. Winter buds: 1/4-1/2 inch long; sharp-pointed; angled with gray-woolly scales.

BARK. Thick; nearly black; deeply furrowed with narrow, scaly ridges; inner bark thick, orange-yellow, very bitter.

WOOD. Important; similar to *Q. rubra.*

SILVICAL CHARACTERS. Moderately intolerant; similar to *Q. rubra;* on dry to good, moist sites.

Southern Red Oak

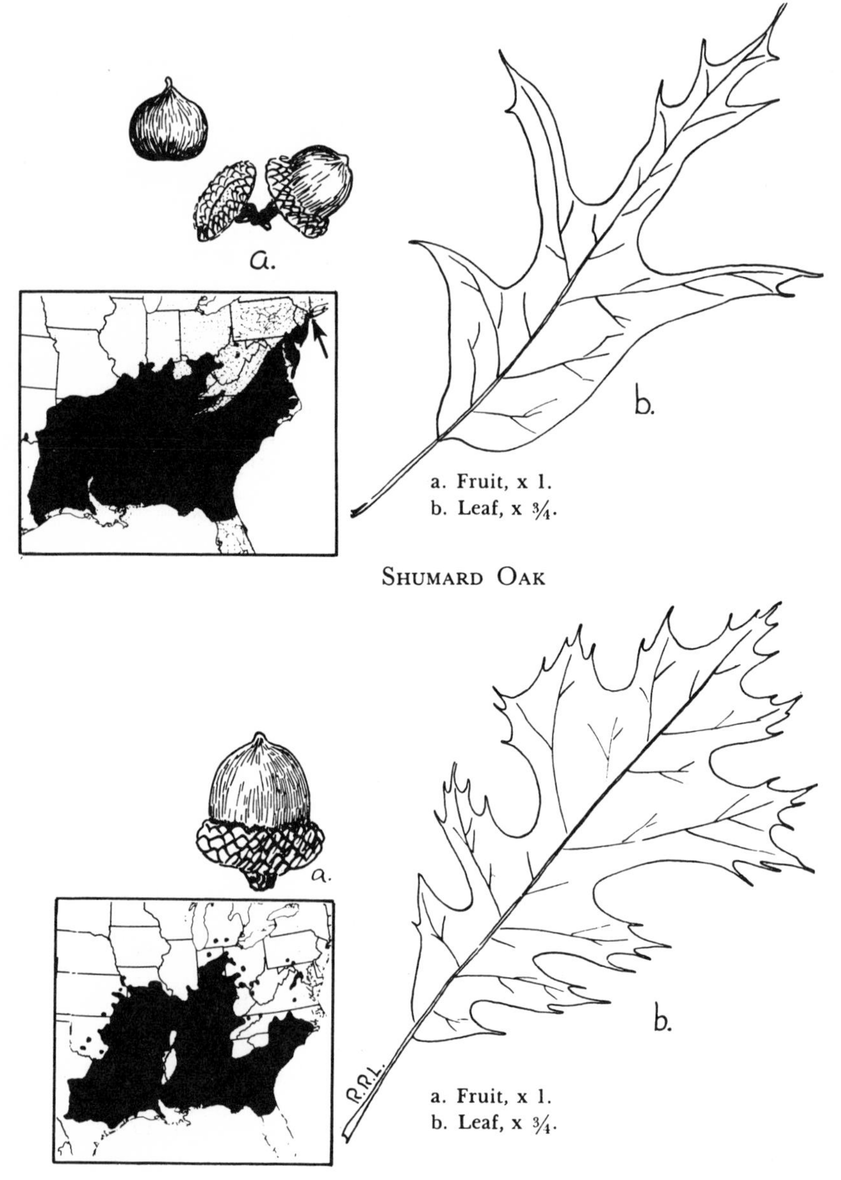

a. Fruit, x 1.
b. Leaf, x ¾.

Shumard Oak

a. Fruit, x 1.
b. Leaf, x ¾.

Southern Red Oak

Quercus falcata Michx. (*Quercus rubra* L. in part)

HABIT. A medium-sized tree 70-80 feet high and 2-3 feet in diameter (max. 120 by 7 1/2 feet); large, rounded crown.

LEAVES. Deciduous; obovate to ovate; 5-10 inches long; lobed in 2 ways often on same tree: (1) shallowly 3-lobed at apex, (2) deeply and irregularly 5-7-lobed with falcate lobes; lobes bristle-tipped; dark green above, gray-green and white- or red-woolly below; petioles 1-1 1/2 inches long, slender.

FRUIT. Maturing in 2 years; sessile or nearly so; acorn 1/2 inch long; orange-brown, subglobose; enclosed at base or to 1/3 length in thin, shallow cup with red-brown, appressed, pale, hairy scales; kernel bitter, yellow.

TWIGS. Stout; red-brown; nearly glabrous. Winter buds: 1/8-1/4 inch long; ovoid; blood-red; hairy; not strongly angled.

BARK. Thick; nearly black; deeply furrowed with broad, scaly ridges; inner bark slightly yellow.

WOOD. Important (especially varieties); similar to *Q. rubra.*

SILVICAL CHARACTERS. Similar to *Q. velutina;* dry to wet sites. The economically important cherrybark oak, *Q. falcata* var. *pagodaefolia* Ell., is widely distributed on well-drained bottomlands and is easily distinguished by more uniformly, 6-11-lobed leaves and gray-black flaky or scaly ridged bark.

* * *

Shumard Oak

Quercus shumardii Buckl.

HABIT. A large tree 80-100 feet high and 3-5 feet in diameter (max. 180 by 8 feet); open, wide-spreading crown.

LEAVES. Deciduous; obovate to oblong; 6-8 inches long; 5-9-lobed with sinuses extending more than halfway to midrib; lobes many-toothed and bristle-tipped; dark green above, paler below and glabrous except for tufts; petiole slender.

FRUIT. Maturing in 2 years; sessile or nearly so; acorn 3/4-1 1/4 inches long, red-brown, oblong-ovoid; enclosed at base in thick, shallow cup with appressed scales; kernel bitter, whitish.

TWIGS. Slender to stout; gray; glabrous. Winter buds: 1/4 inch long; straw-colored; often angled; downy to glabrous.

WOOD. Important; similar to *Q. rubra.*

SILVICAL CHARACTERS. Similar to *Q. velutina;* moist sites. Passing into the variety *texana* (Buckl.) Ashe in southern Oklahoma and central Texas.

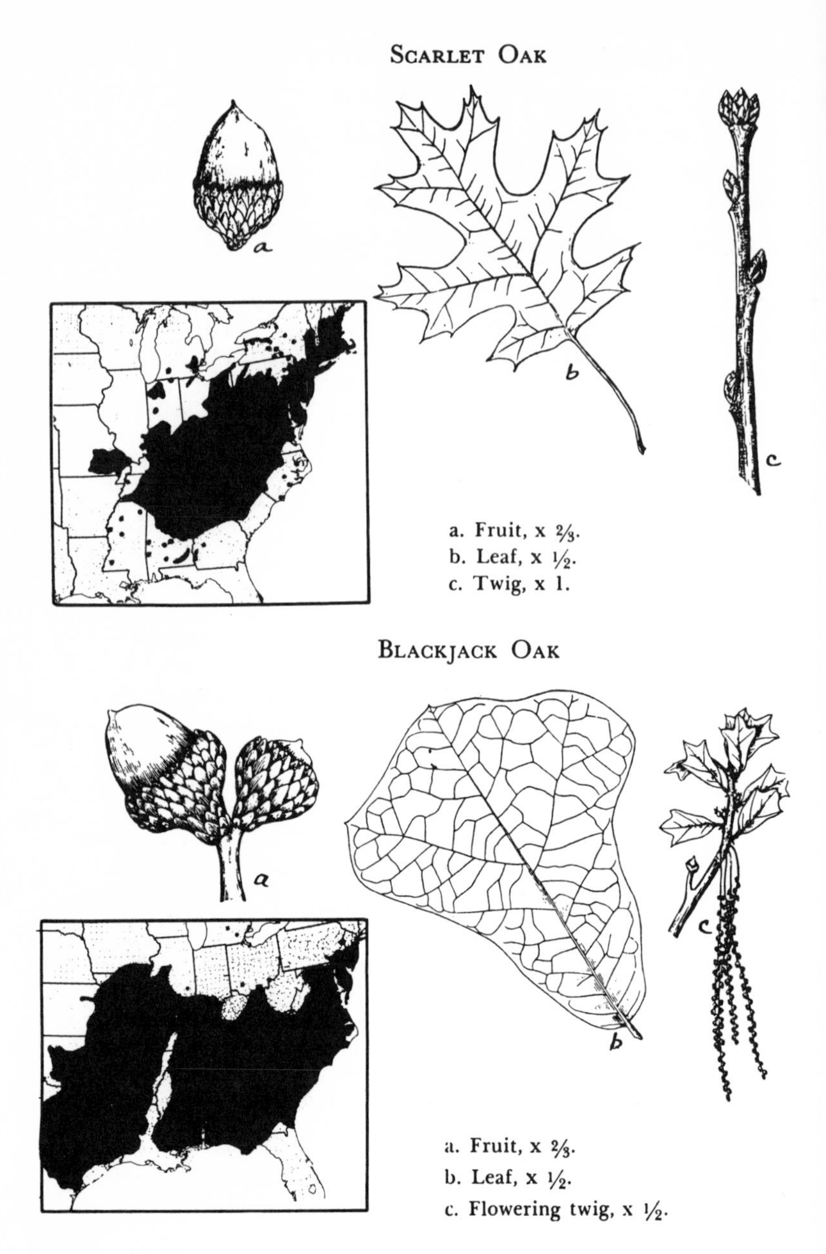

Scarlet Oak

a. Fruit, x ⅔.
b. Leaf, x ½.
c. Twig, x 1.

Blackjack Oak

a. Fruit, x ⅔.
b. Leaf, x ½.
c. Flowering twig, x ½.

Scarlet Oak

Quercus coccinea Muenchh.

HABIT. A medium-sized tree 60-80 feet high and 1-3 feet in diameter (max. 100 by 4 feet); open, rounded crown.

LEAVES. Deciduous; obovate to oval; 3-7 inches long; deeply 5-9-lobed with wide circular sinuses; lobes toothed and bristle-tipped; bright, shiny green above, paler and glabrous below except axillary tufts; turning brilliant scarlet in autumn; petioles 1 1/2-2 1/2 inches long, slender.

FRUIT. Maturing in 2 years; sessile or nearly so; acorn 1/2-1 inch long, red-brown, subglobose, usually with distinctive concentric rings near apex; enclosed for 1/2 length or less in thick cup with lustrous scales; kernel bitter, white.

TWIGS. Slender; smooth; red-brown. Winter buds: 1/8-1/4 inch long; not strongly angled; red-brown; silky hairs on upper half of scales.

BARK. Similar to *Q. velutina,* but inner bark red, not bitter.

WOOD. Important; similar but inferior to *Q. rubra.*

SILVICAL CHARACTERS. Similar to *Q. velutina;* dry, sandy sites.

* * *

Blackjack Oak

Quercus marilandica Muenchh.

HABIT. A small, shrubby tree 20-30 feet high and 1/2-1 foot in diameter (max. 50 by 2 1/2 feet; rounded, contorted crown.

LEAVES. Deciduous; thick; variable, but typically obovate, more or less 3-lobed at apex; 5-7 inches long; lobes entire or toothed, bristle-tipped; dark-green, lustrous above, paler, tawny, scurfy-pubescent below; petioles 1/2-2 inches long, stout.

FRUIT. Maturing in 2 years; sessile or nearly so; acorn 3/4 inch long, yellow-brown, subglobose; enclosed for 1/2 length in bowl-shaped cup with large, loose, red-brown, hairy scales; kernel yellow, bitter.

TWIGS. Stout; green to red-brown. Winter buds: 1/4-1/2 inch long; angled; hairy; rusty-brown; conical.

BARK. Thick; nearly black; divided into rough, square blocks.

WOOD. Unimportant; similar to *Q. rubra.*

SILVICAL CHARACTERS. Similar to *Q. velutina;* poor, dry sites.

Pin Oak

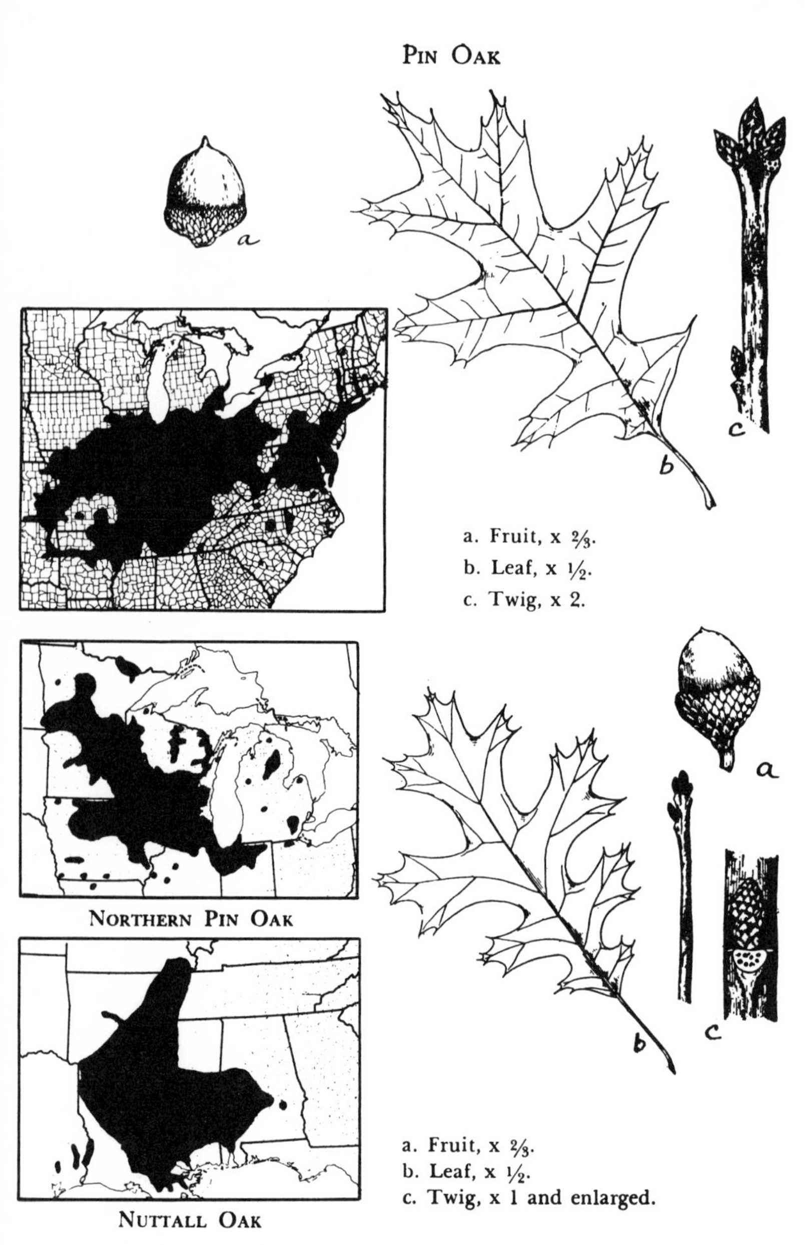

a. Fruit, x 2/3.
b. Leaf, x 1/2.
c. Twig, x 2.

Northern Pin Oak

a. Fruit, x 2/3.
b. Leaf, x 1/2.
c. Twig, x 1 and enlarged.

Nuttall Oak

Pin Oak

Quercus palustris Muenchh.

HABIT. A medium-sized tree 60-80 feet high and 1-3 feet in diameter (max. 120 by 6 feet); pyramidal crown with drooping lower branches that prune poorly.

LEAVES. Deciduous; obovate; 3-6 inches long; deeply 5-7-lobed with wide, rounded sinuses; lobes toothed and bristle-tipped; bright green and lustrous above, paler and glabrous below except axillary tufts; petioles slender.

FRUIT. Maturing in 2 years; sessile or nearly so; acorn 1/2 inch long, hemispherical, light brown; often striate; enclosed at base in thin cup with red-brown, appressed, free-tipped scales; kernel bitter, yellow.

TWIGS. Slender; smooth; green to red-brown; many short pinlike branches. Winter buds: 1/8 inch long; red-brown; shiny; sharp-pointed; angled; ciliate.

BARK. Thick; gray-brown; long, smooth, finally scaly ridged.

WOOD. Important; similar to *Q. rubra.*

SILVICAL CHARACTERS. Similar to *Q. velutina;* on moist sites; a common ornamental; easily transplanted.

* * *

Northern Pin Oak

Quercus ellipsoidalis E. J. Hill

This species is similar to pin oak except for the ellipsoidal acorn 1/2-3/4 inch long enclosed for 1/3-1/2 its length in bowl-shaped cup. From central Michigan and northern Ohio to eastern Iowa and southern Minnesota.

* * *

Nuttall Oak

Quercus nuttallii Palmer (*Quercus palustris* f. *nuttallii* Muller)

This species is similar to pin oak except for its oblong-ovoid acorn 3/4-1 1/4 inches long enclosed for about 1/3 length in a bowl-shaped cup and for yellow axillary hairs on lower surface of the leaf. Bottomlands of Mississippi, Alabama, Tennessee, Missouri, Arkansas, Louisiana, Oklahoma, Texas.

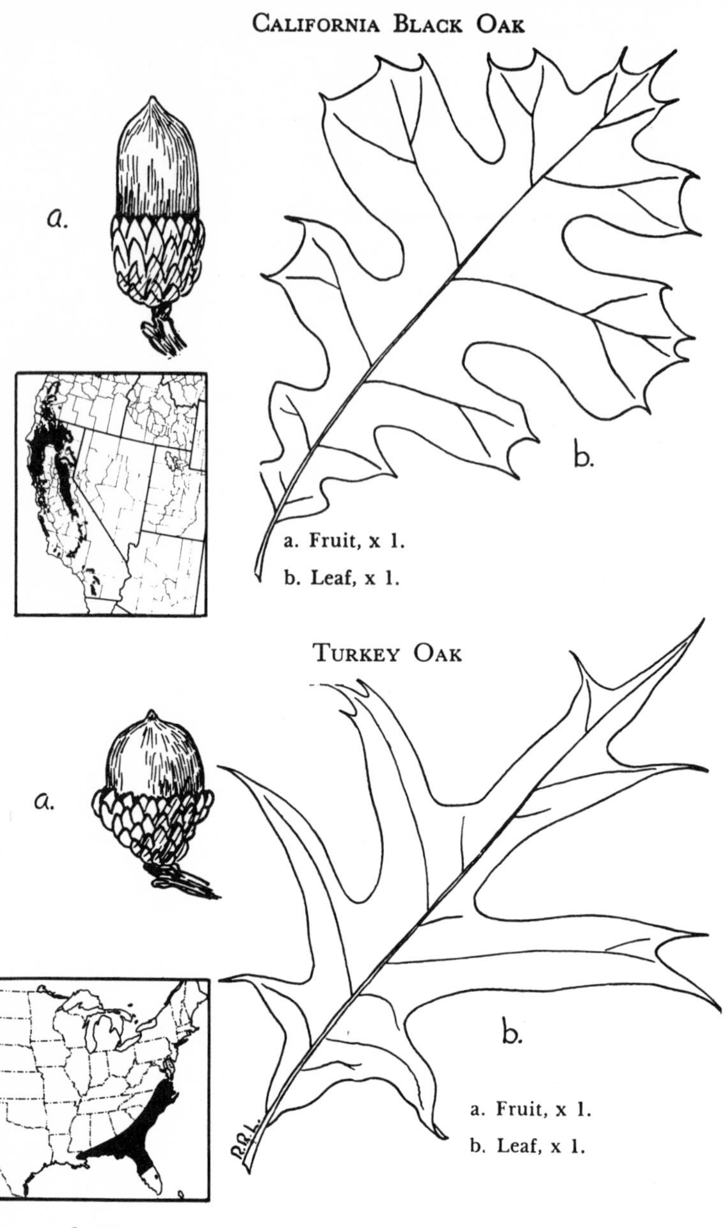

CALIFORNIA BLACK OAK

a. Fruit, x 1.
b. Leaf, x 1.

TURKEY OAK

a. Fruit, x 1.
b. Leaf, x 1.

California Black Oak

Quercus kelloggii Newb.

HABIT. A medium-sized tree 60-90 feet high and 2-4 feet in diameter (max. 130 by 9 feet); large, open, globose crown.

LEAVES. Deciduous; obovate; 3-8 inches long; deeply 5-7-lobed, rounded usually narrow sinuses; lobes usually toothed, bristle-tipped; dark yellow-green above, paler below; usually hairy.

FRUIT. Maturing in 2 years; short-stalked; acorn 1-1 1/2 inches long, ellipsoidal, light chestnut-brown; enclosed for 1/3-3/4 its length in bowl-shaped cup with thin, chestnut-brown, erose-margined scales; inner surface of nut shell woolly.

TWIGS. Slender to rather stout; red-brown. Winter buds: 1/4 inch long; ovoid; chestnut-brown; ciliate on margins.

BARK. Thick; nearly black; divided by deep furrows into wide ridges; light brown and smooth on young stems.

WOOD. Unimportant; similar to *Q. rubra.*

SILVICAL CHARACTERS. Similar to *Q. velutina;* on dry, sandy soils.

* * *

Turkey Oak

Quercus laevis Walt. (*Quercus catesbaei* Michx.)

HABIT. A small tree 20-30 feet high and 1-2 feet in diameter (max. 60 by 2 1/2 feet); open, irregular crown.

LEAVES. Deciduous; variable; distinctive; 3-12 inches (usually about 5); deeply 3-5-lobed; lobes spreading, falcate, bristle-tipped; lustrous yellow-green above, paler below; glabrous except tufts of red hairs in axils of veins; persisting on tree into winter; petioles 1/4-3/4 inches, stout.

FRUIT. Maturing in 2 years; short-stalked; acorn 1 inch long, oval, light brown; enclosed for 1/3-1/2 length in bowl-shaped, hairy cup of loosely imbricated scales; kernel bitter, yellow.

TWIGS. Stout; red-brown; glabrous. Winter buds: 1/2 inch long; narrow; tapering; chestnut-brown; red-hairy.

BARK. Rather thick; nearly black; deeply furrowed with rough, scaly ridges.

WOOD. Unimportant; similar to *Q. rubra.*

SILVICAL CHARACTERS. Similar to *Q. velutina;* dry, sandy soils.

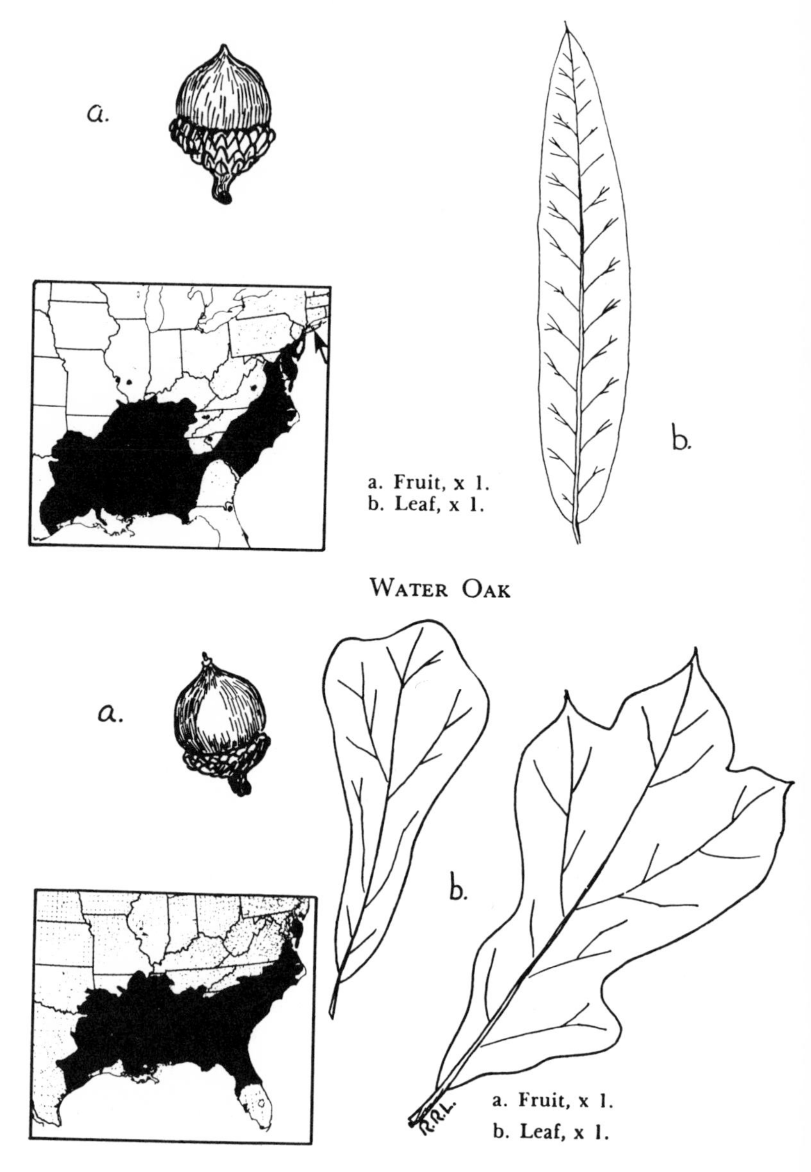
WILLOW OAK
a.
b.
a. Fruit, x 1.
b. Leaf, x 1.
WATER OAK
a.
b.
R.R.L.
a. Fruit, x 1.
b. Leaf, x 1.

Willow Oak

Quercus phellos L.

HABIT. A medium-sized tree 70-100 feet high and 2-4 feet in diameter (max. 130 by 7 feet); dense, rounded crown.

LEAVES. Deciduous; typically linear-lanceolate to ovate-lanceolate; 2-5 inches long; entire or wavy; acute with bristle tip; bright green above, paler and glabrous to hairy below.

FRUIT. Maturing in 2 years; sessile or nearly so; acorn not over 1/2 inch long, green-brown to yellow-brown, hemispherical; enclosed at base by shallow cup with thin, hairy, red-brown scales; kernel bitter, yellow.

TWIGS. Slender; red-brown; glabrous. Winter buds: 1/8-1/4 inch long; sharp-pointed; chestnut-brown; smooth.

BARK. Rather thick; nearly black; broken by deep furrows into rough ridges on old trunks; smooth and red-gray when young.

WOOD. Moderately important; similar to *Q. rubra*.

SILVICAL CHARACTERS. Similar to *Q. velutina;* wet sites; fast growing; commonly planted as ornamental.

* * *

Water Oak

Quercus nigra L.

HABIT. A medium-sized tree 50-70 feet high and 2-3 feet in diameter (max. 125 by 6 feet); rounded, even crown.

LEAVES. Deciduous, though tardily so, as many remain green until late winter; extremely variable in shape, size, margin; commonly spatulate, also obovate to narrow-oblong; margin entire to shallowly 3-lobed at apex to deeply 5-7-lobed; 2-6 inches long; dull blue-green above, paler below; mostly glabrous.

FRUIT. Maturing in 2 years; sessile or nearly so; acorn 1/3-2/3 inch long, nearly black, subglobose; enclosed at base by shallow cup with thin, hairy scales; kernel bitter, bright orange.

TWIGS. Slender; red-brown; glabrous. Winter buds: 1/8-1/4 inch long; angled; sharp; loose, red-brown, hairy scales.

BARK. Rather thin; gray-black; broken by shallow furrows into wide, scaly ridges; smooth when young.

WOOD. Moderately important; similar to *Q. rubra*.

SILVICAL CHARACTERS. Similar to *Q. velutina;* moist to wet sites; reproduces aggressively; common ornamental.

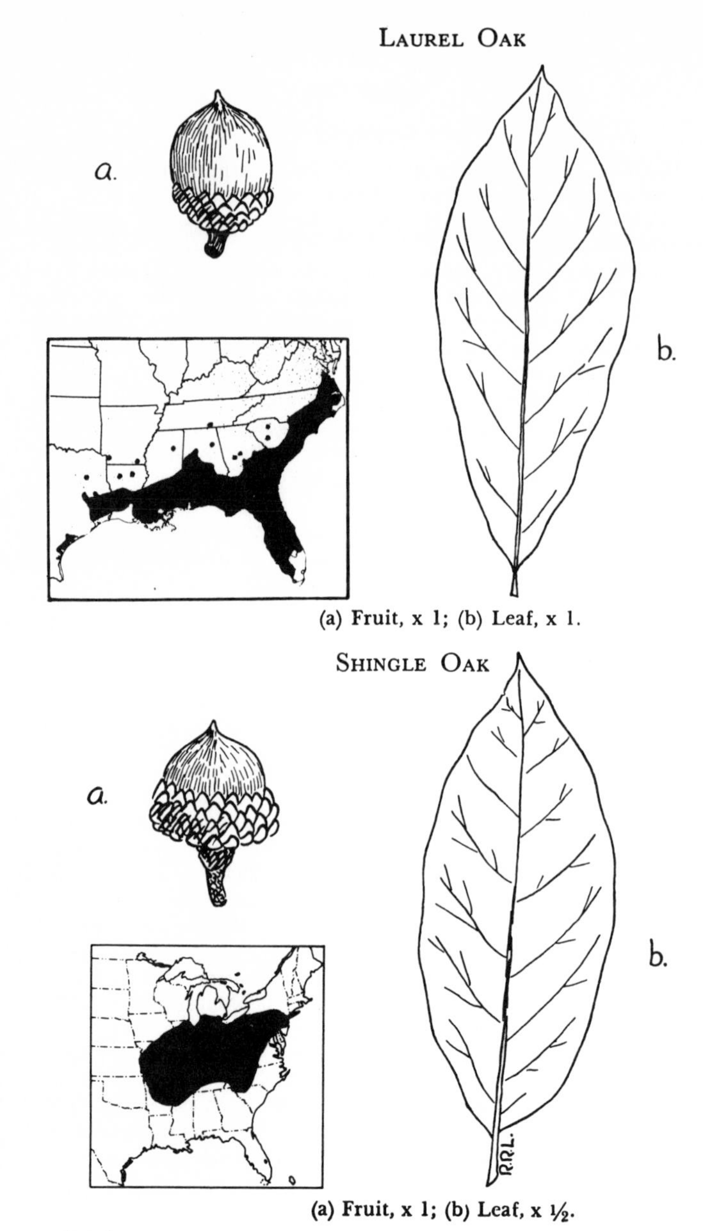

LAUREL OAK

(a) Fruit, x 1; (b) Leaf, x 1.

SHINGLE OAK

(a) Fruit, x 1; (b) Leaf, x ½.

Laurel Oak

Quercus X *laurifolia* Michx.

HABIT. A medium-sized tree 50-70 feet high and 2-3 feet in diameter (max. 100 by 8 feet); dense, rounded crown.

LEAVES. Nearly evergreen, falling in early spring just before new leaves appear; elliptical to oblong-lanceolate; 2-4 inches long; entire or wavy, occasionally 3-lobed at apex; bristle-tipped or rounded at apex; bright green above, paler below; glabrous.

FRUIT. Maturing in 2 years; sessile or nearly so; acorn 1/3-1/2 inch long, nearly black, hemispherical; base enclosed by shallow cup with thin, red-brown, hairy scales; kernel bitter.

TWIGS. Slender; red-brown; glabrous. Winter buds: 1/10-3/16 inch long; ovoid; pointed; tight, bright red scales.

BARK. Thick; nearly black; deeply furrowed with broad, flat ridges; dark brown and smooth or scaly when young.

WOOD. Little used; similar to *Q. rubra.*

SILVICAL CHARACTERS. Recent study indicates laurel oak is a hybrid of *Q. phellos* and *Q. nigra.*

* * *

Shingle Oak

Quercus imbricaria Michx.

HABIT. A medium-sized tree 50-60 feet high and 2-3 feet in diameter (max. 4 feet); narrow, rounded crown.

LEAVES. Deciduous; elliptical to oblong-obovate; 4-6 inches long; lustrous above, paler and white-hairy below.

FRUIT. Maturing in 2 years; usually short-stalked; acorn 1/2-3/4 inch long, chestnut-brown, ovoid; enclosed for 1/3-1/2 length in bowl-shaped cup with thin, red-brown, appressed, hairy scales; kernel bitter.

TWIGS. Slender; green-brown. Winter buds: 1/8-3/16 inch long; conical; chestnut-brown; pubescent.

BARK. Thick; gray-brown; shallowly furrowed and ridged.

SILVICAL CHARACTERS. Similar to *Q. velutina;* moist sites.

* * *

The bluejack oak, *Q. incana* Bartr., is a similar small tree of dry, sandy sites extending from Virginia to Florida west to Texas. Characterized by small elliptical leaves, blue-green above and white-woolly below, and shallow acorn cups.

LIVE OAK

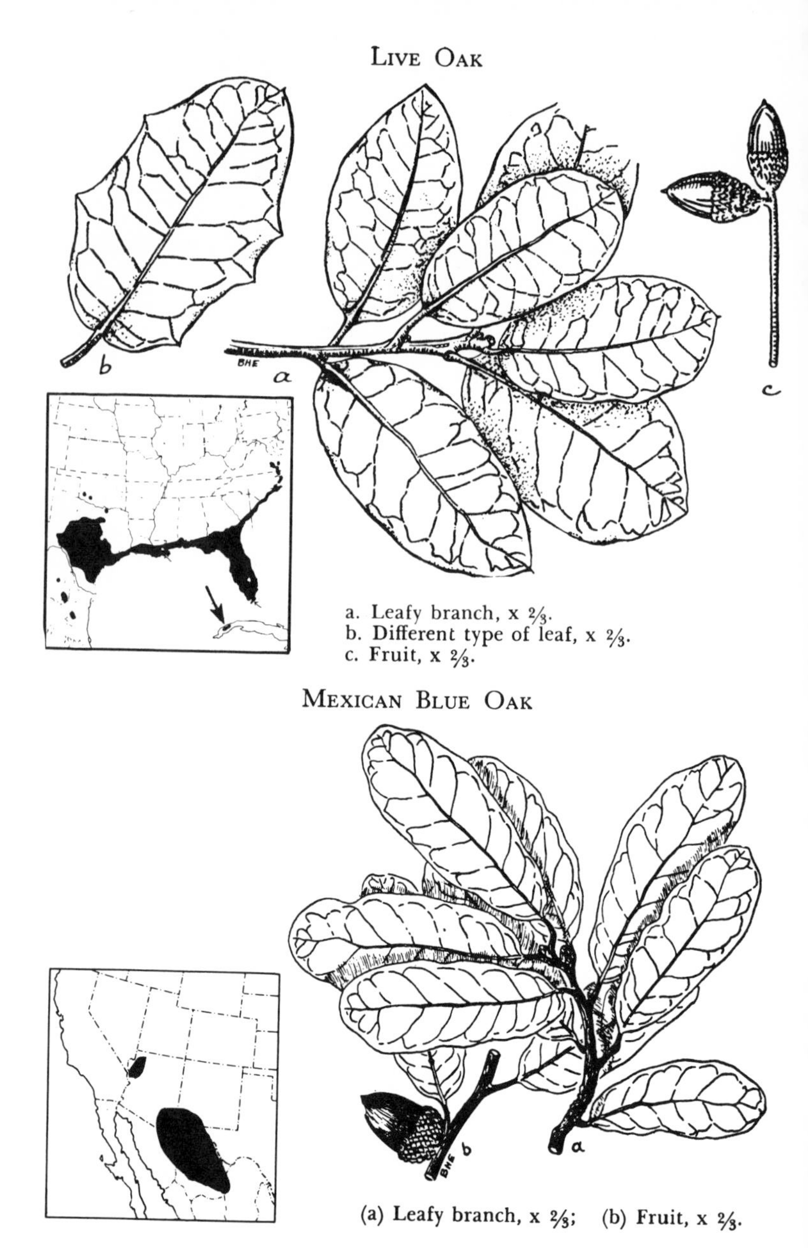

a. Leafy branch, x ⅔.
b. Different type of leaf, x ⅔.
c. Fruit, x ⅔.

MEXICAN BLUE OAK

(a) Leafy branch, x ⅔; (b) Fruit, x ⅔.

FAGACEAE

Live Oak

Quercus virginiana Mill.

HABIT. A medium-sized tree 40-50 feet high and 3-4 feet in diameter (max. 80 by 11 1/2 feet); trunk usually dividing near the ground into several large, horizontal limbs; crown dense, rounded with span of 100-150 feet (max. 168 feet).

LEAVES. Persistent until new leaves appear; elliptical to oblong-obovate; 2-5 inches long; entire and wavy margin (rarely with few teeth); rounded tip; dark green above, paler and somewhat hairy below.

FRUIT. Maturing in one year; stalks 1-5 inches long; acorn 3/4-1 inch long, dark brown, ellipsoidal; enclosed for 1/3-1/2 its length in bowl-shaped cup with thin, red-brown, woolly scales; inner surface of nut shell glabrous; kernel sweet.

TWIGS. Slender; gray-brown; hairy. Winter buds: 1/16 inch long.

BARK. Rather thick; dark red-brown; shallowly furrowed with flat, scaly ridges.

WOOD. The only important live oak; diffuse porous; very heavy, hard, and strong; durable; used for ships, posts, etc.

SILVICAL CHARACTERS. Intolerant; typical of dry sites; long-lived; 3 varieties reach tree size.

* * *

Mexican Blue Oak

Quercus oblongifolia Torr.

HABIT. A shrub or small tree rarely 30 feet high and 18-30 inches in diameter; handsome, rounded crown.

LEAVES. Persistent until new leaves appear; ovate to elliptic; 1-2 inches long; entire or wavy with revolute margins (rarely coarsely toothed); blue-green above, paler and glabrous below.

FRUIT. Maturing in one year; sessile or rarely stalked; acorn 1/2-3/4 inch long, chestnut-brown, ovoid to obovoid; enclosed for 1/3 length in bowllike cup with thin, red-tipped, woolly scales; inner surface of nut shell glabrous; kernel sweet.

TWIGS. Slender; red-gray. Winter buds: 1/16-1/8 inch long; brown.

BARK. Thick; ash-gray; broken into nearly square, platelike scales on old trunks.

SILVICAL CHARACTERS. Intolerant; on dry foothills.

ARIZONA WHITE OAK

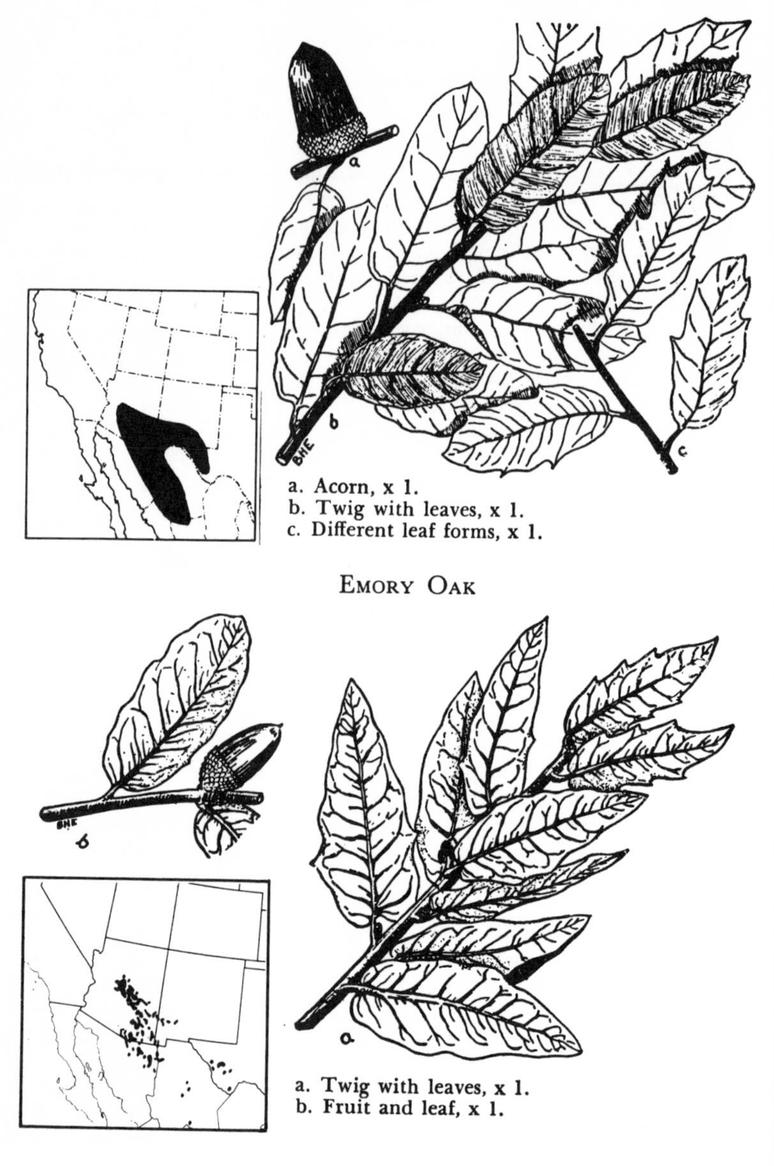

a. Acorn, x 1.
b. Twig with leaves, x 1.
c. Different leaf forms, x 1.

EMORY OAK

a. Twig with leaves, x 1.
b. Fruit and leaf, x 1.

Arizona White Oak

Quercus arizonica Sarg.

HABIT. A shrub or small tree rarely 60 feet high and 3 feet in diameter; handsome, round-topped crown with massive, contorted, nearly horizontal branches.

LEAVES. Persistent until appearance of new leaves; oblong-lanceolate to broadly ovate; 1-4 inches long (most about 2 1/2); entire or with wavy, spiny teeth; revolute margin; dark blue-green above, paler and densely hairy below.

FRUIT. Maturing in one year; sessile or short-stalked; acorn 3/4-1 inch long, chestnut-brown, oblong to obovoid; enclosed for 1/2 its length in deep cup with thick, corky, woolly, red-tipped scales; inner surface of nut shell glabrous; kernel sweet.

TWIGS. Stout; red-brown; red and hairy at first, becoming glabrous. Winter buds: 1/16 inch long; subglobose, brown; ciliate.

BARK. Rather thick; ash-gray; furrowed and scaly-ridged.

SILVICAL CHARACTERS. Intolerant; on dry mountain slopes; most common live oak in the Southwest.

* * *

Emory Oak

Quercus emoryi Torr.

HABIT. A shrub or small tree rarely 60 feet high and 2-5 feet in diameter; round-topped, even crown with stout, drooping branches.

LEAVES. Persistent until appearance of new leaves; oblong-lanceolate; 1-2 1/2 inches long; entire or remotely wavy toothed; leathery; dark green above, paler below; glabrous or slightly hairy.

FRUIT. Maturing in one year; sessile or nearly so; acorn 1/2-3/4 inch long, nearly black, oblong to ovoid; enclosed for 1/3-1/2 its length in bowl-shaped cup with brown, hairy scales; inner surface of nut shell woolly; kernel sweet.

TWIGS. Slender; red and hairy at first, becoming dark brown and glabrous.

BARK. Thick; nearly black; deeply furrowed and scaly-plated.

SILVICAL CHARACTERS. Intolerant; on dry foothills and mountains; very abundant; in pure or mixed stands; coppices freely; acorns important source of human and animal food.

Interior Live Oak

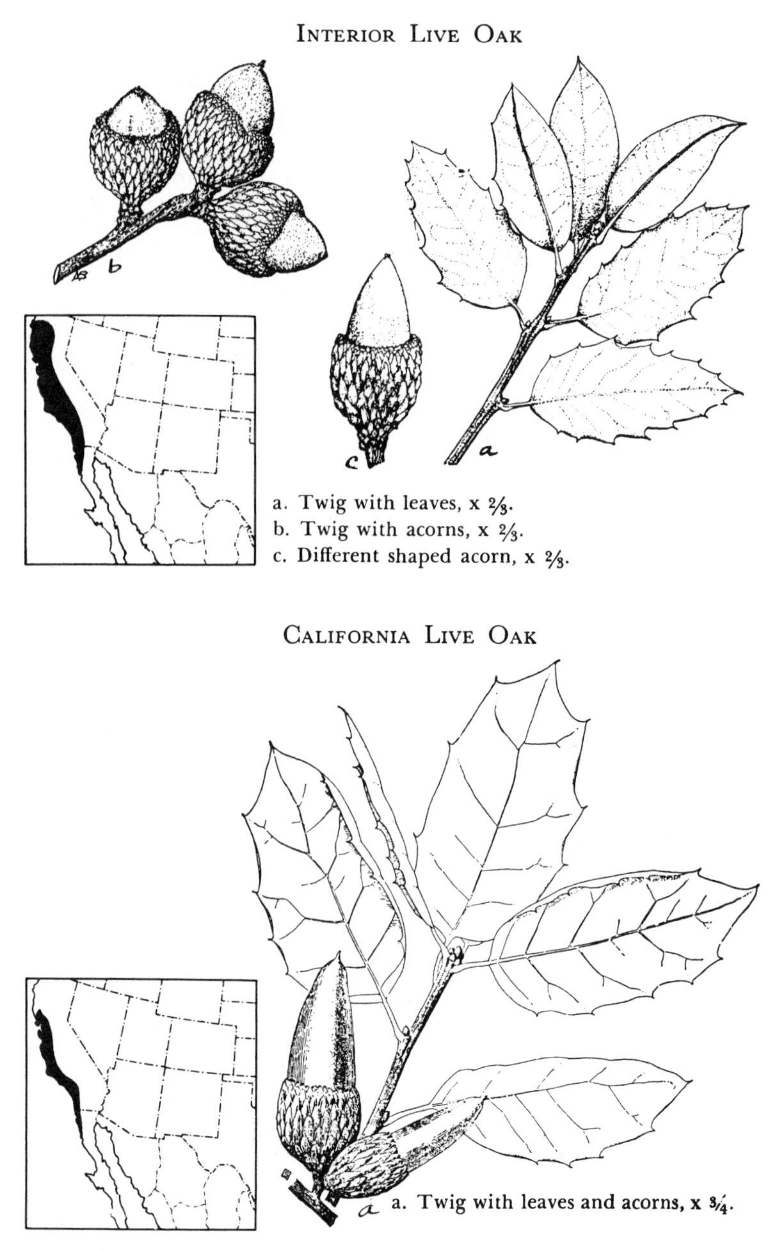

a. Twig with leaves, x ⅔.
b. Twig with acorns, x ⅔.
c. Different shaped acorn, x ⅔.

California Live Oak

a. Twig with leaves and acorns, x ¾.

Interior Live Oak

Quercus wislizenii A. DC.

HABIT. Commonly shrubby though becoming a medium-sized tree 60-80 feet high and 2-4 feet in diameter (max. 100 by 6 1/2 feet); wide, rounded crown with large, spreading branches.

LEAVES. Persistent 2 years; lanceolate to broadly elliptic; 1-3 inches long (mostly about 1 1/4 inches); entire or with spiny teeth; leathery; glabrous; lustrous dark green above, paler below.

FRUIT. Maturing in 2 years; sessile or short-stalked; acorn 1-1 1/2 inches long, chestnut-brown, slender-oblong, pointed; enclosed for 1/2 or less of its length in bowl-shaped cup with thin, closely imbricated, ciliate scales; inner surface of nut shell woolly; kernel bitter.

TWIGS. Slender; dark brown. Winter buds: 1/8-1/4 inch long; ovoid.

BARK. Thick; nearly black; deeply furrowed with scaly ridges.

SILVICAL CHARACTERS. Intolerant; slow-growing; dry sites.

* * *

California Live Oak

Quercus agrifolia Née

HABIT. A medium-sized tree 50-75 feet high and 2-4 feet in diameter (max. 90 by 12 1/2 feet); trunk dividing shortly above ground into several large, horizontal limbs that often rest on the ground; broad crown often 150 feet across.

LEAVES. Persistent until new leaves appear; oval or oblong to suborbicular; 1-3 inches long; margins entire to spiny-toothed, revolute; leathery; lustrous dark green above, paler below; glabrous to hairy.

FRUIT. Maturing in one year; sessile or nearly so; acorn 3/4-1 1/2 inches long, chestnut-brown, slender-conic, pointed; enclosed for 1/3 of length or only at base in bowl-shaped cup with thin, closely imbricated scales; inner surface of nut shell woolly; kernel bitter.

TWIGS. Slender; gray-brown. Winter buds: 1/16 inch long; globose.

BARK. Thick; nearly black; deeply furrowed with scaly ridges.

SILVICAL CHARACTERS. Intolerant; slow-growing; dry sites.

Canyon Live Oak

a. Twig with leaves, x ⅔.

b. Different form of leaves, x ⅔.

c. Acorn, x ⅔.

Silverleaf Oak

a. Twig with leaves and acorns, x ⅔.

Canyon Live Oak

Quercus chrysolepis Liebm.

HABIT. A shrub to medium-sized tree 60-80 feet high and 1-5 feet in diameter (max. 100 by 12 feet); usually dividing above ground into large, horizontal branches; large, spreading crown; forming dense thickets on dry slopes.

LEAVES. Persistent 3-4 years; oblong-ovate to elliptic; 1-4 inches long; mostly entire on old trees and sinuate-dentate on young trees, or both forms appearing together; revolute margins; leathery; bright yellow-green and glabrous above; tawny-tomentose, becoming glabrous and blue-green below.

FRUIT. Maturing in 2 years; sessile or nearly so; acorn 1/2-2 inches long, chestnut-brown, ellipsoidal to ovate; enclosed at base in thick, shallow cup with hairy to golden-woolly scales; inner surface of nut shell woolly; kernel bitter.

TWIGS. Slender; brown to gray; woolly, sometimes becoming smooth. Winter buds: 1/8 inch long; ovoid; acute; chestnut-brown.

BARK. Rather thick (3/4-1 1/2 inches); gray-brown, tinged with red; smooth except for small scales on old trunks.

SILVICAL CHARACTERS. Tolerant when young; long-lived; canyon walls to dry mountain slopes; passes into variety *palmeri* (Engelm.) Sarg. in the Southwest.

* * *

Silverleaf Oak

Quercus hypoleucoides A. Camus (*Quercus hypoleuca* Engelm.)

HABIT. A shrub or small tree 20-30 feet high (rarely 60 feet) and 10-15 inches in diameter; narrow, rounded crown.

LEAVES. Persistent until appearance of new leaves; lanceolate to elliptic; 2-4 inches long; entire or with few coarse teeth near apex; revolute margins; leathery; dark yellow-green and lustrous above, woolly below.

FRUIT. Maturing in 2 years; sessile or short-stalked; acorn 1/2-2/3 inch long, chestnut-brown, ovoid; enclosed for 1/3 of length in thick cup with thin, brown, silvery-hairy scales; inner surface of nut shell woolly; kernel bitter.

TWIGS. Stout; red-brown. Winter buds: 1/8 inch long; brown.

BARK. Rather thick (3/4-1 inch); nearly black; deeply furrowed into broad, thick-scaled ridges.

SILVICAL CHARACTERS. Intolerant; on dry, low mountain sites.

ULMACEAE

The Elms

Characteristics of the Genus *Ulmus* L.

HABIT. Handsome trees often planted for shade or ornamental purposes; irregularly pinnate branching.

LEAVES. Alternate; simple; deciduous; lanceolate to obovate; stipulate; penniveined; simply or more commonly doubly serrate; 2-ranked; mostly unequal and oblique at the base and acuminate at the apex; petioles short.

FLOWERS. Regular; perfect; small; in fascicles or cymes; appearing before the leaves (in a few species not appearing until autumn); on slender, drooping pedicels; calyx 5-9-lobed; corolla absent; 4-6 stamens; ovary usually 1-celled by abortion, 1-ovuled, flattened, deeply 2-lobed style.

FRUIT. A distinctive oblong to suborbicular samara; maturing in spring or autumn of first year; surrounded at base by calyx remnants; seed cavity compressed, surrounded by a thin papery wing commonly notched at the apex and tipped with the remnants of the persistent style.

TWIGS. Slender to stout; somewhat zigzag; corky wings on some species; pith solid, round; leaf scars with 3 somewhat depressed bundle scars. Winter buds: terminal absent; lateral with overlapping scales.

BARK. Thick and furrowed; often with layers of 2 colors.

WOOD. Moderately important; heavy; hard; strong; tough; ring-porous; not durable; difficult to split; specialized uses.

SILVICAL CHARACTERS. Rather tolerant trees; fast-growing; wide-spreading, lateral roots.

GENERAL. This genus contains about 20 species of trees scattered through the Northern Hemisphere; many are among our most important shade and ornamental trees. Six species are native to North America. Five exotic species commonly planted in this country are included in the key; of these *U. pumila*, widely grown in the central and western states, has been naturalized from Kansas to Minnesota.

KEY TO THE SPECIES OF ELMS

1. Leaves doubly serrate, unequal at base.
   2. Some of twigs with corky wings; fruit hairy, at least on margins; leaves not over 4 inches long.
      3. Buds, young twigs pale-hairy; spring fruit, ovoid to suborbicular, hairy and ciliate, seed cavity not pronounced; leaves smooth above, hairy below . . . . *U. thomasii*, Rock Elm, p. 219
      3. Buds and young twigs glabrous or nearly so; fruit oblong, seed cavity pronounced; southern.
         4. Leaves 2-4 inches, acuminate, glabrous above, slightly hairy along veins below; fall fruit, fringed margin with long white hairs; Kentucky, Illinois to Georgia, west to Oklahoma . . . . *U. serotina*, September Elm, p. 217
         4. Leaves 1-2 1/2 inches; fruit covered, long white hairs.
            5. Leaves acuminate, coarsely doubly toothed, smooth above, hairy below; spring fruit; Illinois, Virginia to Oklahoma, south . . . . *U. alata*, Winged Elm, p. 217
            5. Leaves rounded or acute at apex, unequally doubly toothed, almost singly toothed; roughened above, hairy below; fall fruit; Arkansas to Mississippi, west to Louisiana, Oklahoma, Texas . . . . *U. crassifolia*, Cedar Elm, p. 217
   2. Twigs without corky ridges; fruit appearing in spring; seed cavity distinct.
      6. Fruit hairy or ciliate; buds glabrous or with rusty hairs; leaves usually 4-7 inches long, rough on top.
         7. Buds with rusty hairs; fruit hairy over seed cavity; leaves rough on both surfaces; bark of trunk dark red-brown . . . . *U. rubra*, Slippery Elm, p. 219
         7. Buds nearly glabrous; fruit hairy only on margin; leaves rough above; bark of trunk with alternating gray and brown layers . . . . *U. americana*, American Elm, p. 217
      6. Fruit glabrous; buds pale-hairy or glabrous; European ornamentals.
         8. Twigs hairy; leaves rough above; petioles 1/8-1/4 inch.
            9. Leaves 2-3 inches; petioles 1/6-1/4 inch; bark gray, furrowed; seed near apex of deeply notched fruit . . . . *U. procera* Salisb., English Elm
            9. Leaves 3-6 1/2 inches long; petioles 1/8 inch; bark dark, remaining smooth many years; seed near center of slightly notched fruit . . . . *U. glabra* Huds., Wych Elm
         8. Twigs glabrous; leaves 2-3 inches; smooth above; petioles 1/4-1/2 inch. . . . . *U. carpinifolia* Gled., Smooth-leaved Elm
1. Leaves singly or nearly singly serrate, nearly equal at base, 3/4-3 inches long; fruit glabrous; Asiatic ornamentals.
   10. Leaves 3/4-2 inches long, predominantly rounded to acute at apex; fall fruit, ovate, notched at apex . . . . *U. parvifolia* Jacq., Chinese Elm
   10. Leaves 1-3 inches long, predominantly acuminate to acute at apex; spring fruit, suborbicular, closed notch . . . . *U. pumila* L., Siberian Elm

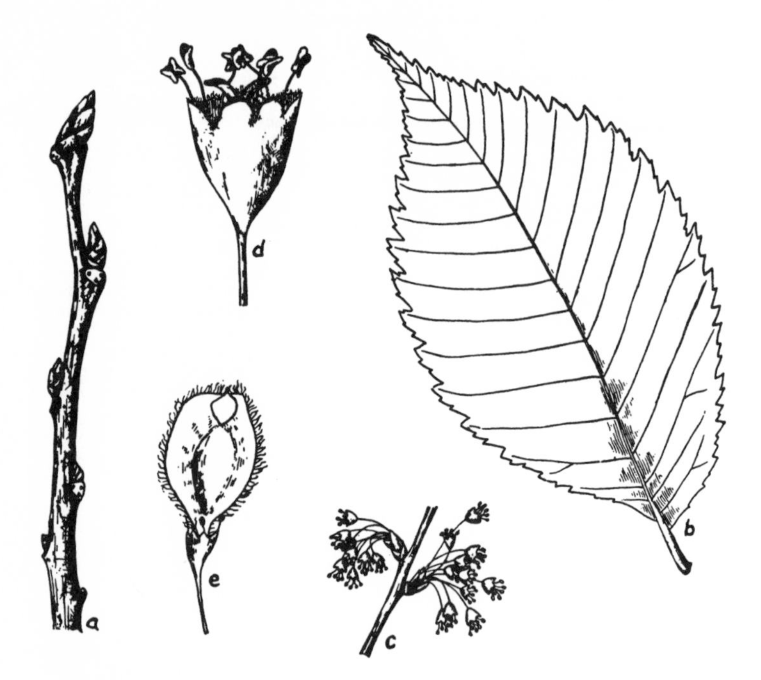

a. Winter twig, x 2.

b. Leaf, x ½.

c. Flowering branchlet, x ½.

d. Flower, enlarged.

e. Fruit, x 2.

ULMACEAE

American Elm. White Elm

Ulmus americana L.

HABIT. A handsome tree 75-100 feet high and 3-6 feet in diameter (max. 120 by 10 feet); open-grown trees commonly dividing near ground into several ascending limbs forming a vase-shaped crown; perhaps our most common ornamental species.

LEAVES. Oblong-obovate to oval; 4-6 inches long; coarsely doubly serrate; thick and firm; base conspicuously oblique; dark green and more or less rough above, paler below.

FRUIT. A smooth, oval to ovate samara 1/2 inch long, hairy on margins and deeply notched at apex; seed cavity distinct; long-stalked fascicles; maturing in late spring.

TWIGS. Slender; round; red-brown; hairy at first, becoming glabrous. Winter buds: lateral 1/4 inch long; chestnut-brown; acute; glabrous or with scales hairy-fringed.

BARK. Thick; ash-gray; variable, but typically diamond-shaped by furrows; alternate light and dark layers in outer bark.

WOOD. The most important of the elms.

SILVICAL CHARACTERS. Moderately tolerant; shallow-rooted; typical of moist sites, but hardy and widely planted as a shade tree; seriously threatened by Dutch Elm disease; with mixed hardwoods.

* * *

Three unimportant elms are found in the Southeast; all are characterized by corky wings on some twigs and by small leaves. They are distinguished in the key.

1. Winged elm, *Ulmus alata* Michx., a common small tree that flowers and fruits in spring; characterized by doubly serrate leaves 1-2 1/2 inches long and hair-covered fruit 1/3 inch long with a distinct seed cavity.
2. Cedar elm, *Ulmus crassifolia* Nutt., flowers and fruits in the fall; characterized by nearly singly-toothed leaves 1-2 1/2 inches long with rounded to acute apex and hair-covered fruit 1/3-1/2 inch long with a distinct seed cavity.
3. September elm, *Ulmus serotina* Sarg., flowers and fruits in the fall; characterized by acuminate leaves 2-4 inches long and fruit ciliate with white hairs.

SLIPPERY ELM

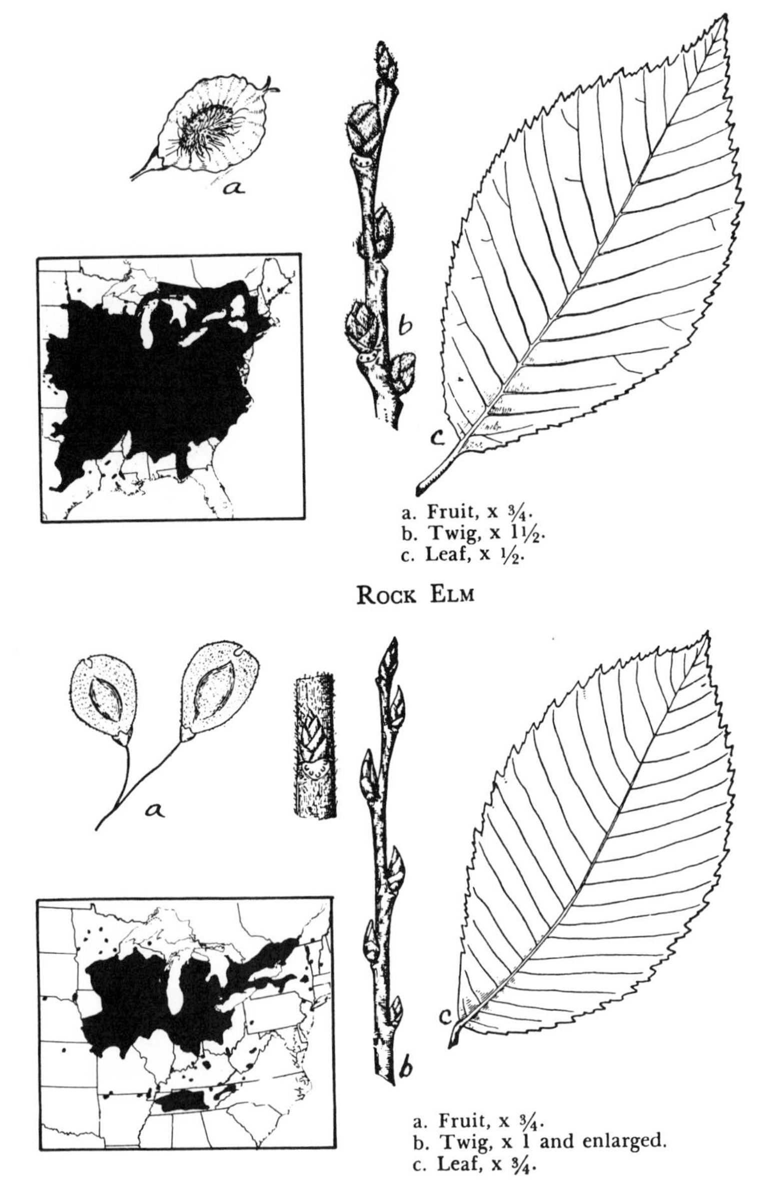

a. Fruit, x 3/4.
b. Twig, x 1 1/2.
c. Leaf, x 1/2.

ROCK ELM

a. Fruit, x 3/4.
b. Twig, x 1 and enlarged.
c. Leaf, x 3/4.

Slippery Elm

Ulmus rubra Muhl. (*Ulmus fulva* Michx.)

HABIT. A medium-sized tree 40-70 feet high and 1-2 feet in diameter (max. 5 feet); clear trunk; rounded, open crown with spreading branches and ascending twigs.

LEAVES. Oblong-obovate to oval; 4-7 inches long; coarsely doubly serrate; thick and firm; base conspicuously oblique; dark green and rough above; paler and rough below.

FRUIT. A suborbicular samara about 3/4 inch long; seed cavity distinct, brown-woolly; wings and margins smooth; apex entire or nearly so; in short-stalked fascicles; maturing in spring.

TWIGS. Rather stout; ash-gray; rough. Winter buds: lateral 1/4 inch long; nearly black, hairy.

BARK. Thick; dark red-brown; nearly parallel furrows; without alternate light and dark layers; inner bark mucilaginous.

SILVICAL CHARACTERS. Moderately tolerant; wide lateral roots; with mixed hardwoods; varied sites.

* * *

Rock Elm. Cork Elm

Ulmus thomasii Sarg. (*Ulmus racemosa* Thomas)

HABIT. A handsome medium-sized tree 60-80 feet high and 2-3 feet in diameter (max. 100 by 5 1/2 feet); trunk persisting into crown; narrow crown with drooping lower branches.

LEAVES. Obovate to oval; 2 1/2-5 inches long; coarsely doubly serrate; thick and firm; base nearly equal; dark green and usually glabrous above, paler and slightly hairy below.

FRUIT. An obovate to oval samara 3/4-1 inch long, hairy all over; seed cavity indistinct; apex shallowly notched; margins ciliate; in racemose cymes on long slender stalks; maturing in spring.

TWIGS. Rather stout; red-brown; usually developing corky wings. Winter buds: lateral 1/4 inch long; brown; acute; downy-ciliate.

BARK. Thick; dark gray-brown; deeply and irregularly furrowed; alternate light and dark layers in outer bark.

SILVICAL CHARACTERS. Intermediate in tolerance; wide lateral roots; with mixed hardwoods; varied sites; poor reproduction.

ULMACEAE

The Hackberries

Characteristics of the Genus *Celtis* L.

HABIT. Shrubs or trees often planted for shade or ornamentals; irregularly pinnate branching.

LEAVES. Alternate; simple; lanceolate to ovate; acute or acuminate at apex; often oblique at base; serrate or entire margins; membranaceous or subcoriaceous; deciduous; mostly long-petiolate; 2-ranked; stipules thin, caducous, fugacious, enclosing leaf in bud.

FLOWERS. Regular; polygamo-monoecious or rarely monoecious; minute; staminate in fascicles toward base of twig; above these the pistillate or perfect, solitary or in few-flowered fascicles; pedicellate; appearing soon after unfolding of leaves; calyx 4-5-lobed, green-yellow, deciduous; corolla absent; stamens as many as calyx lobes and opposite them; ovary 1-celled, ovoid, sessile, green, and lustrous.

FRUIT. Subglobose or ovoid drupe; tipped with remnants of style; thick firm skin; thin, pulpy flesh; nutlet bony, thick-walled, reticulate-pitted; ripening in autumn, often remaining long after leaves fall. Seed: filling cavity in nutlet.

TWIGS. Slender; round; unarmed or spinose; pith round and usually finely chambered at nodes. Winter buds: terminal absent; lateral small, appressed, scaly; branchlets prolonged by an upper lateral bud.

BARK. Usually gray and smooth, sometimes with conspicuous, corky, warty excrescences.

WOOD. Rather heavy; fairly hard; not strong; odorless; ring-porous; moderately important; often sold as elm.

GENERAL. This genus consists of about 70 species scattered through the north temperate and tropical regions; 5 tree species and several varieties have been listed for the United States; native species often disfigured by gall-making insects distorting the buds and producing broomlike clusters of branchlets; this genus is in a confused state and needs revision; distributions have not been accurately determined.

KEY TO THE SPECIES OF HACKBERRIES

1. Fruit dark purple, on pedicels somewhat longer than leaf petioles; leaves 2 1/2-4 inches long, usually sharply serrate with numerous teeth, green on both surfaces, without conspicuous reticulate veinlets on lower surface . . . . *C. occidentalis*, Hackberry, p. 223
1. Fruit orange to red-brown to yellow; leaves entire or sparingly toothed.
   2. Fruit on pedicels much longer than leaf petioles; leaves 1 1/2-3 inches long, conspicuous reticulate veinlets below.
      3. Leaves broadly ovate, green below; fruit yellow to orange-red . . . . *C. reticulata*, Netleaf Hackberry, p. 225
      3. Leaves oblong-ovate, pale below; fruit dark red-brown; Texas . . . . *C. lindheimeri*, Lindheimer Hackberry, p. 223
   2. Fruit on pedicels shorter or but little longer than leaf petioles; leaves without conspicuous reticulate veinlets below.
      4. Leaves mostly oblong-lanceolate, long-acuminate, 2-5 inches long; fruit yellow to orange-red . . . . *C. laevigata*, Sugarberry, p. 225
      4. Leaves ovate-lanceolate, short-acuminate, 1 1/2-2 1/2 inches long; fruit dark orange-red to red-purple . . *C. tenuifolia*, Georgia Hackberry, p. 223

* * *

Planertree. Waterelm

Planera aquatica Gmel.

A small, monotypic tree growing in swampy sites on the coastal plain from North Carolina to northern Florida west to Texas and north in the Mississippi River Valley to southern Illinois. Characterized by elmlike leaves, 2-3 inches long, but crenate-serrate with gland-tipped teeth and peculiar prickly drupe 1/3 inch long, covered with odd plates and processes.

* * *

Florida Trema

Trema micrantha (L.) Blume

A small tree growing from central Florida to Central and South America. Characterized by 2-ranked, ovate, cordate, finely serrate, persistent leaves 3-4 inches long, dark green above and pale tomentose below; stout, hoary, tomentose twigs; fruit a yellow-brown drupe 1/6-1/5 inch long crowned with persistent style. West Indies trema, *T. lamarckiana* (R. & S.) Blume, reaches southern Florida.

Hackberry

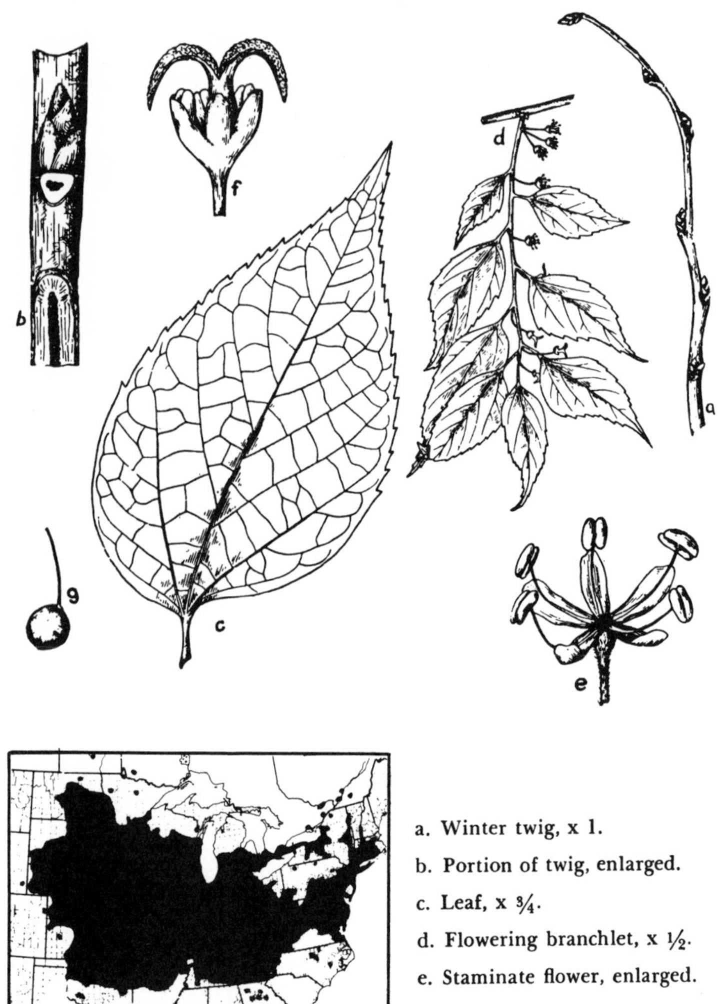

a. Winter twig, x 1.

b. Portion of twig, enlarged.

c. Leaf, x ¾.

d. Flowering branchlet, x ½.

e. Staminate flower, enlarged.

f. Pistillate flower, enlarged.

g. Fruit, x 1.

Hackberry

Celtis occidentalis L.

HABIT. Occasionally shrubby, but often rather large tree 100-120 feet high and 1-2 feet in diameter (max. 130 by 5 feet); crown rounded, large, spreading branches.

LEAVES. Ovate to ovate-lanceolate; 2 1/2-4 inches long; long-acuminate apex; obliquely rounded base; coarsely serrate or rarely almost entire; rather thin; light, dull green and rough above, slightly paler and pilose along veins below; turning light yellow in autumn; petioles 1/4-1/2 inch long.

FRUIT. 1/3 inch in diameter; subglobose; ovoid, or obovoid; dark purple; stems 1/2-3/4 inch long; thick, tough skin; dark orange-colored flesh; oblong, pointed, light brown nutlet with prominently reticulated pit.

TWIGS. Slender; ridged; light brown, becoming darker; pubescent or glabrous; marked by pale, oblong lenticels; pith often finely chambered at nodes. Winter buds: 1/4 inch long; ovoid; pointed; pubescent; chestnut-brown.

BARK. Rather thick (1-1 1/2 inches); dark brown; smooth, or more or less roughened by irregular wartlike excrescences or by long ridges.

SILVICAL CHARACTERS. Intermediate in tolerance; rapid growing; lateral roots; the largest and handsomest of the genus; frequently planted in West because of drouth resistance; adapted to variety of sites, doing best on moist, rich soils; stunted and scraggly on poor, dry sites.

* * *

Lindheimer Hackberry

Celtis lindheimeri Engelm.

A small tree native only to Texas and quite similar to netleaf hackberry. Identifying characters are included in the key.

* * *

Georgia Hackberry

Celtis tenuifolia Nutt.
[*Celtis pumila* var. *georgiana* (Small) Sarg.]

A shrub or small tree found on hills from Pennsylvania to Indiana, Missouri, eastern Kansas and south to Oklahoma, Louisiana, northern Florida. Identifying characters are included in the key.

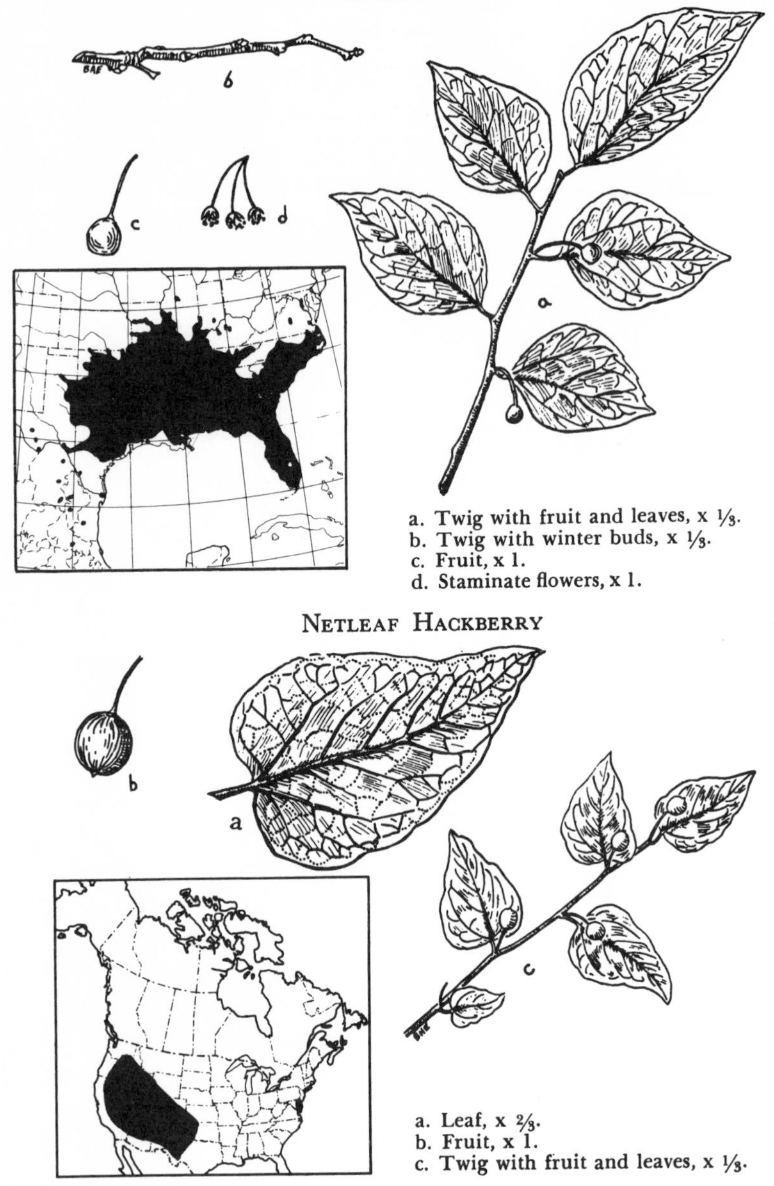

Sugarberry

a. Twig with fruit and leaves, x 1/3.
b. Twig with winter buds, x 1/3.
c. Fruit, x 1.
d. Staminate flowers, x 1.

Netleaf Hackberry

a. Leaf, x 2/3.
b. Fruit, x 1.
c. Twig with fruit and leaves, x 1/3.

Sugarberry. Sugar Hackberry

Celtis laevigata Willd. (*Celtis mississippiensis* Bose)

HABIT. A medium-sized tree 60-80 feet high and 2-3 feet in diameter (max. 5 feet); crown rounded, spreading branches.

LEAVES. Oblong-lanceolate; 2-5 inches long; acuminate apex; obliquely rounded or cordate at base; entire or irregularly and sparingly toothed; thin; light green and smooth or slightly rough above, smooth below, with veinlets not conspicuous; petioles 1/4-1/2 inch long.

FRUIT. 1/4 inch long; subglobose; dark orange-red; on pedicels shorter or slightly longer than leaf petioles.

BARK. Rather thin (1/3-2/3 inch thick); pale gray; covered with prominent wartlike excrescences.

SILVICAL CHARACTERS. Intermediate in tolerance; rapid growing; lateral roots; site varies from moist stream banks to the variety *texana* (Scheele) Sarg. on dry rocky bluffs in West.

* * *

Netleaf Hackberry

Celtis reticulata Torr. (*Celtis douglasii* Planch.) (*Celtis laevigata* var. *brevipes* Sarg.)

HABIT. A shrub or small tree rarely 30 feet high and 1 foot in diameter; crown open with stout, ascending branches.

LEAVES. Broadly ovate; 1 1/4-3 inches long; acute or acuminate tips; obliquely rounded at base; margins entire or with a few coarse teeth; thick; dark green and rough or smooth above, paler below with conspicuous reticulate veinlets; petioles 1/8-1/2 inch long.

FRUIT. 1/4 inch long; subglobose; orange-red to yellow; on pedicels longer but not twice as long as leaf petioles.

BARK. Thick; red-brown to ash-gray; rough with prominent, short, projecting ridges.

SILVICAL CHARACTERS. Moderately intolerant; on dry, rocky hillsides; a confused species including many forms previously considered distinct.

Red Mulberry

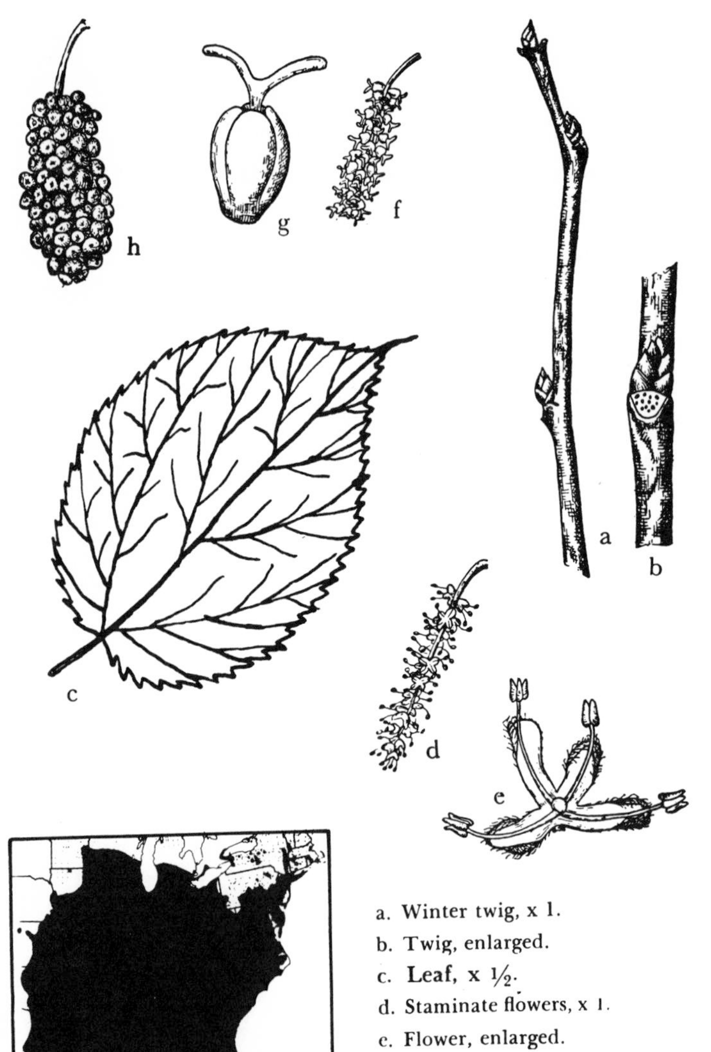

a. Winter twig, x 1.
b. Twig, enlarged.
c. Leaf, x ½.
d. Staminate flowers, x 1.
e. Flower, enlarged.
f. Pistillate flowers, x 1
g. Flower, enlarged.
h. Fruit, x 1.

MORACEAE

Red Mulberry

Morus rubra L.

HABIT. A small tree with milky juice 20-40 feet high and 8-15 inches in diameter (max. 70 by 3 1/2 feet); dense, round-topped crown.

LEAVES. Alternate; simple; suborbicular; 3-5 inches long; unlobed or deeply 1-3 lobed; margins coarsely serrate; thin; deciduous; dark blue-green, smooth, or rough above, and paler, more or less hairy below, with all veinlets hairy; stipulate.

FLOWERS. Monoecious or dioecious; minute; appearing with leaves; dense spikes; calyx 4-lobed, green; corolla absent.

FRUIT. Drupaceous; juicy; enclosed in thickened, berrylike calyx and united into a multiple fruit (syncarp) resembling a blackberry; 1-1 1/4 inches long; purple; oblong; edible.

TWIGS. Slender; brown; smooth; leaf scars with many bundle scars. Winter buds: terminal absent; lateral ovoid; acute; lustrous; 1/4 inch long; light brown.

BARK. Thin; dark brown tinged with red; scaly and furrowed, bark of roots yellow.

WOOD. Rather heavy and hard; ring-porous; dark; unimportant.

SILVICAL CHARACTERS. Tolerant; rich, moist sites; taproot. White mulberry, extensively cultivated in United States in the past as food for silkworms, has become naturalized.

* * *

The 2 native and 3 naturalized mulberries can be separated by the following characters:

1. Fruits ovoid to cylindric; buds with 3-6 outer scales . . . . *Morus*
   2. Leaves usually rough above, hairy below, usually undivided on old shoots, 2 1/2-6 inches long.
      3. Leaves truncate or slightly heart-shaped at base; fruit dark purple at maturity; native . . . . *M. rubra* L., Red Mulberry
      3. Leaves deeply heart-shaped at base, rarely lobed; fruit dark red; Asiatic . . . . *M. nigra* L., Black Mulberry
   2. Leaves smooth above, nearly glabrous below, variously lobed.
      3. Leaves 2 1/2-7 inches long, lustrous; fruit white, pink, or violet; Chinese . . . . *M. alba* L., White Mulberry
      3. Leaves 1-2 inches long; fruit nearly black; Arizona to west Texas . . . . *M. microphylla* Buckl., Texas Mulberry
1. Fruits globose, orange-red; buds with 2-3 outer scales; leaves 0-3-lobed, 3-8 inches long, velvety-hairy below; Asiatic . . . . *Broussonetia papyrifera* (L.) Vent., Paper-mulberry

OSAGE-ORANGE

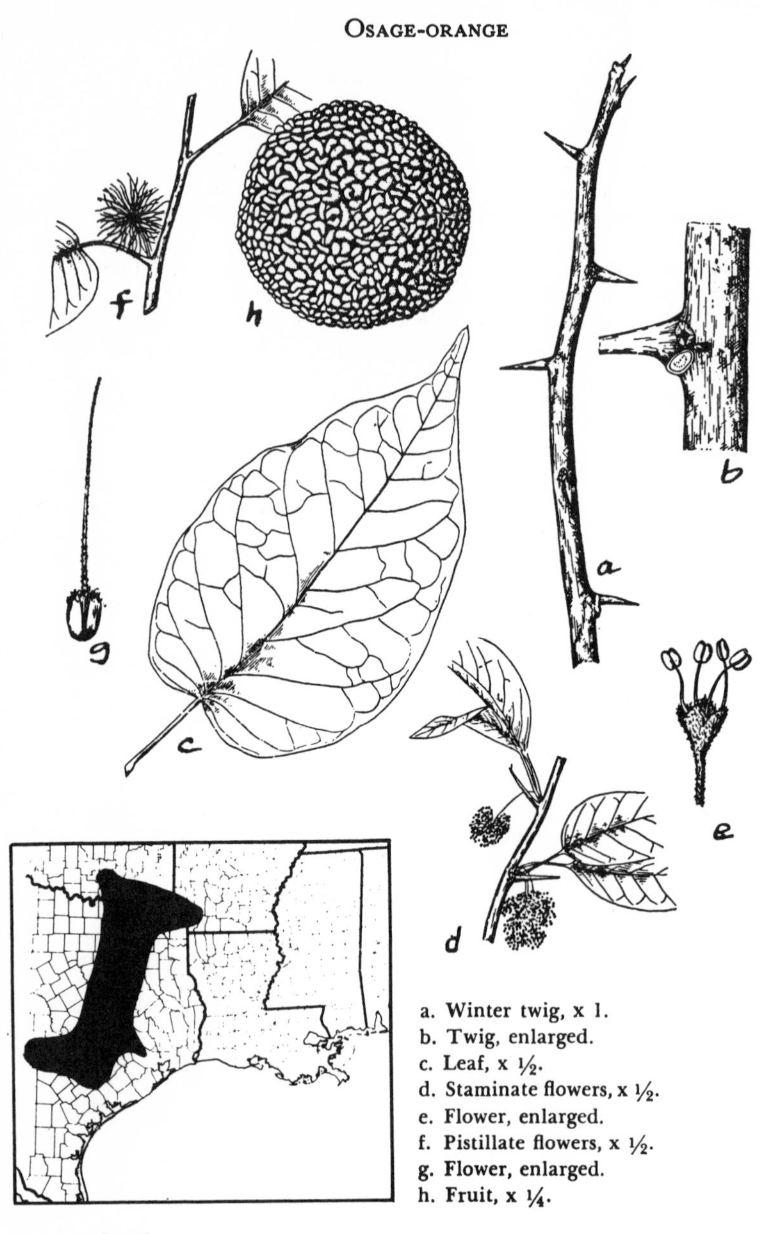

a. Winter twig, x 1.
b. Twig, enlarged.
c. Leaf, x ½.
d. Staminate flowers, x ½.
e. Flower, enlarged.
f. Pistillate flowers, x ½.
g. Flower, enlarged.
h. Fruit, x ¼.

MORACEAE

Osage-orange

Maclura pomifera (Raf.) Schneid. (*Toxylon pomiferum* Raf.)

HABIT. A small tree with milky juice, 20-30 feet high and 1-2 feet in diameter (max. 65 by 6 feet); rounded open crown with few large, curving branches; extensively planted.

LEAVES. Alternate; simple; ovate to oblong-lanceolate; 3-5 inches long; long-pointed apex; entire; thick and firm; deciduous; dark green and shining above; paler and smooth below.

FLOWERS. Dioecious or monoecious; minute; appearing after the leaves; male in dense racemes; female in dense, globose heads; calyx 4-lobed, hairy; corolla absent.

FRUIT. Drupaceous; numerous small drupes crowded and grown together into a multiple fruit resembling an orange; 4-5 inches in diameter; pale green; containing a bitter, milky juice; becoming woody.

TWIGS. Stout; orange-brown; armed with straight, stout, axillary thorns about 1/2 inch long and with spurlike lateral branchlets. Winter buds: terminal absent; lateral small, globular, brown, depressed, and partially embedded in the bark.

BARK. Thin; dark orange-brown; furrowed; flat ridges.

WOOD. Hard and heavy; tough; durable; ring-porous; bright orange; used for bows; a yellow dye is extracted from the roots.

SILVICAL CHARACTERS. Intolerant; a monotypic genus with a small natural distribution, but now widely planted throughout much of the United States.

* * *

Fig

Ficus L.

Two species of fig are native to southern Florida, *Ficus aurea* Nutt., the Florida strangler fig, and *Ficus laevigata* Vahl, the shortleaf fig. *Ficus carica* L. of western Asia is often cultivated for fruit and sometimes found as an escape in the southern states.

MAGNOLIACEAE

The Magnolias

Characteristics of the Genus *Magnolia* L.

HABIT. Handsome trees with straight stems and round-topped, pyramidal crowns; frequently planted as ornamentals.

LEAVES. Simple; alternate; deciduous or persistent; margins entire; unlobed; sometimes auriculate at base; mostly large and thin and very conspicuous.

FLOWERS. Perfect; appearing after the leaves; large and usually showy; terminal and solitary; sepals 3; petals 6-15, in series of 3, white or yellow to greenish; stamens and pistils spirally arranged, numerous.

FRUIT. A large, conelike aggregate of spirally arranged follicles; each follicle 1-2-seeded, the seed drupelike with a scarlet, fleshy outer coat and suspended from the cone by a long slender thread at maturity.

TWIGS. Stout; round; aromatic and somewhat bitter tasting; pith homogeneous or diaphragmed between nodes; round; conspicuously marked by large leaf scars with numerous bundle scars and narrow encircling stipular rings. Winter buds: terminal present, large, single outer scale; bud scales large, membranaceous stipules adnate to the base of the petiole and deciduous with the unfolding of each successive leaf.

BARK. Ash-gray or brown; smooth or scaly.

WOOD. Rather light and soft; diffuse-porous; light, yellow-brown heartwood; unimportant except for cucumbertree, the wood of which is often sold as yellow-poplar.

SILVICAL CHARACTERS. Intolerant; fairly rapid growth; short-lived; deep, wide root systems; typical of moist forest sites; in mixture with other hardwoods.

GENERAL. About 35 species of magnolia are native to Asia and North America, with 8 species native to the United States. Numerous native and Asiatic species are commonly planted through the southern part of the United States as ornamentals for their showy flowers, which in some species are over a foot in diameter.

KEY TO THE SPECIES OF MAGNOLIA

1. Leaves evergreen, persistent 2 years, thick and leathery, 5-8 inches long, densely red-hairy below; fruit, buds, and twigs densely red or white-hairy; flowers white, fragrant, 7-8 inches across . . . . *M. grandiflora* L., Southern Magnolia, p. 233
1. Leaves deciduous (in Sweetbay semievergreen in the south); parts not densely red-hairy.
   2. Leaves 4-10 inches long; flowers 2-4 inches across.
      3. Leaves obovate-spatulate, auriculate at base; buds, twigs, and lower leaf surface glabrous; flowers creamy white; rare; coastal plain; Georgia and Florida west to Louisiana . . . . *M. pyramidata* Bartr., Pyramid Magnolia
      3. Leaves elliptic to oblong-obovate; buds and lower leaf surfaces usually white-hairy.
         4. Leaves oblong-ovate to elliptic, 6-10 inches long; flowers greenish or yellow; styles deciduous from follicles of fruit . . . . *M. acuminata* L., Cucumbertree, p. 233
         4. Leaves oblong-lanceolate to elliptic, 4-6 inches long, nearly evergreen; coastal swamps from Massachusetts to Pennsylvania, south to southern Florida and west to Texas; north in river valleys to Tennessee and Arkansas . . . . *M. virginiana* L., Sweetbay
   2. Leaves 10-30 inches long; flowers 5 or more inches across; styles persistent on follicles of fruit.
      5. Leaves, buds, and fruit glabrous.
         6. Leaves obovate-lanceolate, wedge-shaped at base, 18-20 inches long; petioles 1-1 1/2 inches long; terminal bud about 1 inch, covered with glaucous bloom; flowers white, malodorous; Pennsylvania to Ohio, Missouri, Oklahoma, and south . . . *M. tripetala* L., Umbrella Magnolia
         6. Leaves obovate-spatulate, auriculate at base, 10-12 inches long; flowers pale yellow, fragrant; Virginia and Kentucky to Alabama and Georgia . . . . *M. fraseri* Walt., Fraser Magnolia
      5. Leaves, buds, and fruit hairy; flowers white, fragrant; petioles 3-4 inches long; buds 1 3/4-2 inches long, covered with thick, white tomentum.
         7. Leaves silvery-hairy on lower surface, obovate, heart-shaped at base, 20-30 inches long; fruit ovoid to globose; North Carolina to Kentucky and Arkansas, south to Florida and Louisiana . . . . *M. macrophylla* Michx., Bigleaf Magnolia
         7. Leaves hairy on midrib of lower surface, obovate-spatulate, 16-22 inches long; fruit ovoid-cylindric; western Florida and Texas . . . . *M. ashei* Weatherby, Ash Magnolia

* * *

Florida Anise-Tree

Illicium floridanum Ellis

An ornamental shrub or small tree with persistent, entire, elliptical, long, acuminate leaves; showy, perfect flowers; distinctive, buttonlike fruit that is a flattened aggregate of follicles. Coastal plain from northwest Florida to southeastern Louisiana. Yellow Anise-tree, *I. parviflorum* Michx. is local in southeastern Florida.

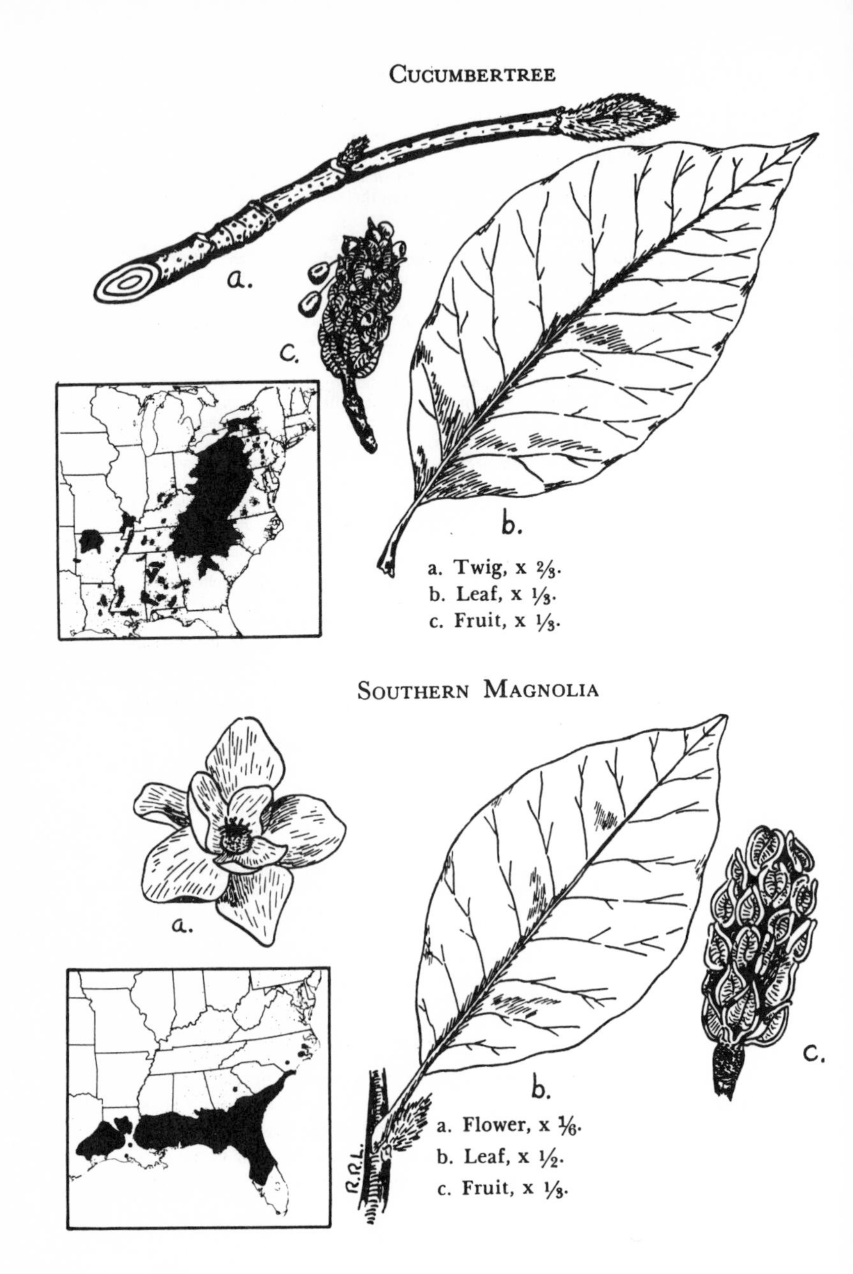

CUCUMBERTREE

a. Twig, x ⅔.
b. Leaf, x ⅓.
c. Fruit, x ⅓.

SOUTHERN MAGNOLIA

a. Flower, x ⅙.
b. Leaf, x ½.
c. Fruit, x ⅓.

MAGNOLIACEAE

Cucumbertree. Cucumber Magnolia

Magnolia acuminata L.

HABIT. A medium-sized tree 70-90 feet high and 3-4 feet in diameter (max. 100 by 6 feet); clear, straight bole; pyramidal crown with small branches.

LEAVES. Deciduous; broadly elliptical to ovate; 6-10 inches long; acute to acuminate at apex; rounded to broadly wedge-shaped at base; margin entire or slightly repand; thin; bright yellow-green and glabrous above; paler and glabrous to hairy below.

FLOWERS. Yellow-green; 2-3 inches long; not showy.

FRUIT. Conelike aggregate of follicles; cylindric to ovoid; 2-3 inches long; glabrous. Seed: 1/2 inch long; red; suspended on slender, white thread.

TWIGS. Moderately stout; lustrous. Winter buds: terminal 1/2-3/4 inch long, with white silvery hairs; lateral smaller.

BARK. Thin; dark brown; shallowly furrowed with narrow, scaly ridges.

GENERAL. The rare and local variety *cordata* (Michx.) Sarg., with yellow flowers and hairy twigs, ranges from central North Carolina to central Georgia.

* * *

Southern Magnolia. Evergreen Magnolia

Magnolia grandiflora L.

HABIT. A medium-sized tree 60-80 feet high and 2-3 feet in diameter (max. 135 by 4 1/2 feet); clear, straight bole; pyramidal crown with small spreading branches.

LEAVES. Persistent 2 years; narrowly oval to ovate; 5-8 inches long; acute to acuminate at apex; wedge-shaped at base; margin entire; thick and leathery; bright green and lustrous above; at maturity red-woolly below.

FLOWERS. White; showy and fragrant; 6-8 inches in diameter; 6-12 large petals.

FRUIT. Conelike aggregate of follicles; ovoid; 3-4 inches long; red-woolly. Seed: 1/2 inch long; red; suspended on slender, white thread.

TWIGS. Stout; red-woolly. Winter buds: terminal 1-1 1/2 inches long, white- or red-woolly; lateral smaller.

BARK. Thin; gray to light brown; on old stems scaly with small scales rarely over 1 inch in length.

Yellow-poplar

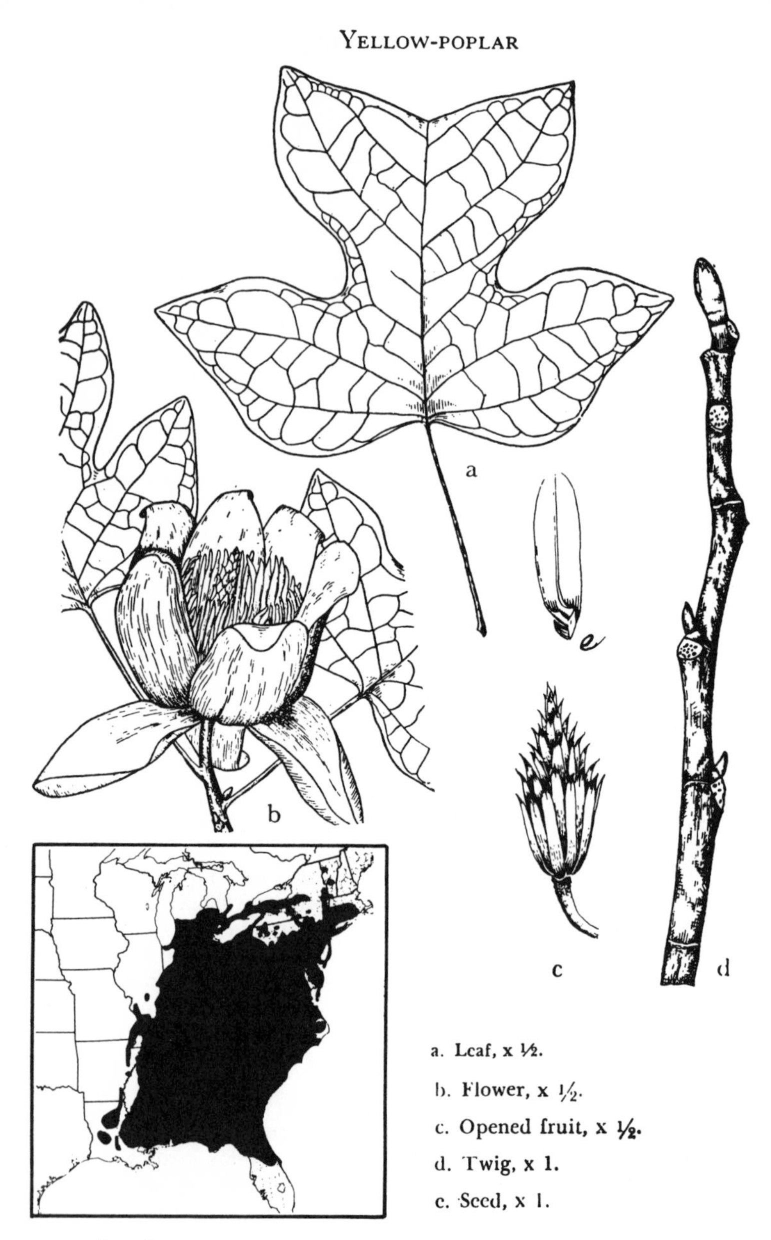

a. Leaf, x ½.

b. Flower, x ½.

c. Opened fruit, x ½.

d. Twig, x 1.

e. Seed, x 1.

MAGNOLIACEAE

Yellow-poplar. Tuliptree

Liriodendron tulipifera L.

HABIT. A large, handsome tree 80-100 feet high and 4-6 feet in diameter (max. 198 by 12 feet); clear, straight bole; open, oblong, or conical crown with small branches.

LEAVES. Simple; alternate; deciduous; suborbicular; 4-6 inches long; characteristically 4-lobed resembling a tulip in outline; lobes entire-margined; truncate at base and truncate or notched at apex; glabrous; lustrous dark green above; paler below; petioles slender, 5-6 inches long; stipules large, conspicuous.

FLOWERS. Perfect; appearing after the leaves; yellow-green; cup-shaped; 1 1/2-2 inches long; 3 sepals; 6 petals in 2 rows; stamens and pistils numerous and spirally arranged around a central axis.

FRUIT. A large (2 1/2-3 inches long) erect, conelike aggregate of spirally arranged samaras; each samara 1 1/2 inches long, 4-angled, terminally winged, deciduous from the slender, more or less persistent, central axis.

TWIGS. Rather stout; lustrous red-brown; stipular scars encircling twig; bitter; pith round, diaphragmed; leaf scars round, with several bundle scars. Winter buds: terminal present, about 1/2 inch long, flattened, dark red or green; valvate stipular scales; only 2 outer scales visible, the entire bud resembling a duck's bill.

BARK. Smooth and dark green on young stems; becoming thick, ash-gray, furrowed, with rough, rounded ridges; inner bark bitter.

WOOD. Highly important, valuable hardwood species; rather light and soft; light yellow to dark brown heartwood; easily worked; used for plywood, furniture, interior finish, and construction work.

SILVICAL CHARACTERS. Very intolerant; fast-growing; reaching maturity at 200-250 years; deep, wide-spreading roots; in mixture with other hardwoods; typical of moist forest sites.

GENERAL. Only 2 species of this genus are now in existence, one a native of central China, *Liriodendron chinense* Sarg. The native species attains the greatest height of any of our broadleaf species and perhaps the greatest diameter.

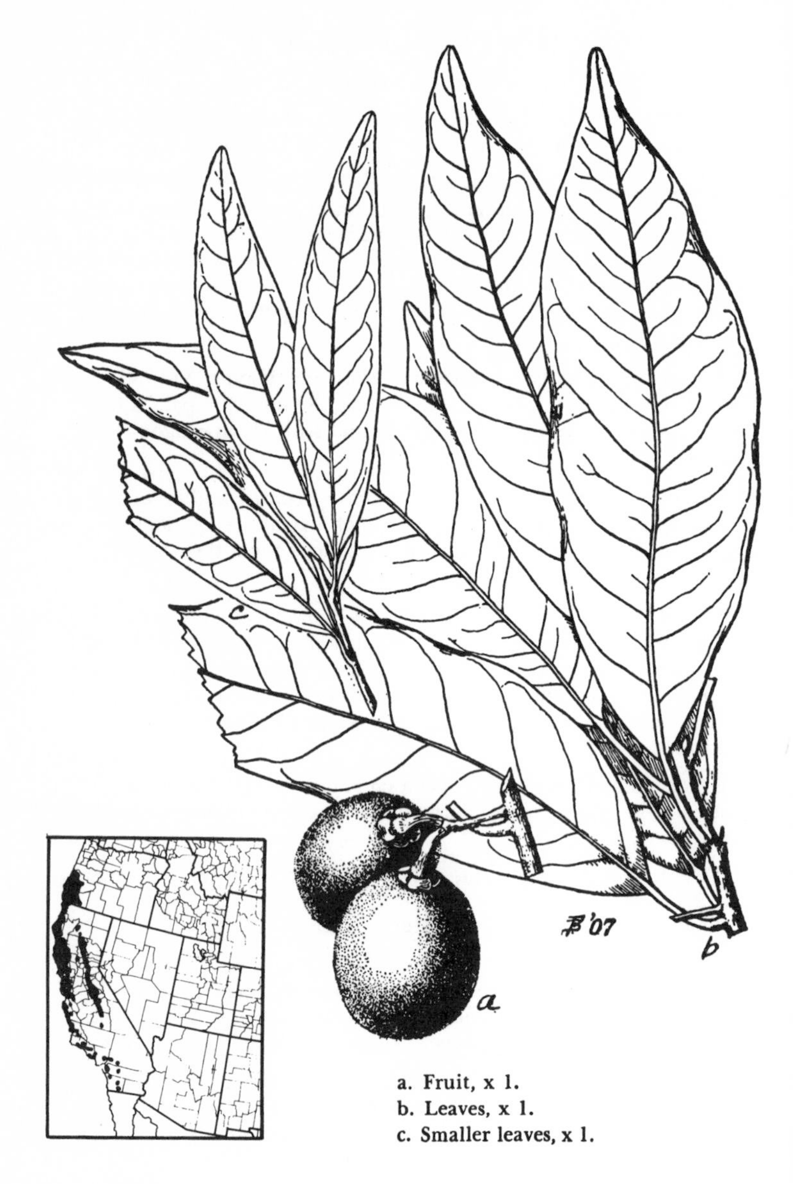

a. Fruit, x 1.
b. Leaves, x 1.
c. Smaller leaves, x 1.

LAURACEAE

California-laurel. Oregon-myrtle

Umbellularia californica (Hook. and Arn.) Nutt.

HABIT. A small to medium-sized tree 20-80 feet high and 1-3 feet in diameter (max. 175 by 12 feet); broad, rounded, dense crown; trunk often divided; shrubby on dry sites.

LEAVES. Alternate; simple; lanceolate to elliptical; 2-5 inches long; thick and leathery; entire; glabrous; persistent 2-6 years; spicy-scented; dark green and lustrous above; paler and dull below.

FLOWERS. Perfect; small; in umbels; appearing before the new leaves; calyx yellow-green, 6-parted; corolla absent.

FRUIT. An olivelike, acrid, yellow-green, drupelike berry about 1 inch in diameter; maturing in 1 year; surrounded at base by enlarged lobes of calyx; seed ovoid, light brown.

TWIGS. Slender; yellow-green; glabrous; aromatic. Winter buds: terminal present, minute, naked.

BARK. Rather thin; dark brown; tight scales; smooth and gray-green on branches.

WOOD. Rather hard and heavy; diffuse-porous; fine-textured; lustrous light brown; expensive and highly valued for furniture, finish, and novelties; sold as myrtle wood.

SILVICAL CHARACTERS. Moderately tolerant; deep, widespreading, fleshy roots; on varied sites, but best growth on moist bottomlands; a monotypic genus.

* * *

ANNONACEAE

Pawpaw

Asimina triloba (L.) Dunal

This distinctive shrub or small tree is scattered throughout the eastern United States. It is characterized by a bananalike fruit, a cylindric, yellow-green, edible berry 3-5 inches long containing numerous seeds; simple, alternate, deciduous, entire, membranaceous leaves 10-12 inches long and 4-6 inches wide; large, perfect flowers, 3 sepals and 6 petals; rusty-pubescent young twigs, becoming glabrous; small, red-woolly buds; thin, dark brown bark, containing whitish blotches and wartlike protuberances. In the southeast about eight shrubby species are recognized with smaller leaves and fruits.

Sassafras

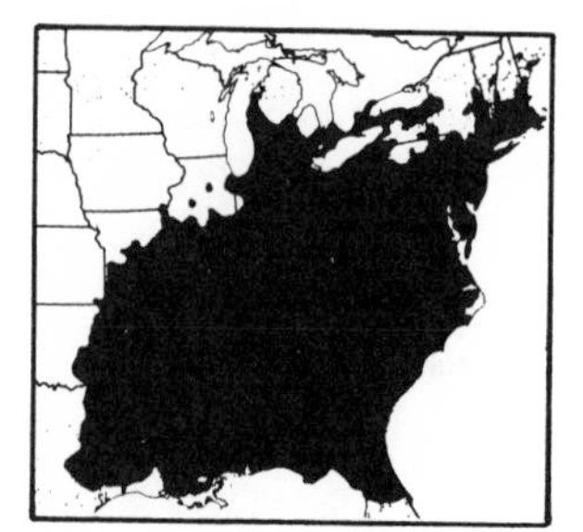

a. Pistillate flowers, x 1/2.
b. Fruit, x 1/2.
c. Staminate flowers, x 1/2.
d. Leaves, x 1/2.

Redbay

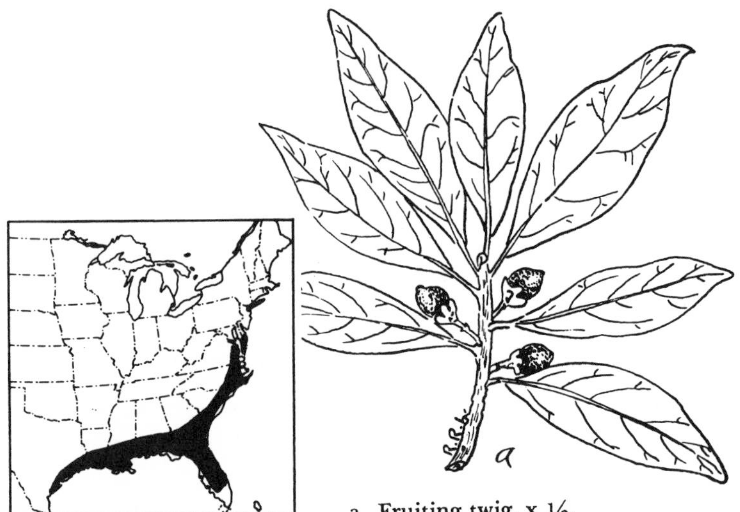

a. Fruiting twig, x 1/2.

LAURACEAE

Sassafras

Sassafras albidum (Nutt.) Nees [*Sassafras variifolium* (Salisb.) Ktze.]

HABIT. A shrub to medium-sized tree 20-50 feet high and 1-2 feet in diameter (max. 90 by 5 feet); flat, open crown.

LEAVES. Alternate; simple; deciduous; oval to obovate; aromatic; three forms on same tree, entire, 1-lobed or 3-lobed; 3-6 inches long; thin; bright green above; glabrous and glaucous below.

FLOWERS. Dioecious; small; in racemes; yellow-green; with the leaves; corolla absent.

FRUIT. Subglobose, lustrous, blue drupelike berry, 1/3 inch long; borne on bright red, club-shaped pedicel 1 1/2-2 inches long; maturing in one year.

TWIGS. Rather stout; glabrous; glaucous; yellow-green; aromatic. Winter buds: terminal present, 1/3 inch long; green, 3-4-scaled.

BARK. Thick; red-brown; furrowed on old trunks.

SILVICAL CHARACTERS. Very intolerant; fast-growing; short-lived; indicator of poor sites; oil of sassafras distilled from roots and bark.

* * *

Redbay

Persea borbonia (L.) Spreng.

HABIT. A small to medium-sized tree, rarely 60 feet high; rounded, dense crown with stout, erect branches.

LEAVES. Alternate; simple; persistent; elliptic to lanceolate; 2-4 inches long; entire; thick; aromatic; nearly glabrous; bright green and lustrous above; paler and glaucous below.

FLOWERS. Perfect; small; yellow; panicles; no corolla.

FRUIT. Oblong, dark blue, lustrous drupelike berry, 1/2 inch long; on glabrous red peduncles, 1/2-1 inch long; calyx persistent.

TWIGS. Slender; glabrous; dark green. Winter buds: terminal present, naked, 1/4 inch long, red-woolly.

BARK. Thin; dark red; furrowed and scaly.

SILVICAL CHARACTERS. Intermediate in tolerance; wet swampy sites.

GENERAL. The swampbay, *P. palustris* (Raf.) Sarg., now considered a form of redbay but differs, having lower leaf surfaces and twigs brown-woolly, peduncles 1 1/2-2 inches long, fruit about 3/4 inch. The avocado, *P. americana* Mill, is naturalized in southern Florida. *P. humilis* Nash is a shrubby species reported from central Florida and Texas.

Sweetgum

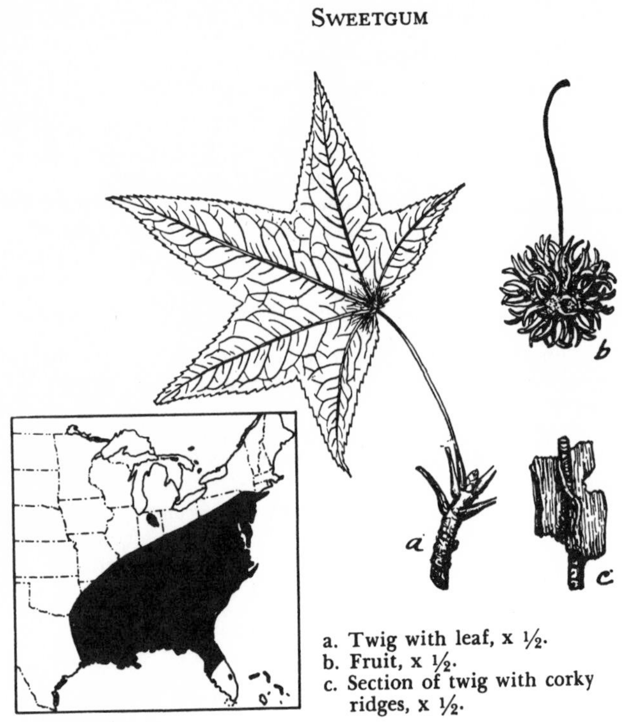

a. Twig with leaf, x ½.
b. Fruit, x ½.
c. Section of twig with corky ridges, x ½.

Witch-hazel

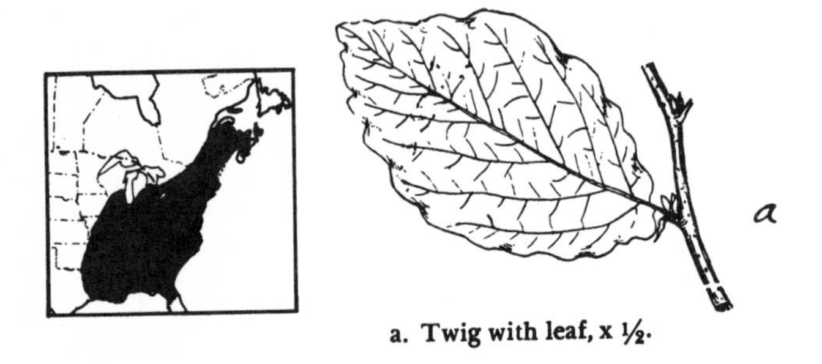

a. Twig with leaf, x ½.

HAMAMELIDACEAE

Sweetgum. Redgum

Liquidambar styraciflua L.

HABIT. A medium-sized to large tree 80-120 feet high and 2-4 feet in diameter (max. 200 by 7 feet); small, oblong crown with small branches; long, clear, frequently buttressed bole.

LEAVES. Alternate; simple; deciduous; star-shaped; deeply and palmately 5-7-lobed; 4-7 inches long; lobes toothed and acuminate; lustrous and bright green above; paler below and glabrous except for axillary tufts of hairs; stipulate.

FLOWERS. Monoecious; both male and female in heads; male in terminal racemes 2-3 inches long without calyx or corolla; female in solitary, long-stalked heads, minute calyx, and 2-celled ovary; appearing with leaves.

FRUIT. A multiple, globose fruit 1-1 1/2 inches in diameter, of woody, 2-celled, beaked capsules; maturing in one year but persistent through the winter; 2 seeds in each capsule, each about 3/8 inch long with a short terminal wing.

TWIGS. Slender to stout; round or slightly angled; frequently developing corky wings during the second year; green to red-brown; pith angled and homogeneous. Winter buds: terminal present, 1/4-1/2 inch long, ovate, orange-brown, scaly.

BARK. Thick; gray-brown; deeply furrowed, with narrow, rounded and somewhat scaly ridges.

WOOD. Very important, exceeded only among hardwoods by oaks in volume of timber cut; diffuse-porous; fairly hard and heavy; heartwood dark red-brown and often figured; used for veneer, lumber, furniture, and containers.

SILVICAL CHARACTERS. Very intolerant; on rich moist bottomlands or swampy sites; wide, shallow, lateral roots; reproduction aggressive; storax gum, for drugs and soaps, obtained from bark. This genus is composed of four species, two in Asia, one in central America.

* * *

Witch-hazel

Hamamelis virginiana L.

Witch-hazel is a distinctive shrub or small tree found through the eastern United States. It is characterized by conspicuous, perfect, yellow, autumnal flowers; woody, 2-valved capsules from which the black seeds are forcibly ejected; simple, alternate, deciduous, oval leaves, 4-6 inches long, which are crenately toothed or lobed and smooth; and naked buds, 1/4-1/2 inch long.

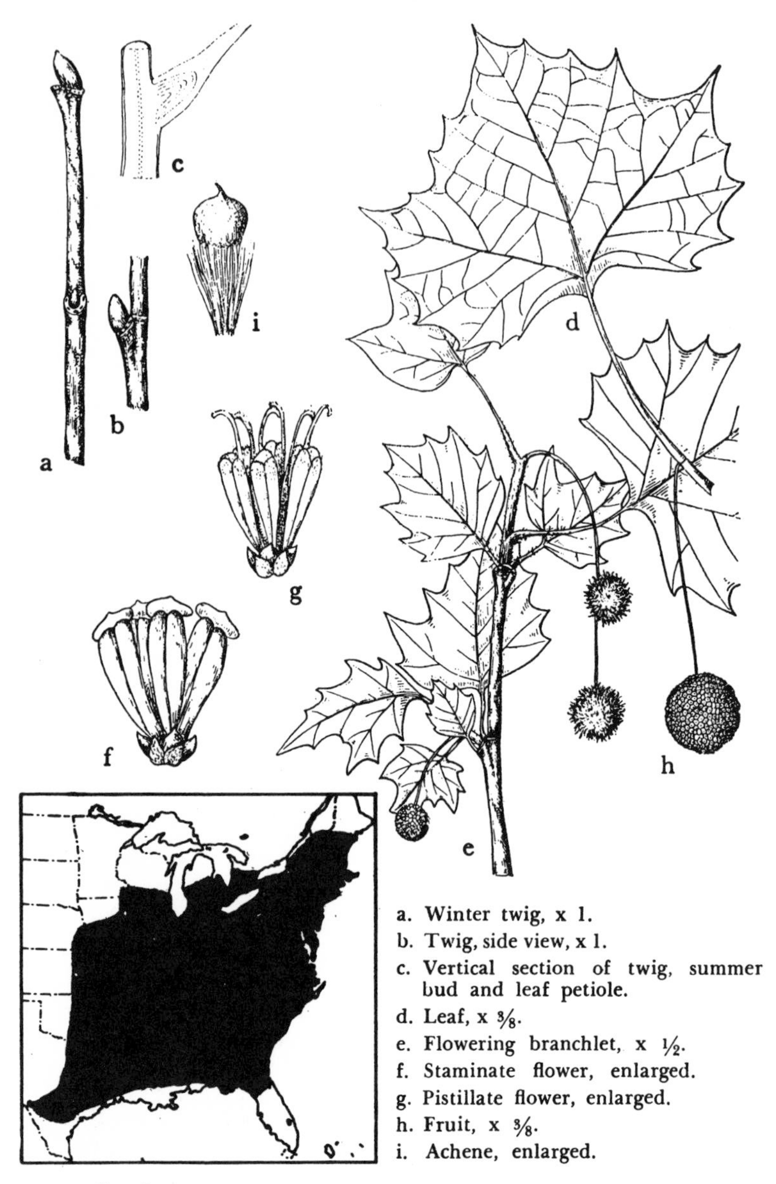

a. Winter twig, x 1.
b. Twig, side view, x 1.
c. Vertical section of twig, summer bud and leaf petiole.
d. Leaf, x 3/8.
e. Flowering branchlet, x 1/2.
f. Staminate flower, enlarged.
g. Pistillate flower, enlarged.
h. Fruit, x 3/8.
i. Achene, enlarged.

PLATANACEAE

American Sycamore. Planetree
Platanus occidentalis L.

HABIT. A large tree 70-100 feet high and 3-8 feet in diameter (max. 175 by 14 feet); with yellow-poplar the largest of eastern hardwoods; long, clear bole; open crown.

LEAVES. Alternate; simple; deciduous; broadly ovate; 3-5-lobed; 4-7 inches long; lobe margins coarsely toothed; thin and firm; bright green and smooth above; paler and hairy along the veins below; petioles 2-3 inches long; stipules leaflike.

FLOWERS. Monoecious; both male and female in heads with minute individual flowers, each containing 3-8 sepals and petals; appearing with the leaves.

FRUIT. A multiple, globose fruit 1-1 1/4 inches in diameter; of elongated, obovoid achenes with spur at apex and ring of erect hairs around base; maturing in one year; borne singly on stems 3-6 inches long.

TWIGS. Stout; round; orange-brown; lustrous; encircled by stipular scars; pith round, homogeneous. Winter buds: terminal absent; lateral 1/4-3/8 inch, conical, brown, single visible scale, formed in petiole base.

BARK. Red-brown and scaly near base; thin and characteristically mottled on upper trunk by exfoliating outer layers that expose lighter colored inner layers.

WOOD. Intermediate in importance; diffuse-porous with broad rays; fairly hard and heavy; heartwood light red-brown; used for boxes, cooperage, vehicles, furniture, ties, and fuel.

SILVICAL CHARACTERS. Very intolerant; rapid growth; on moist sites; shallow, lateral roots; withstands smoke.

* * *

A monotypic family with a single genus and 6 or 7 species. The following key distinguishes between 3 native and 2 widely planted species:

1. Heads of fruit solitary or in pairs; leaves not deeply lobed.
    2. Leaf lobes broader than long; fruit heads usually single; bark creamy-white to brown . . . . *P. occidentalis* L., American Sycamore
    2. Leaf lobes about as long as broad; 2 (rarely 3) fruit heads per stem; bark olive-green . . . . *P.* X *acerifolia* Willd., London Planetree
1. Heads of fruit usually 3 or more; leaves deeply lobed.
    2. Leaves mostly 5-7-lobed, glabrous to hairy below; Eurasian . . . . *P. orientalis* L., Oriental Planetree
    2. Leaves woolly below; native and western.
        3. Arizona and New Mexico; leaves 5-7-lobed . . . . *P. wrightii* S. Wats., Arizona Sycamore
        3. California; leaves 3-5-lobed . . . *P. racemosa* Nutt., California Sycamore

ROSACEAE

The Apples

Characteristics of the Genus *Malus* Mill.

HABIT. Shrubs to medium-sized trees, the native species having little or no importance; broad, rounded, open crowns.

LEAVES. Alternate; simple; deciduous, or rarely half-evergreen; toothed and in native species more or less lobed; stipules free from the petioles, early deciduous; petiolate.

FLOWERS. Regular; perfect; in short terminal racemes; on spurlike, sometimes spinescent branches; appearing with or after the leaves; calyx 5-lobed, sometimes persistent and erect on fruit; petals white to red, showy; stamens 15-50; ovary inferior, usually with 5 carpels.

FRUIT. A fleshy pome with papery carpels joined at their apex; indehiscent; important as food in some species. Seed: 1-2 in each cell; ovoid; acute; chestnut-brown.

TWIGS. Slender to stout; round; frequently spiny or with spinescent spur shoots; fruit spurs roughened by leaf scars with 3 bundle scars. Winter buds: terminal present, small, scaly, obtuse.

BARK. Thin; scaly; gray to red-brown; often fissured.

WOOD. Heavy; hard; fine-textured; diffuse-porous; heartwood red-brown; used to some extent for tool handles.

GENERAL. This taxonomically difficult genus contains about 25 species scattered through the Northern Hemisphere. The 6 species native to North America are unimportant; however, several Eurasian species are widely cultivated for their fruits or flowers. The common apple, *M. pumila* Mill., was introduced in Colonial times and has escaped and become naturalized. Many authors combine *Malus* with *Pyrus*, the name for apple becoming *Pyrus malus* L.

* * *

Common Pear

Pyrus communis L.

This Eurasian species is widely cultivated for its fruit and is a common escape in North America. In addition to its distinctive "pear-shaped" pomaceous fruit, it differs from the closely related apple in having more lustrous, thinner, and less hairy leaves.

KEY TO THE SPECIES OF APPLE

1. Leaves never lobed, rolled up in bud; naturalized.
   2. Calyx persistent on fruit; leaves hairy below; fruit 1 inch or more in diameter . . . . *Malus pumila* Mill., Apple, p. 247
   2. Calyx deciduous from fruit; leaves glabrous; fruit 1/3-1/2 inch in diameter . . . . *M. baccata* (L.) Borkh., Siberian Crab Apple
1. Leaves more or less lobed, at least on vigorous shoots, folded in bud; native.
   3. Eastern; calyx persistent on green (rarely yellow) waxy fruit.
      4. Leaves and calyx woolly; Great Plains, Wisconsin, and Minnesota to Louisiana and Texas . . . . . *M. ioensis* (Wood) Britt., Prairie Crab Apple
      4. Mature leaves and calyx not woolly.
         5. Leaves on vigorous shoots distinctly and deeply lobed.
            6. Leaves rounded or wedge-shaped at base; northern and central states . . . . *M. coronaria* (L.) Mill., Sweet Crab Apple, p. 247
            6. Leaves heart-shaped at base; North Carolina to Alabama . . . . *M. glabrata* Rehd., Biltmore Crab Apple
         5. Leaves on vigorous shoots, slightly lobed.
            7. Fruit about 2 inches broad, much broader than high, with deep cavity at top and base; Ohio to West Virginia, south to North Carolina and Georgia . . *M.* X *platycarpa* Rehd., Bigfruit Crab Apple
            7. Fruit about 1 inch broad and long, with shallow cavities. Virginia to Illinois, south to Florida and Louisiana . . . . *M. angustifolia* (Ait.) Michx., Southern Crab Apple
   3. West Coast; calyx deciduous on yellow to red nonwaxy fruit; Alaska to California . . . . *M. diversifolia* (Bong.) Roem., Oregon Crab Apple

* * *

Torrey Vauquelinia

Vauquelinia californica (Torr.) Sarg.

This is a shrub or small tree rarely 20 feet high, native to southern Arizona, southwestern New Mexico, and northern Mexico. It is characterized by simple, persistent, lanceolate leaves 1 1/2-3 inches long, with remote, small, glandular teeth; small, regular, perfect, white flowers borne in leafy, woolly panicles; and a woody, ovoid, capsular, 5-celled fruit about 1/4 inch long, which is woolly, subtended by the remnants of the flower, and long persistent on the branches.

Sweet Crab Apple

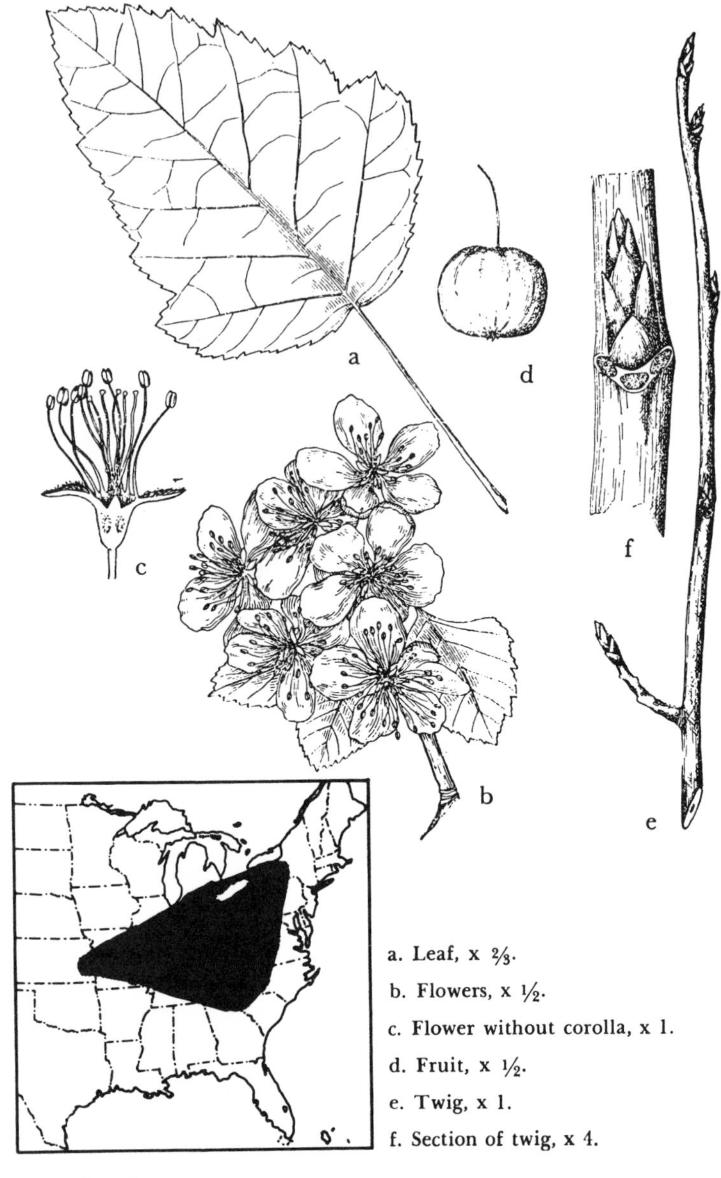

a. Leaf, x ⅔.

b. Flowers, x ½.

c. Flower without corolla, x 1.

d. Fruit, x ½.

e. Twig, x 1.

f. Section of twig, x 4.

ROSACEAE

Sweet Crab Apple

Malus coronaria (L.) Mill. (*Pyrus coronaria* L.)

HABIT. A bushy shrub or small tree 15-30 feet high and 1/2-1 foot in diameter; crown broad, rounded, and bushy.

LEAVES. Ovate to ovate-oblong; 3-4 inches long; 3-lobed on vigorous shoots; sharply and deeply toothed; thin and smooth; dark green above; paler below; petioles long and slender.

FLOWERS. Handsome; fragrant; 1 1/4-2 inches across; white to rose-colored; in 5-6-flowered umbellike racemes; appearing after the leaves.

FRUIT. Pome depressed-globose; 1-1 1/2 inches in diameter; pale to yellow-green; waxy; fragrant; flesh very tart; ripening in late autumn, often remaining on tree until spring.

TWIGS. Rather stout; developing short, stout, often spinelike spurs; red-brown. Winter buds: terminal 1/8-1/4 inch long, scaly, bright red.

BARK. Thin; red-brown; divided by shallow fissures into broad, scaly ridges.

SILVICAL CHARACTERS. Intolerant; rich, moist sites along streams or in thickets along roads and fences; fibrous roots; planted as an ornamental for its showy flowers.

* * *

Apple

Malus pumila Mill. (*Pyrus malus* L.)

This species, native to Eurasia, is the parent of most of our cultivated apples and has escaped in many parts of North America. It is stated that over 3,000 varieties of this species have been developed by pomologists. The apple is a small to medium-sized tree that can be identified from its unlobed, crenate-serrate to entire, thick leaves that are hairy below and rolled up in the bud.

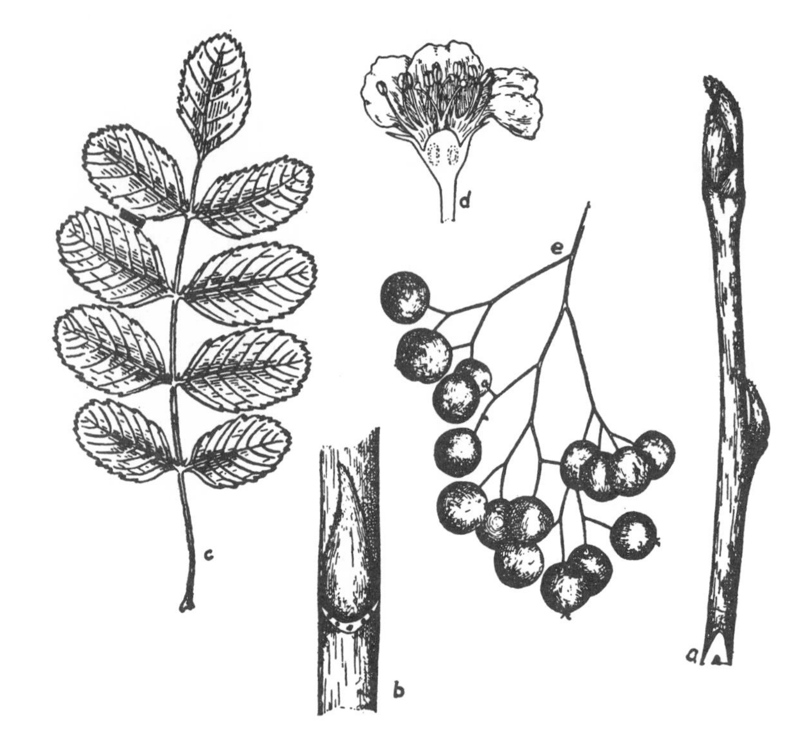

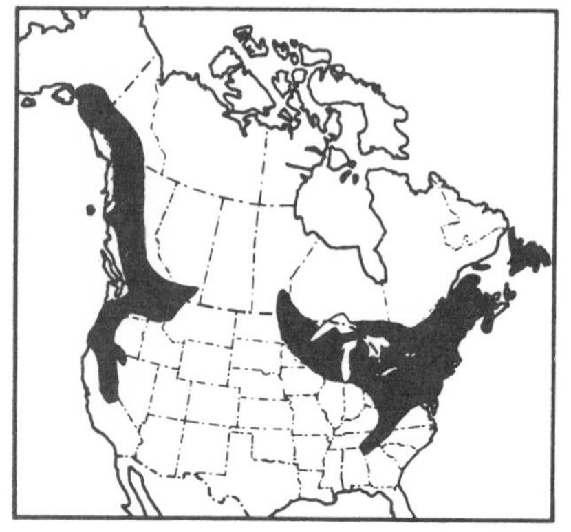

a. Winter twig, x 1.

b. Portion of twig, enlarged.

c. Leaf, x ½.

d. Vertical section of flower, enlarged

e. Portion of a fruiting cyme, x 1.

Eastern: *S. americana.*

Western: *S. sitchensis.*

American Mountain-ash

Sorbus americana Marsh. [*Pyrus americana* (Marsh.) DC.]
(*Sorbus sitchensis* Roem.)

HABIT. A shrub or small tree, seldom 30 feet high and 12 inches in diameter; crown round-topped, handsome; spreading, slender branches.

LEAVES. Alternate; pinnately compound in native species; 6-8 inches long; 7-17 mostly sessile leaflets; oblong-oval to lanceolate; 2-4 inches long and 1/2-1 inch wide; serrate; glabrous; dark green above; pale below; deciduous.

FLOWERS. Regular; perfect; small; in broad, flat cymes 3-5 inches across; on short, stout pedicels; appearing after leaves; 5 petals, cream-white, 1/4 inch in diameter.

FRUIT. Berrylike pome; subglobose; 1/4-1/2 inch in diameter; bright orange-red; thin, acrid flesh; papery carpels. Seed: 1 or 2 in each cell; 1/8 inch long; ovoid; brown.

TWIGS. Stout; round; red-brown and pubescent, becoming dark brown and glabrous; marked by large pith, large leaf scars, and oblong lenticels. Winter buds: terminal 1/4-3/4 inch long, acute, dark red, pilose, gummy exudation.

BARK. Thin (1/8 inch); light gray; smooth or slightly roughened by scales; inner bark fragrant.

WOOD. Light; soft; weak; close-grained; diffuse-porous; heartwood pale brown; sapwood thick and lighter colored.

SILVICAL CHARACTERS. Intolerant; slow-growing; short-lived, abundant seeder; fibrous roots; prefers moist sites, but grows well on rocky hillsides; often cultivated.

GENERAL. The taxonomy in this genus is confused. The related western species, *S. sitchensis* Roem., is usually a shrub; distribution is included in map. The key below separates native and naturalized species reaching tree size:

1. Winter buds densely covered with long white hairs; leaflets 1-2 inches long, oblong; widely planted and naturalized through the northern United States and Canada . . . . *S. aucuparia* L., European Mountain-ash
1. Winter buds glabrous or slightly hairy; leaflets 1 1/2-4 inches long.
   2. Winter buds glabrous or slightly white-hairy; leaflets usually lanceolate and acuminate . . . . *S. americana* Marsh., American Mountain-ash
   2. Winter buds slightly red-hairy; leaflets oblong, short-pointed; distribution similar to *S. americana* but extending to southern Greenland . . . . *S. decora* Schneid., Showy Mountain-ash

Downy Serviceberry

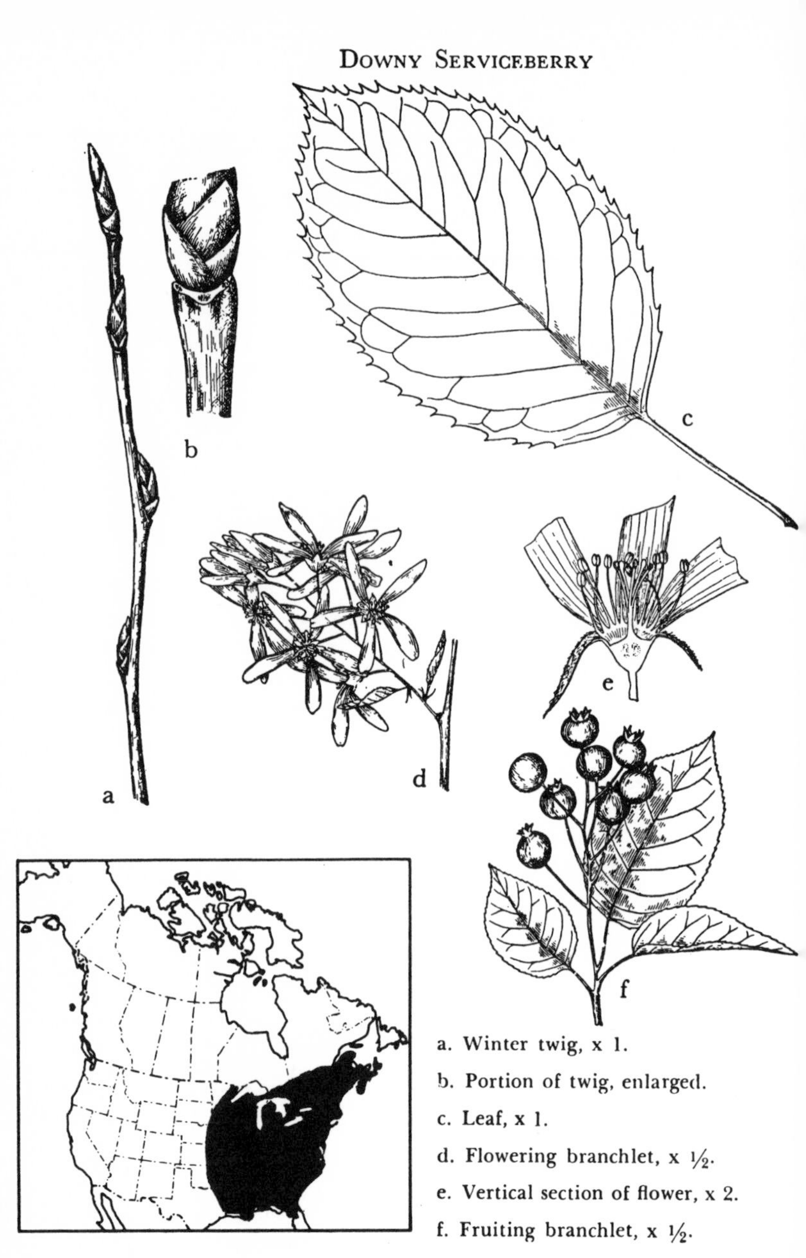

a. Winter twig, x 1.

b. Portion of twig, enlarged.

c. Leaf, x 1.

d. Flowering branchlet, x ½.

e. Vertical section of flower, x 2.

f. Fruiting branchlet, x ½.

ROSACEAE

Downy Serviceberry. Shad Bush

Amelanchier arborea (Michx.) Fern. (*Amelanchier canadensis* auth.)

HABIT. A shrub or small tree 25-40 feet high and 8-14 inches in diameter (max. 70 by 2 feet); crown open.

LEAVES. Alternate; simple; oblong-ovate to oval or ovate; 2-4 inches long; acute at apex; finely serrate; dark green and glabrous above; pale below; deciduous; petioles slender.

FLOWERS. Regular; perfect; in erect, crowded racemes; appearing with leaves; calyx tube persistent on fruit; corolla white, 1/2-3/4 inch long, of 5, strap-shaped petals.

FRUIT. Berrylike pome; globose; 1/4-1/2 inch in diameter; dark red to purple; more or less covered with glaucous bloom; flesh sweet; open at summit. Seed: 5-10; dark chestnut-brown.

TWIGS. Slender; round; red-brown to dark gray; narrow leaf scars, 3 bundle scars. Winter buds: terminal 1/4-1/2 inch long, conical, acute or acuminate, chestnut-brown.

BARK. Thin (1/8 inch); light brown, tinged with red; smooth or slightly furrowed with scaly ridges.

WOOD. Heavy; hard; close-grained; diffuse-porous; heartwood light brown; sapwood thick, light; unimportant.

SILVICAL CHARACTERS. Tolerant when young; abundant seeder; deep fibrous roots; capable of sprouting repeatedly; moist valleys and borders of streams to dry mountain slopes.

GENERAL. The taxonomy within this genus is confused, relatively minor characters being used to separate species. Three species, keyed out below, commonly form small trees and the following shrubby species have been reported as reaching tree size: *A. alnifolia* Nutt., *A. interior* Niel., *A. sanguinea* (Pursh) DC., and *A. utahensis* Koehne.

1. Leaves coarsely toothed, usually only above the middle, with about 7-13 teeth per inch, rounded at apex; top of ovary woolly; buds ovoid . . . . *A. florida* Lindl., Pacific Serviceberry
1. Leaves finely toothed nearly to base with about 15-30 teeth per inch, pointed at apex; top of ovary glabrous; buds lanceolate.
   2. Leaves woolly when young; flowers in woolly racemes; fruit dry, on pedicels 1/2-1 inch long . . . . *A. arborea* (Michx.) Fern., Downy Serviceberry
   2. Leaves nearly glabrous from first; racemes glabrous; fruit sweet and juicy, on pedicels 1-2 inches long; Newfoundland to Minnesota, south to Kansas and Alabama . . . . *A. laevis* Wieg., Allegheny Serviceberry

Black Hawthorn

a. Twig with leaves and flowers, x 1.

b. Twig with fruit, x 1.

c. Diagram of flower, x 2.

ROSACEAE

The Hawthorns

Characteristics of the Genus *Crataegus* L.

HABIT. Deciduous shrubs or small trees; usually spiny; crown generally rounded, wide-spreading.

LEAVES. Alternate; simple; deciduous; usually serrate and often more or less lobed; membranaceous to coriaceous; stipules persistent until autumn or deciduous in spring, small to leaflike, often bright-colored; petiolate.

FLOWERS. Regular; perfect, in few- or many-flowered terminal corymbs; pedicellate; calyx 5-lobed, tubular, persistent on fruit or deciduous; corolla 5-petaled, white, inserted on edge of disk lining calyx tube; stamens 5-25; ovary inferior, of 1-5 carpels connate at base, with 2 ovules in each cell; styles as many as carpels.

FRUIT. Small, variously colored pome with 1-5 bony, 1-seeded nutlets; flesh usually dry and mealy; generally open or concave at apex.

TWIGS. Round; rigid; more or less zigzag; generally armed with stiff, sharp thorns; marked by oblong lenticels and small leaf scars. Winter buds: terminal usually present, small, globose, scaly, lustrous brown.

BARK. Dark red to gray; scaly or shallowly furrowed.

WOOD. Heavy; hard; tough; close-grained; diffuse-porous; heartwood red-brown; sapwood thick, light-colored; unimportant; used for tool handles, canes, turned articles.

SILVICAL CHARACTERS. Intolerant; reproduction aggressive; growth slow.

GENERAL. An extremely large and complex genus containing over a thousand species, the identification of which presents great difficulties even to the specialist; hybridizing is common. The number of stamens, which is variable but usually within constant limits, and the color of the anthers appear to be the most satisfactory characters for the identification of species. Some 149 species have been listed as trees in the United States and Alaska; of these *C. douglasii* Lindl., the black hawthorn, which is common through the West, has been pictured on the opposite page as being typical of the genus.

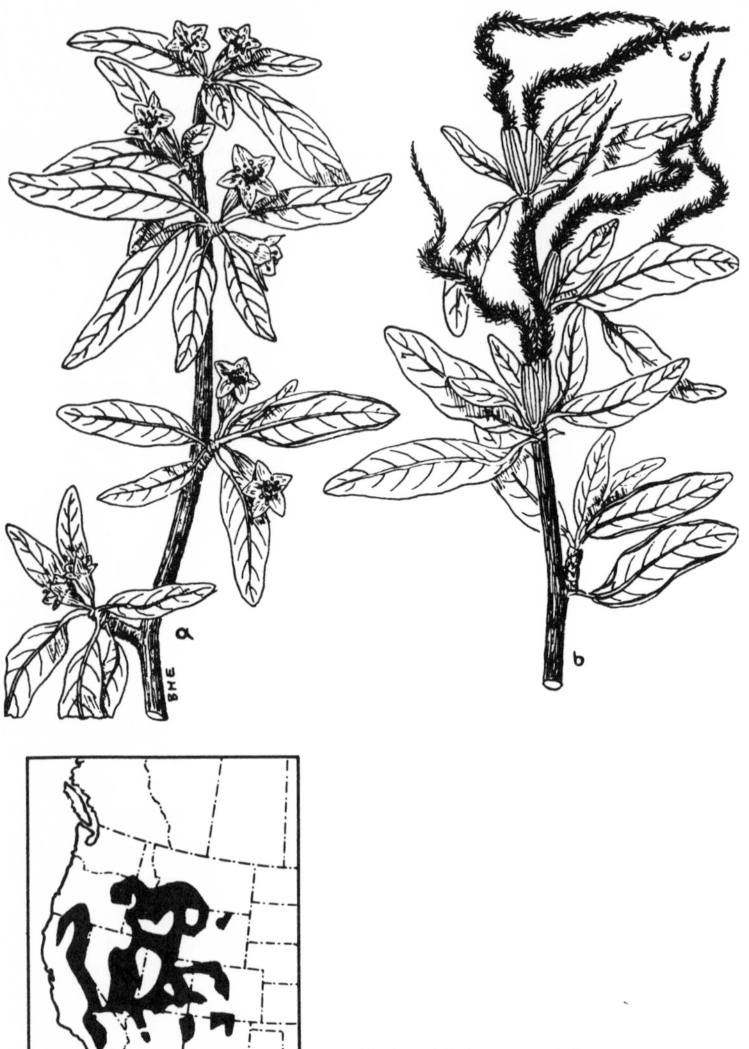

a. Twig with leaves and flowers, x 1.

b. Twig with fruit, x 1.

Curlleaf Mountain-mahogany

Cercocarpus ledifolius Nutt.

HABIT. A shrub or small tree rarely 40 feet high and 30 inches in diameter; trunk short and crooked; crown round and compact with stout, more or less crooked, spreading branches.

LEAVES. Alternate; simple; lanceolate to lanceolate-elliptic; 1/2-1 inch long; acute; entire; coriaceous; revolute margins; dark green and lustrous above; pale and tomentulose below; resinous; persistent for 2 years; short-petiolate; stipules minute, deciduous.

FLOWERS. Regular; perfect; solitary; inconspicuous; sessile in axils of leaves; calyx tube long, hoary-tomentose, whitish; corolla absent; stamens 15-30, in 2-3 rows, free, anthers hairy; ovary inferior, composed of 1 carpel and ovule.

FRUIT. Linear-oblong, coriaceous achene 1/4 inch long, enclosed in persistent calyx tube 1/2 inch long; chestnut-brown; covered with long hairs; tipped with persistent, hairy, elongated style 2-3 inches long. Seed: solitary; linear; acute.

TWIGS. Stout; round; rigid; red-brown and pubescent at first, becoming dark brown or silver-gray and glabrous; spurlike lateral branches. Winter buds: minute, scaly, pubescent.

BARK. Thick on old trunks (1 inch); red-brown; hard; firm; furrowed and scaly.

WOOD. Exceedingly heavy and hard; brittle; close-grained; diffuse-porous; warping badly; heartwood clear red or dark brown; sapwood rather thin, yellow; unimportant.

SILVICAL CHARACTERS. Intolerant; rather long-lived; slow-growing; on dry, gravelly, windswept slopes.

GENERAL. Three other shrubby species occasionally reach tree size; the tree species are keyed out below:

1. Leaves entire, lanceolate, leathery, persistent; flowers solitary or in 2's or 3's . . . . *C. ledifolius* Nutt., Curlleaf Mountain-mahogany
1. Leaves usually somewhat toothed, elliptic to broadly obovate, deciduous.
   2. Leaves entire or slightly toothed near apex; flowers solitary or in 2's or 3's; Texas to Arizona and south . . . . *C. breviflorus* A. Gray., Hairy Mountain-mahogany
   2. Leaves with numerous teeth above the middle; flowers in many-flowered racemes; Oregon and California to Arizona and Mexico . . . . *C. betuloides* Nutt., Birchleaf Mountain-mahogany

ROSACEAE

The Cherries, Plums, and Peaches

Characteristics of the Genus *Prunus* L.

HABIT. Shrubs or usually small trees, only a few species reaching sizes of commercial importance.

LEAVES. Alternate; simple; deciduous or persistent; usually serrate, rarely entire; stipules free from petiole, early deciduous; petiolate.

FLOWERS. Regular; perfect or rarely dioecious; solitary or in terminal or axillary racemes, corymbs, or umbels; appearing from separate buds with, before, or after leaves; calyx 5-lobed, tubular; corolla 5-petaled, usually white, deciduous, stamens usually 15-20; ovary inserted in bottom of calyx tube, inferior or superior, 1-celled, 2-ovuled.

FRUIT. Thin dry, or thick fleshy, 1-seeded drupe; stone bony, smooth or rugose; indehiscent; important as food in several species. Seed: filling cavity of nut; suspended; thin-coated, pale brown.

TWIGS. Slender or stout; round; astringent; often with spinescent spur branches; red to brown; marked by lenticels, stipular scars, usually by small, elevated, horizontal leaf scars with 3 bundle scars. Winter buds: terminal present or absent; lateral nearly equal in size; scales imbricated, the inner accrescent and often colored.

BARK. Astringent; gray to dark brown; plated or scaly.

WOOD. From light to heavy and hard; fine-textured; diffuse-porous; durable; heartwood light or dark brown, often reddish; sapwood lighter colored; a few species important for timber.

GENERAL. About 30 species of *Prunus* are native to North America, 18 being arborescent at times. Seven introduced species, four of which are highly important fruit trees, have become naturalized in parts of the United States and Canada. These species, included in the key, are: *P. avium* (L.) L., Mazzard cherry; *P. cerasus* L., sour cherry; *P. domestica* L., garden plum; *P. insititia* L., Damson plum; *P. mahaleb* L., Mahaleb cherry; *P. persica* (L.) Batsch., peach; and *P. spinosa* L., sloe or blackthorn. The almond, *P. amygdalus* Batsch., and the apricot, *P. armeniaca* L., while not naturalized, are commonly planted in warmer regions. As with other cultivated fruits, these have been improved until there are numerous forms or varieties. The commonly planted ornamental flowering almond, *P. triloba* Lindl., is a shrub or small tree.

KEY TO THE SPECIES OF CHERRIES, PLUMS, AND PEACHES

1. Fruit usually over 1/2 inch in diameter and usually slightly 2-lobed by ventral groove.
   2. Terminal bud present; fruit and ovary hairy; naturalized . . . . *P. persica*, Peach, p. 259
   2. Terminal bud absent; fruit and ovary glabrous . . . . Plums
      3. Leaves rolled up in bud; flowers 1 or 2; naturalized.
         4. Twigs glabrous; fruit 1 inch across . . . . *P. domestica*, Garden Plum, p. 265
         4. Twigs hairy or woolly; fruit about 1/2 inch across.
            5. Flowers single; buds round . . . . *P. spinosa* L., Sloe
            5. Flowers in 2's; buds ovate . . . . . *P. insititia*, Damson Plum, p. 265
      3. Leaves folded in bud (except Klamath plum); flowers 3 to many; native.
         6. West Coast; leaves round-ovate, rolled in bud . . . . *P. subcordata* Benth., Klamath Plum
         6. East of Pacific Coast; leaves ovate to lanceolate.
            7. Leaves with sharp teeth, dull dark green above.
               8. Leaves lanceolate to elliptic; fruit purple, glaucous.
                  9. Fruit stone turgid; calyx hairy; Connecticut to North Carolina . . . . *P. alleghaniensis* Port, Allegheny Plum
                  9. Fruit stone flattened; calyx glabrous; North Carolina to Texas and Arkansas . . . . *P. umbellata* Ell., Flatwoods Plum
               8. Leaves oblong to obovate; fruit red to yellow; stone flattened.
                  10. Leaves hairy below; twigs hairy; central and southern . . . . *P. mexicana* Wats., Mexican Plum
                  10. Leaves nearly glabrous below; twigs glabrous; Canada to Florida, west to the Rocky Mountains . . . . *P. americana*, American Plum, p. 265
            7. Leaves with dull, rounded teeth.
               11. Leaves dull green, elliptic to obovate, coarsely or doubly toothed; stone flattened . . . . *P. nigra*, Canada Plum, p. 265
               11. Leaves lustrous, thin, ovate to lanceolate, finely toothed; stones turgid; southern Indiana to Kansas and south.
                   12. Leaves 2 1/2-6 inches long; fruit 1/4-1 inch; calyx glandular.
                       13. Leaves oblong to obovate . . . . *P. hortulana* Bail, Hortulan Plum
                       13. Leaves elliptic to lanceolate . . . . *P. munsoniana* W. & H., Wildgoose Plum
                   12. Leaves 1-2 inches long; fruit 1/2 inch; calyx lobes without glands . . . . *P. angustifolia* Marsh., Chickasaw Plum
1. Fruit seldom 1/2 inch across, not lobed; leaves folded in bud; terminal bud present . . . . Cherries
   14. Flowers solitary or few, in umbels or corymbs.
       15. Calyx persistent on fruit; naturalized.
           16. Leaves thin, 2 1/2-5 inches long, hairy below at least along veins; fruit sweet . . . . *P. avium*, Mazzard Cherry, p. 259
           16. Leaves semileathery, 2-3 inches long, glabrous; fruit sour . . . . *P. cerasus* L., Sour Cherry, p. 259
       15. Calyx deciduous from fruit.
           17. Leaves orbicular to broad-ovate; naturalized . . . . *P. mahaleb* L., Mahaleb
           17. Leaves narrower; native.
               18. Leaves oblong-obovate; Montana and British Columbia to Arizona and New Mexico . . *P. emarginata*, Bitter Cherry, p. 263

18. Leaves oblong-lanceolate; Newfoundland to British Columbia, south to Georgia, Iowa, Colorado . . . . *P. pensylvanica*, Pin Cherry, p. 263

14. Flowers 12 or more in elongated racemes.
    19. Leaves deciduous; flowers in terminal, usually leafy racemes.
        20. Calyx deciduous from fruit; leaves oblong-oval to obovate, with spreading teeth; entire area . . . . *P. virginiana*, Common Chokecherry, p. 261
        20. Calyx persistent on fruit; leaves oblong-lanceolate with incurved teeth . . . . *P. serotina*, Black Cherry, p. 261
    19. Leaves persistent; flowers in axillary, leafless racemes.
        21. Southeast; leaves entire or rarely with a few small spiny teeth.
            22. Leaves elliptic to ovate; southern Florida . . . . *P. myrtifolia* Urb., Myrtle Laurelcherry
            22. Leaves oblong-lanceolate; North Carolina to Texas . . . . *P. caroliniana* Ait., Carolina Laurelcherry
        21. California.
            23. Leaves spiny toothed . . . *P. ilicifolia* Dietr., Hollyleaf Cherry
            23. Leaves entire or minutely toothed. . . . . *P. lyonii* Sarg., Catalina Cherry

* * *

Christmasberry

Photinia arbutifolia Lindl. (*Heteromeles arbutifolia* Roem.)

An evergreen shrub or small tree native to lower mountain slopes in California and Lower California, widely planted as an ornamental because of its lustrous evergreen leaves and clusters of red fruits that resemble those of mountain-ash. This tree is characterized by alternate, simple, thick, glossy, oblong to elliptic, serrate, persistent leaves; small, white, perfect flowers in terminal panicles; and fruit a red, ovoid, clustered pome about 1/3 inch long.

* * *

Cliffrose. Quininebush

Cowania mexicana D. Don

This shrub or small tree rarely 25 feet high is found on dry slopes from Nevada to Colorado and south into Mexico. It is characterized by odd, simple, alternate leaves, 3-5-lobed, 1/3-1/2 inch long, thick, dark green above and woolly below, and tardily deciduous to persistent; perfect, regular, solitary, showy flowers, which are pale yellow to white and about one inch across; from which develop 5-12 achenes, each about 1/4 inch long and tipped with a 2-inch white, hairy style.

ROSACEAE

Peach
Prunus persica (L.) Batsch (*Amygdalus persica* L.)

This small Asiatic tree with its many improved varieties produces the peaches of commerce. It has escaped from cultivation, mostly along roads or fences. The peach differs from other subgenera of *Prunus* in having large, grooved, hairy fruits; 1-2 sessile flowers with hairy ovaries; and scaly buds with the terminal present and separate lateral flower buds.

* * *

Almond
Prunus amygdalus Batsch

A small Asiatic tree cultivated for the edible kernels of the stone in the drupaceous fruit. This species is restricted to warm climates. This tree closely resembles the peach, differing from it in that the fruit becomes dry and splits open at maturity. The commonly planted flowering almond, *P. triloba* Lindl., is a handsome, pink-flowered shrub or small tree native to China. The desert almond, *P. fasciculata* Gray, is a shrub or rarely a small tree native to the Southwest.

* * *

Sour Cherry
Prunus cerasus L.

This native of southeastern Europe and Asia is widely naturalized as an escape through much of North America. It is important as the source of numerous varieties of canning cherries.

* * *

Sweet Cherry. Mazzard
Prunus avium L.

Native to Europe and Asia, but widely naturalized through North America, this species is widely cultivated for its garden cherries.

* * *

Apricot
Prunus armeniaca L.

This small to medium-sized tree, native to western Asia, is widely planted for its distinctive fruits. It closely resembles the peach in its hairy, grooved fruit, but differs from it in having no terminal bud and solitary axillary buds, and in having broad-ovate leaves rolled up in the bud.

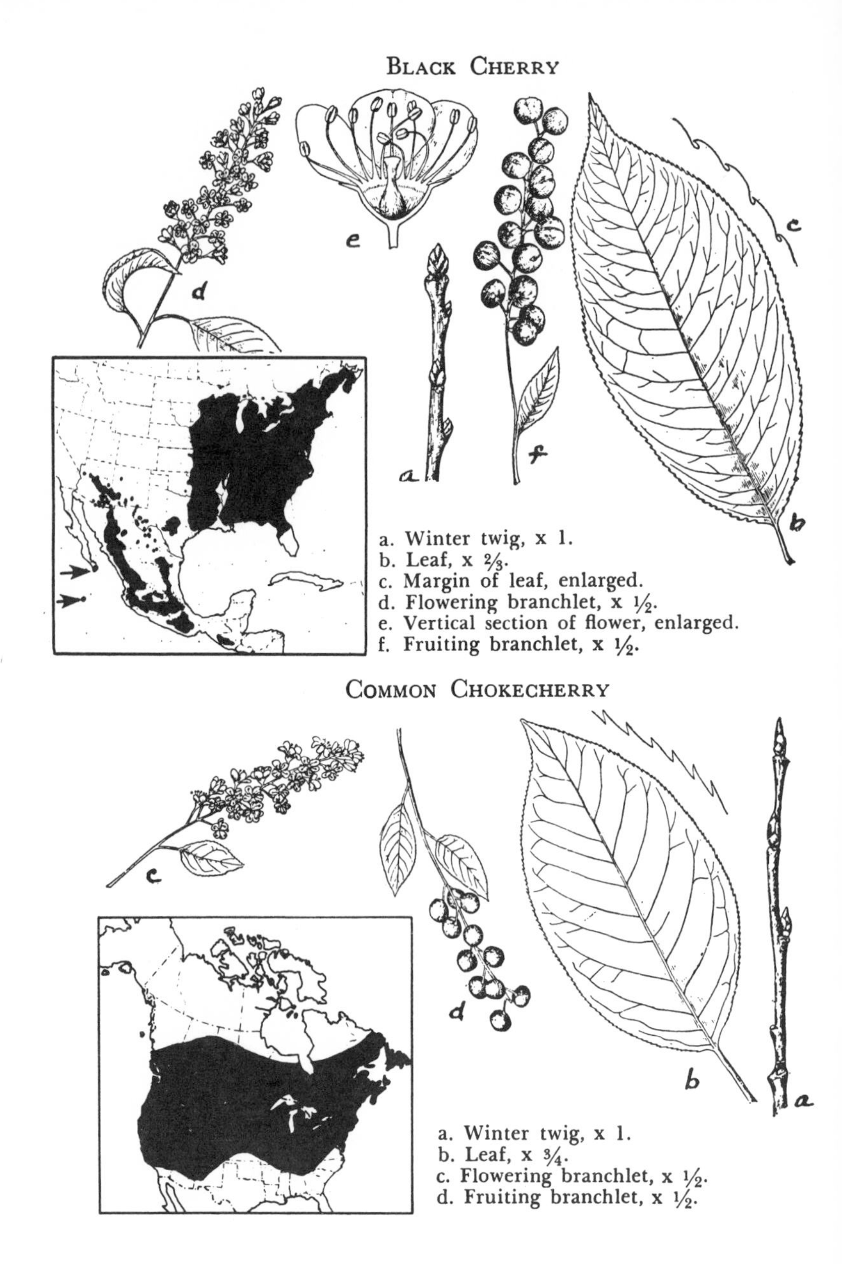

BLACK CHERRY

a. Winter twig, x 1.
b. Leaf, x ⅔.
c. Margin of leaf, enlarged.
d. Flowering branchlet, x ½.
e. Vertical section of flower, enlarged.
f. Fruiting branchlet, x ½.

COMMON CHOKECHERRY

a. Winter twig, x 1.
b. Leaf, x ¾.
c. Flowering branchlet, x ½.
d. Fruiting branchlet, x ½.

ROSACEAE

Black Cherry

Prunus serotina Ehrh.

HABIT. A medium-sized tree 40-60 feet high and 1-3 feet in diameter (max. 100 by 5 feet); clear bole and narrow, oblong crown.

LEAVES. Oval to oblong-lanceolate; thick; acuminate or acute at apex; finely serrate with incurved callous teeth; dark green and lustrous above; paler, with red-brown hairs along base of midrib below; 2-6 inches long; petioles short, usually with 2 glands near blade.

FLOWERS. In many-flowered, loose racemes, 4-5 inches long; appearing with the leaves; calyx cup-shaped, persistent on fruit; petals 5, white, about 1/8 inch long.

FRUIT. Globose; in racemes; 1/3-1/2 inch in diameter; nearly black when ripe; flesh juicy, edible, slightly bitter.

TWIGS. Slender; round; glabrous; red-brown; bitter and aromatic; short spur shoots common on old growth. Winter buds: terminal 1/4 inch long; ovoid, scaly, light brown.

BARK. Thin; smooth and red-brown with horizontal lenticels on young stems; nearly black and exfoliating into distinctive, persistent scales with upturned edges on old trunks.

WOOD. Moderately important; rather heavy and hard; diffuse-porous; heartwood red-brown; a prized and valuable furniture wood.

SILVICAL CHARACTERS. Intermediate in tolerance; typical of rich, moist sites; usually in mixed stands; long taproot.

* * *

Common Chokecherry

Prunus virginiana L.

This shrub or small tree is one of the most widely distributed in North America. It closely resembles black cherry, but can be distinguished by usually red fruit without persistent calyx; leaves thin, with sharply serrate, spreading teeth; smooth or slightly fissured bark. The western forms are designated as either western chokecherry, *P. virginiana* var. *demissa* (Nutt.) Torr., or black chokecherry, *P. virginiana* var. *melanocarpa* (A. Nels.) Sarg.

Pin Cherry

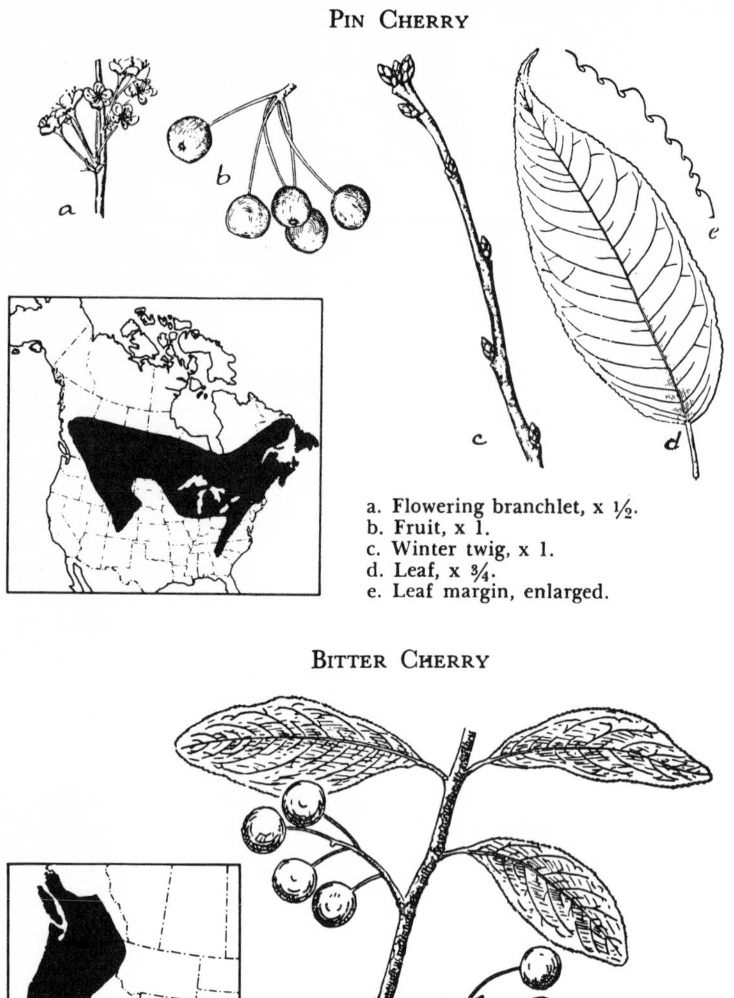

a. Flowering branchlet, x ½.
b. Fruit, x 1.
c. Winter twig, x 1.
d. Leaf, x ¾.
e. Leaf margin, enlarged.

Bitter Cherry

a. Twig with leaves and fruit, x ¾.

Pin Cherry. Wild Red Cherry

Prunus pensylvanica L.

HABIT. A shrub or small tree rarely 30-40 feet high and 18-20 inches in diameter; short trunk; crown narrow, rounded or flat-topped, with slender, horizontal branches.

LEAVES. Obovate to oblong-lanceolate; acuminate or acute at apex; sharply and coarsely serrate, with incurved teeth; glabrous; bright green and lustrous above, paler below; 3-4 inches long and 1-2 inches broad; turning bright yellow in autumn; petiole slender, glabrous, or pilose.

FLOWERS. In 2-5-flowered, sessile umbels; on slender pedicels nearly 1 inch long; appearing when leaves half grown; calyx tube glabrous, marked by conspicuous orange band in mouth of throat; corolla cream-white, 1/2 inch in diameter.

FRUIT. Globose; in 2-5-fruited umbels; 1/4-1/3 inch in diameter; on slender pedicels; light red; thick-skinned; flesh thin, quite sour; occasionally made into jelly; stone oblong, thin-walled, pointed, 3/16 inch long, ridged.

TWIGS. Slender; round; puberulous and light red at first, becoming glabrous and bright to dull red; marked by orange-colored, raised lenticels; bark easily separable from green inner bark; lateral branchlets short, spurlike. Winter buds: terminal 1/8 inch long, acute, ovoid, bright red-brown.

BARK. Thin (1/3-1/2 inch); red-brown, marked by orange-colored bands of lenticels; smooth or scaly.

WOOD. Light; soft; close-grained; heartwood light brown; sapwood thin, yellow; unimportant.

SILVICAL CHARACTERS. Intolerant; fast-growing; short-lived; the abundant reproduction often completely taking over burned areas.

* * *

Bitter Cherry

Prunus emarginata Dougl.

This cherry closely resembles pin cherry in its flowers and fruit; it can be distinguished by its oblong-obovate, obtuse leaves and its more western distribution where it extends from British Columbia to Montana and south to California, Arizona, and New Mexico.

American Plum

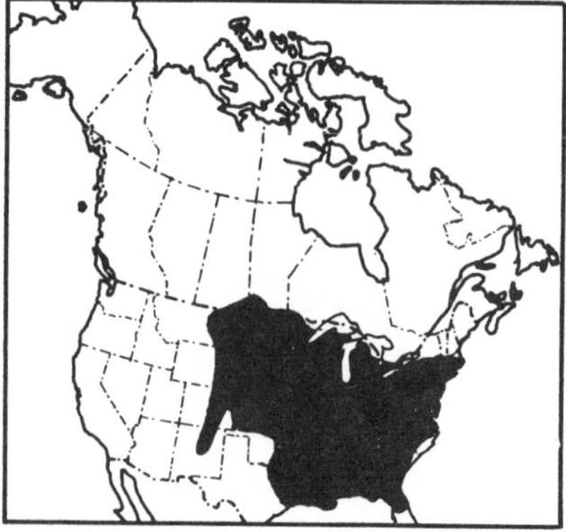

a. Winter twig, x 1.

b. Portion of twig, enlarged.

c. Leaf, x ½.

d. Flowering branchlet, x ½.

e. Vertical section of flower, x 1.

f. Fruiting branchlet, x ½.

American Plum

Prunus americana Marsh.

HABIT. A shrub or small tree 25-30 feet high and rarely 12 inches in diameter; trunk short, usually dividing near the ground; crown broad, with many spreading branches; usually spreading by shoots from the roots into dense thickets.

LEAVES. Oval to slightly oblong-oval, or sometimes obovate; acuminate at apex; sharply and often doubly serrate; thick and firm; dark green above, pale and glabrous below; 3-4 inches long and 1 1/2-1 3/4 inches wide; petiole slender, 1/2-3/4 inch long.

FLOWERS. In 2-5-flowered umbels; on slender, glabrous pedicels 1/2-2/3 inch long; appearing before or with leaves; ill-scented; calyx tube narrow, bright red without, green and pubescent within; corolla white, 1 inch in diameter.

FRUIT. Subglobose; about 1 inch in diameter; red and often spotted at maturity; thick-skinned; nearly free from bloom; flesh bright yellow, juicy, acid; used for jellies; stone oval, rounded at apex, 3/4-1 inch long.

TWIGS. Slender; glabrous; bright green at first, becoming orange-brown and marked by minute, circular, raised lenticels; sometimes spiny-tipped. Winter buds: terminal absent; lateral 1/8-1/4 inch long; acute, chestnut-brown.

BARK. Up to 1/2 inch thick; dark brown, tinged with red; outer layer forming persistent plates.

WOOD. Heavy; hard; close-grained; strong; heartwood dark brown, tinged with red; sapwood thin, lighter colored.

SILVICAL CHARACTERS. Intolerant; on moist bottomlands to banks of intermittent streams on dry uplands and mountain slopes.

* * *

Canada Plum

Prunus nigra Ait.

The Canada plum is very similar to the American plum and often classed as a variety of it. It differs in having leaves with small, rounded teeth and biglandular petioles. Canada plum ranges from New Brunswick to southern Manitoba and south to New York, Indiana, and North Dakota. The garden plum, *P. domestica* L., while native to Europe and Asia, has escaped through much of North America, as has the Bullace or Damson plum, *P. insititia* L.

FABACEAE (LEGUMINOSAE)

Characteristics of the Pea Family

This important family contains about 500 genera and over 15,000 species among which are many of our important food plants. The family is characterized by its distinctive fruit, known as a legume, as well as by alternate, usually compound leaves and regular or distinctive papilionaceous flowers.

The genera growing in North America and reaching tree size are keyed out on the opposite page; important, common, or distinctive species are described on succeeding pages. The several unimportant forms occasionally reaching tree size in the southern fringe of the United States are briefly described below:

Smokethorn, *Dalea spinosa* A. Gray, in Arizona and southeastern California is a desert shrub with minute or no leaves, green, spiny twigs, and a small 1-seeded legume.

Southwestern coralbean, *Erythrina flabelliformis* Kearney, in southern Arizona and New Mexico is characterized by 3 foliolate leaves and showy red flowers. The eastern coralbean, *E. herbacea* L., from North Carolina to Texas is usually a perennial herb but becomes a tree in Florida. The South American *E. crista-galli* L. is cultivated and escaped in the Gulf states.

Kidneywood, *Eysenhardtia polystachya* (Ortega) Sarg., is a shrubby Mexican border form characterized by small leaflets and a legume only 1/2 inch long. *E. angustifolia* Pen. is a small tree in Texas.

Littleleaf leadtree, *Leucaena retusa* Benth., of southern Texas and New Mexico differs from Lysiloma in having a linear legume 6-10 inches long on a peduncle 3-5 inches long. Three similar species, *L. glauca* Benth., *L. greggii* S. Wats., and *L. pulverulenta* Benth., form small trees in southern Texas and Mexico.

Lysiloma, *Lysiloma microphylla* Benth., is a rare shrub or small tree in Arizona characterized by bipinnately compound, persistent leaves and a linear-oblong, compressed legume 5-8 inches long. *L. bahamensis* Benth. is found in southern Florida.

Tesota, *Olneya tesota* A. Gray, is a beautiful spiny small tree of the Arizona-California desert, with purple flowers and a compressed, glandular-hairy legume 2-2 1/2 inches long.

Paradise poinciana, *Poinciana gilliesii* Hook., is an ill-scented shrub or small tree native to South America but naturalized in southern Texas, New Mexico, and Arizona. It is characterized by large, showy, flowers with long-exserted red stamens and pistil. *P. mexicana* (Gray) Rose, is found in southern Texas and Mexico.

KEY TO GENERA REACHING TREE SIZE

1. Leaves bipinnately compound (3-4 pinnate in some genera and sometimes only pinnate in *Gleditsia*); flowers regular or nearly so.
    2. Flowers in globose heads or cylindrical spikes; leaflets mostly less than 1/2 inch long; shrubby; south Florida or Mexican border.
        3. Twigs armed with spines; legumes terete or compressed.
            4. Legume dehiscent; stamens 10, united filaments . . . . *Pithecellobium* Mart., Blackbead, p. 281
            4. Legume indehiscent; filaments free except at base.
                5. Flowers in heads or spikes, yellow or white, stamens more than 10; petioles not spiny or glandular at apex. *Acacia* Mill., Acacia, p. 269
                5. Flowers in cylindric spikes, greenish-white, stamens 10; petioles minutely glandular at apex and tipped with small spinescent rachis . . . . *Prosopis* L., Mesquite, p. 271
        3. Twigs unarmed; legume compressed, dehiscent; flowers in heads, whitish, or pink.
            6. Stamens numerous, long-exserted; flowers pink; planted in Southeast. . . . . *Albizia* Dur., Mimosa, p. 271
            6. Stamens 10-20; flowers white; western.
                7. Stamens 12-20, exserted, filaments united into a tube; legume broad . . . . *Lysiloma* Benth., Lysiloma, p. 266
                7. Stamens 10, inserted, filaments free; legume linear . . . . *Leucaena* Benth., Leadtree, p. 266
    2. Flowers in racemes; stamens 10 or less, filaments free.
        8. Flowers large and showy, yellow, perfect; leaflets less than 1/2 inch long; southwestern shrubs, rarely small trees.
            9. Stamens red and long-exserted; legume broad and flat; naturalized . . . . *Poinciana* L., Poinciana, p. 266
            9. Stamens inserted; native.
                10. Legume linear, terete; rachis of leaf spinescent . . . . *Parkinsonia* L., Parkinsonia, p. 279
                10. Legume oblong, compressed; rachis of leaf not spinescent . . . . *Cercidium* Tulasne, Paloverde, p. 279
        8. Flowers not showy, greenish-white, polygamous or dioecious; leaflets over 1/2 inch long; eastern trees, but widely planted.
            11. Leaves 1-3 feet long with leaflets 2-2 1/2 inches long; legume thick and woody, seeds 3/4 inch . . . *Gymnocladus* Lam., Coffeetree, p. 275
            11. Leaves 1/2-1 foot long with leaflets 1/2-1 1/2 inches long; legume leathery with seeds 1/3 inch long . . . *Gleditsia* L., Honeylocust, p. 277
1. Leaves simple or pinnately compound; flowers papilionaceous.
    12. Leaves simple.
        13. Leaves heart-shaped; flowers in fascicles, red, with free stamens; spineless . . . . *Cercis* L., Redbud, p. 273
        13. Leaves oblong or absent; flowers in racemes, blue—stamens, 9 united anthers; twigs reduced to spines; Ariz., Calif.. *Dalea* Juss., Dalea, p. 266
    12. Leaves pinnately compound.
        14. Leaflets 1-4 inches long.
            15. Leaves 3-foliate; flowers red . . . *Erythrina* L., Coralbean, p. 266
            15. Leaves many-foliate; flowers white, yellow, or blue.
                16. Leaflets 3-4 inches long; flowers white, in panicles; legume compressed . . . . *Cladrastis* Raf., Yellowwood, p. 275
                16. Leaflets 1-2 1/2 inches long.
                    17. Twigs with stipular spines; legumes compressed, not constricted . . . . *Robinia* L., Locust, p. 283
                    17. Twigs unarmed; legumes terete, constricted . . . . *Sophora* L., Sophora, p. 281
        14. Leaflets less than 1 inch long.
            18. Leaves with glandular dots. *Eysenhardtia* H.B.K., Kidneywood, p. 266
            18. Leaves without glandular dots.
                19. Flowers fascicled, yellow . . . *Caragana* Lam., Pea Tree, p. 281
                19. Flowers racemose, purple . . . . . *Olneya* Gray, Tesota, p. 266

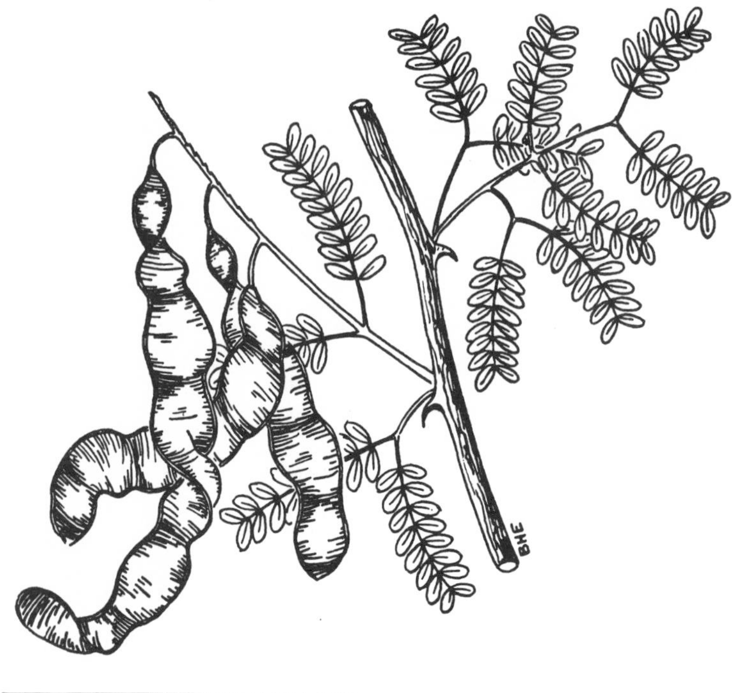

Twig with leaves and fruit, x 1.

Catclaw Acacia. Paradise Flower

Acacia greggii Gray

HABIT. A spiny shrub or small tree rarely 30 feet high and 10-12 inches in diameter; trunk short; crown irregular.

LEAVES. Alternate; evenly doubly pinnate; 1-3 pairs of pinnae, each with 4-5 pairs of oblong, obtuse, thick, pubescent leaflets 1/16-1/4 inch long; persistent; petiole short, glandular.

FLOWERS. Nearly regular; perfect or polygamous; fragrant; in dense, pubescent spikes; calyx 5-lobed, puberulous; corolla 5-petaled, bright yellow, 1/8 inch long; stamens numerous, exserted, 1/4 inch long; ovary stalked and hairy.

FRUIT. Linear-oblong, flat, much curved and contorted, indehiscent, light brown legume 2-6 inches long and 1/2-3/4 inch wide; contracted between the seeds; valves thin and membranaceous. Seed: dark brown; lustrous; 1/4 inch long.

TWIGS. Slender; angled; puberulous or glabrous; red-brown; armed with stout, broad, recurved, infrastipular spines 1/4 inch long, giving tree its common name.

BARK. Thin (1/8 inch); furrowed and scaly; light gray-brown; astringent.

WOOD. Very heavy; hard; strong; close-grained; durable; ring-porous; heartwood red-brown; sapwood thin, light yellow.

SILVICAL CHARACTERS. Lower Sonoran zone; intolerant; vigorous reproducer; thrives in driest and poorest soils; on dry mesas, plains, and in low canyons.

Acacia with more than 400 species is widely distributed throughout the warmer regions of the world. Seven other species of Acacia may reach tree size in the Southwest:

* * *

1. *A. amentacea* DC., with long spines, yellow flowers in spikes, and a flattened arcuate legume; southwestern Texas.
2. *A. constricta* var. *paucispina* Woot., with long spines, yellow flowers in capitate heads, and a terete legume; Texas to Arizona.
3. *A. emoryana* Benth., with short spines, flowers in short spikes, and flat, hairy legume with thickened sutures.
4. *A. farnesiana* (L.) Willd., with spines, yellow flowers in capitate heads, and a terete legume; Texas.
5. *A. millefolia* S. Wats., with slender or no spines, cream-colored flowers in racemes, and a very flat pod; rare in southern Arizona and Mexico.
6. *A. tortuosa* Willd., like (5) but longer pod (3-5 inches); Texas.
7. *A. wrightii* Benth., yellow flowers in spikes; short spines; Texas.

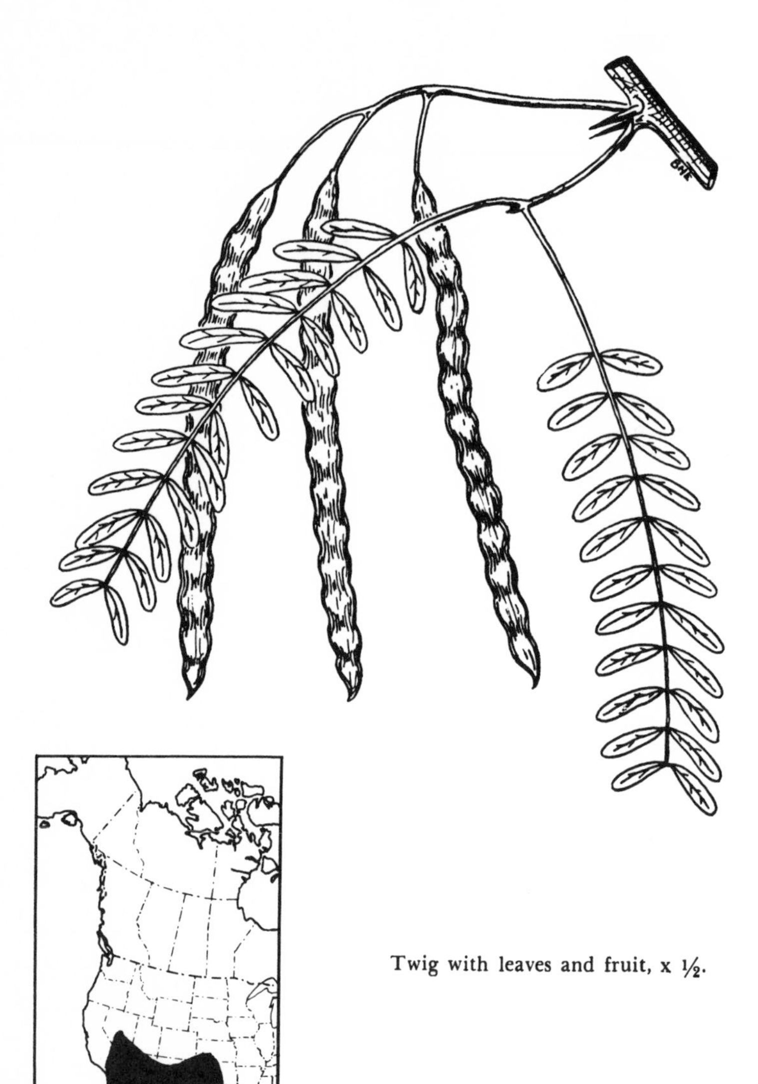

Twig with leaves and fruit, x ½.

Mesquite

Prosopis juliflora (Sw.) DC. [*Prosopis chilensis* (Mol.) Stuntz.]

HABIT. A spiny shrub or small tree rarely 20-50 feet high and 1-4 feet diameter; crown loose and straggling.

LEAVES. Alternate; evenly, doubly (rarely 3-4) pinnate; pinnae with 12-30 linear to linear-oblong, small, deciduous, glabrous leaflets 1/2-2 inches long; petioles spine-tipped.

FLOWERS. Nearly regular; perfect; fragrant; minute; in spikes 1 1/2-4 inches long; green-white; stamens 10.

FRUIT. Linear, flat to subterete, indehiscent, yellowish, straight or falcate legume 4-9 inches long and 1/4-1/2 inch wide; edible. Seed; oblong; compressed; light brown; 1/4 inch long.

TWIGS. Slender; smooth; usually with spines 1/2-2 inches long. Winter buds: terminal absent; lateral small, brown.

BARK. Thick; dark red-brown; furrowed and scaly.

WOOD. Very heavy; hard; close-grained; durable; ring-porous; heartwood dark brown or red; sapwood yellow.

SILVICAL CHARACTERS. Intolerant; long-lived; adapted to desert sites by huge taproot descending 40-50 feet; abundant seeder; aggressive; confined to areas below 6,000 feet.

GENERAL. This species extends from Venezuela through Mexico. Three varieties are shown in the southwestern United States. Screwbean mesquite, *Prosopis pubescens* Benth., ranging from Texas to California north into Nevada and Utah, is readily distinguished by its thick, linear, indehiscent, pale yellow legume 1-2 inches long, twisted by 12-20 turns into a narrow, straight spiral.

* * *

Mimosa. Silktree

Albizia julibrissin Durazz.

A handsome ornamental, widely planted and escaped from Maryland to Indiana and south. The powder puff appearance of the pink flowers in crowded heads and the bipinnate leaves with many tiny leaflets (10-25 pinnae, each with 40-60 leaflets) are distinctive.

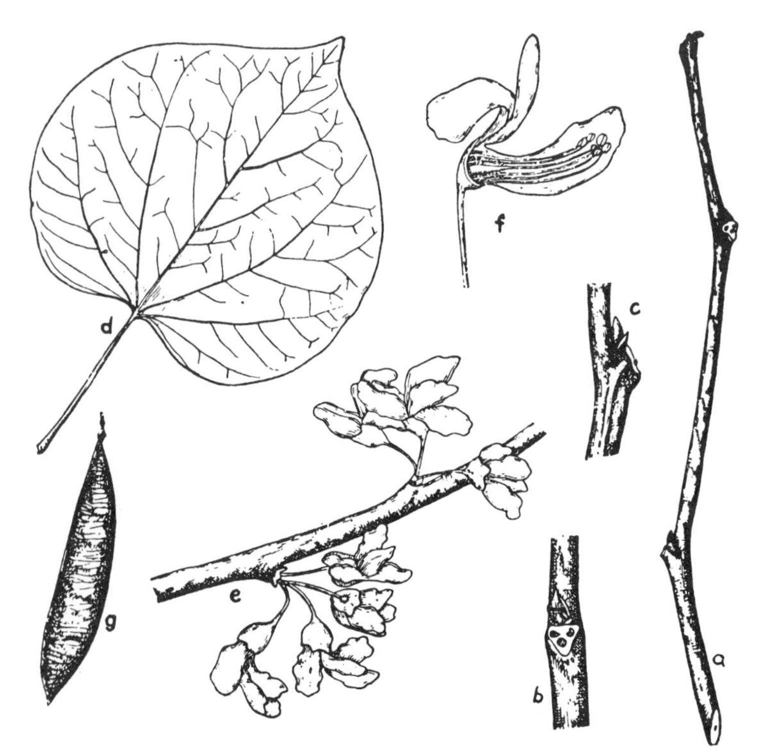

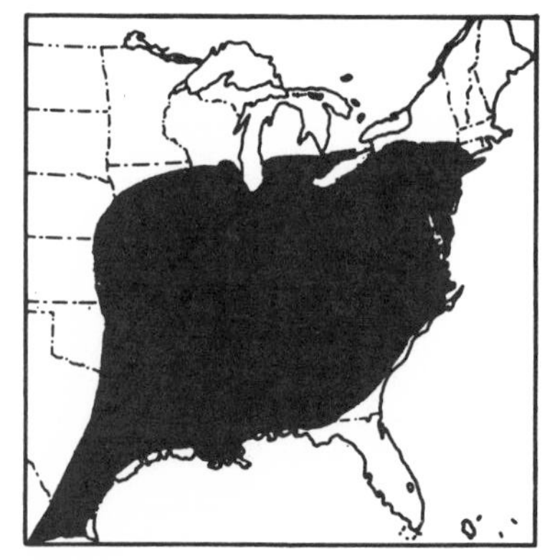

a. Winter twig, x 1.

b. Portion of twig, front view, enlarged.

c. Portion of twig, side view, enlarged.

d. Leaf, x ⅔.

e. Flowering branchlet, x 1.

f. Vertical section of flower, enlarged.

g. Fruit, x ½.

FABACEAE

Eastern Redbud

Cercis canadensis L.

HABIT. Usually a shrub, but not infrequently a small tree up to 40 feet high and 2 1/2 feet in diameter; frequently planted as an ornamental.

LEAVES. Alternate; simple; broad-ovate to reniform; 3-5 inches in diameter; apex acute; base cordate; margins entire; glabrous; deciduous; petioles long, slender; stipules small, membranaceous.

FLOWERS. Irregular (subpapilionaceous); perfect; in simple fascicles; appearing before the leaves; calyx short, top-shaped, purple, 5-toothed, persistent; corolla 5-petaled, rose-colored; stamens 10, free, inserted in 2 rows on margin of thin disk, persistent; ovary short-stalked; ovules numerous in 2 ranks.

FRUIT. Stalked, flat, oblong, russet-brown legume 2-3 1/2 inches long; tipped with remnant of style. Seed: 1/4 inch long; ovoid or oblong; compressed; red-brown.

TWIGS. Slender; round; unarmed; marked by numerous pale lenticels and elevated, often fringed leaf scars with 3 bundle scars. Winter buds: terminal absent; axillary small, scaly, obtuse, chestnut-brown.

BARK. Thin; gray; smooth or becoming scaly on old trunks.

WOOD. Heavy; hard; not strong; ring-porous; heartwood dark red-brown; sapwood white, thin; unimportant.

SILVICAL CHARACTERS. Tolerant; reproduction vigorous; typical of bottomlands and stream borders; a variety recognized in the Southwest is *texensis* (S. Wats.) Hopkins, the Texas redbud.

* * *

California Redbud

Cercis occidentalis Torr.

This similar species is usually shrubby but has been reported as a tree in California and in the Grand Canyon. Its range is from southern Utah and Nevada to California and Arizona and it differs from the eastern redbud in having leaves 2-3 inches long which are obtuse or emarginate at the apex.

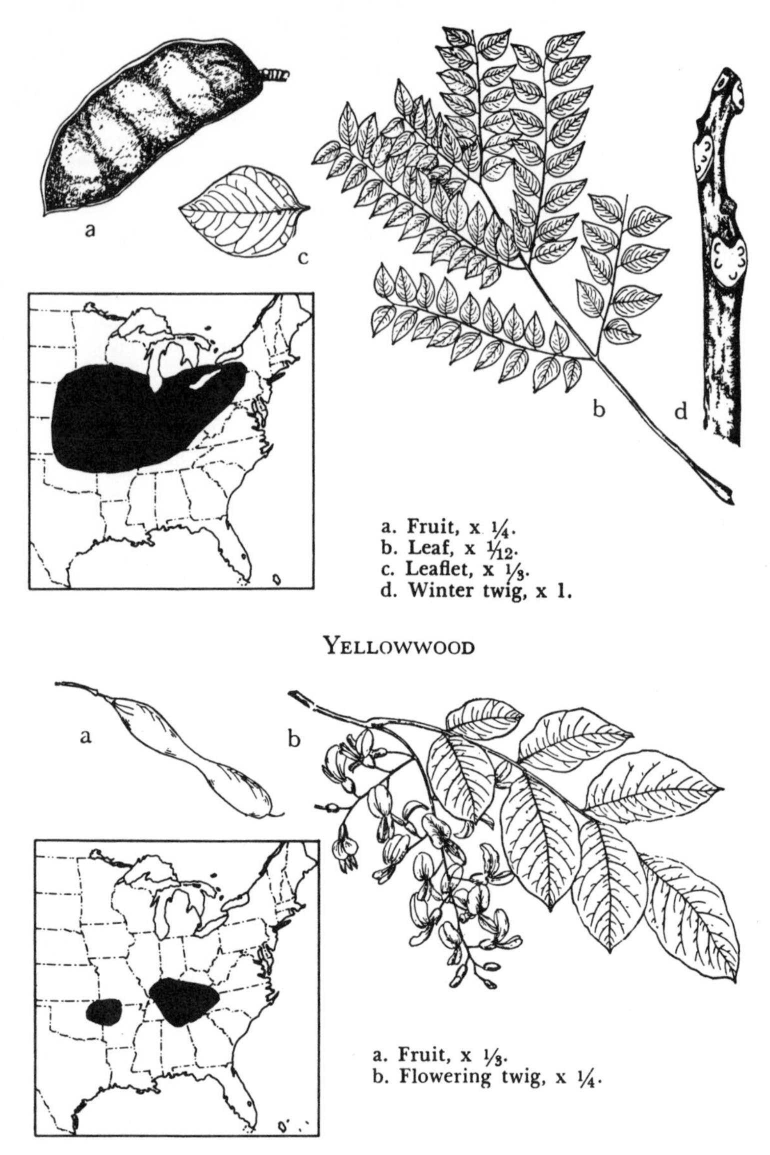

KENTUCKY COFFEETREE

a. Fruit, x 1/4.
b. Leaf, x 1/12.
c. Leaflet, x 1/3.
d. Winter twig, x 1.

YELLOWWOOD

a. Fruit, x 1/3.
b. Flowering twig, x 1/4.

Kentucky Coffeetree

Gymnocladus dioicus (L.) K. Koch

HABIT. A medium-sized to large tree 50-75 feet high and 2-3 feet in diameter (max. 110 by 4 feet); open, pyramidal crown.
LEAVES. Alternate; doubly pinnate with 40 or more leaflets; 1-3 feet long; leaflets 2-2 1/2 inches long, ovate, acute, entire, short-stalked, glabrous, dark green above and paler below; appearing late in the spring; deciduous.
FLOWERS. Nearly regular; dioecious; in racemes; calyx tubular, hairy; petals 5, greenish-white; stamens 10; ovary hairy; appearing after the leaves.
FRUIT. A turgid, woody, short-stalked, red-brown legume 4-10 inches long and 1 1/2-2 inches wide; containing sugary pulp between the 6-9 seeds. Seed: ovoid, 3/4 inch long, with thick bony coat, used as substitute for coffee in early days.
TWIGS. Very stout; unarmed; coated at first with short, dense, red hairs; roughened by large, pale, conspicuous leaf scars. Winter buds: terminal absent; lateral small, depressed in craters, brown, silky-hairy, with 2 in the axil of each leaf.
BARK. Smooth and brown to gray on limbs; becoming 3/4-1 inch thick; gray; fissured; distinctive reflexed scales.
WOOD. Unimportant; heavy and hard; ring-porous; reddish and closely resembling honey locust; sapwood very thin, white.
SILVICAL CHARACTERS. Intolerant; fast-growing; rare throughout its range; a handsome tree often planted as an ornamental.

* * *

Yellowwood

Cladrastis lutea (Michx.) K. Koch

This medium-sized tree is rare and unimportant except for occasional use as an ornamental. It is characterized by alternate, deciduous, pinnately compound leaves with 5-11 leaflets, each 3-4 inches long, obovate, entire, acute, and yellow-green; by perfect, showy, white, papilionaceous flowers; by a glabrous, flat, short-stalked, linear legume 3-4 inches long containing 4-6 flat, brown seeds; by slender, glabrous twigs containing naked lateral buds enclosed in the hollow base of the petiole; and by smooth, thin, gray bark.

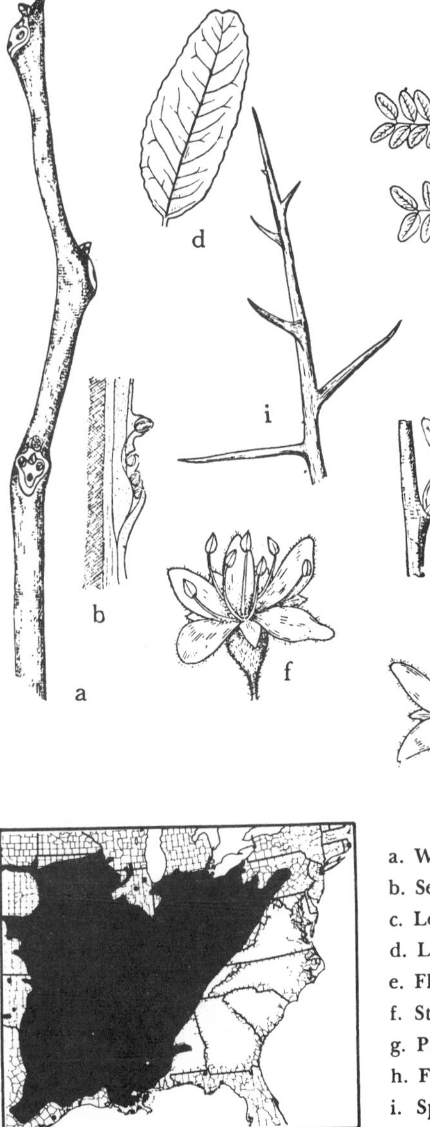

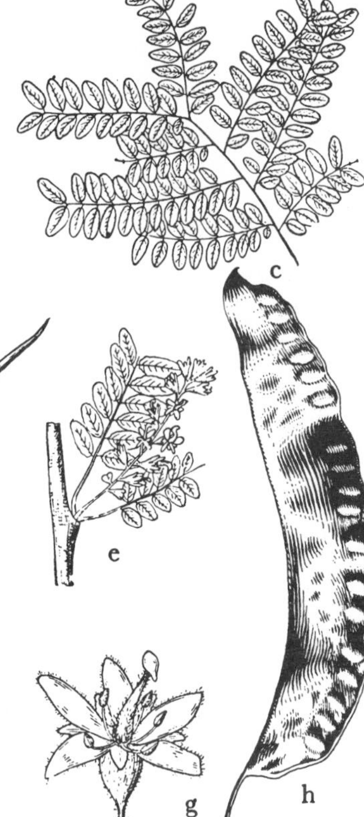

a. Winter twig, x 1.
b. Section through lateral buds.
c. Leaf, x 1/6.
d. Leaflet, x 1.
e. Flowering branchlet, x 1/2.
f. Staminate flower, enlarged.
g. Pistillate flower, enlarged.
h. Fruit, x 1/4.
i. Spine from trunk, x 1/2.

FABACEAE

Honeylocust

Gleditsia triacanthos L.

HABIT. A medium-sized tree 50-75 feet high and 2-3 feet in diameter (max. 140 by 5 feet); open, spreading crown; trunk and branches usually armed with clusters of straight or branched spines several inches long.

LEAVES. Alternate; pinnately or bipinnately compound with both types often on the same tree; 6-12 inches long; leaflets 1/2-1 1/2 inches long, lanceolate-oblong, remotely crenulate, thin, glabrous, lustrous dark green above and yellow-green below; deciduous.

FLOWERS. Regular; polygamous; small; in axillary racemes; calyx 3-5-lobed; petals 3-5, greenish-white; stamens 3-10; ovary 1-celled, woolly; appearing after the leaves.

FRUIT. A strap-shaped, red-brown, usually twisted legume 12-18 inches long and 1 inch wide; containing 12-14 dark brown, oval seeds, each about 1/3 inch long.

TWIGS. Slender to rather stout; usually with straight or 3-branched spines 2-8 inches long; green to red-brown. Winter buds: terminal absent; lateral minute, nearly submerged in leaf scar, brownish, 3 or more superposed.

BARK. Smooth and gray on limbs; becoming 1/2-3/4 inch thick, gray to nearly black, and broken by longitudinal fissures into plates or scaly ridges.

WOOD. Of slight importance; heavy and hard; durable; red-brown; ring-porous; used for veneer and fuel.

SILVICAL CHARACTERS. Intolerant; fast-growing; hardy; prefers rich, moist sites, but hardy on arid, sandy locations; the thornless variety *inermis* (Pursh) Schneid., is commonly planted in cities.

* * *

Waterlocust

Gleditsia aquatica Marsh.

This similar species is native to the coastal region from North Carolina to Texas and up the Mississippi Valley to southern Indiana. It is readily distinguished by its oval, oblique, 1-3-seeded legume, which is 1-2 inches long. The Texas honeylocust, *Gleditsia* X *texana* Sarg., is a hybrid between waterlocust and honeylocust and is characterized by a legume 4-5 inches long. It has been reported from Indiana to Texas.

Blue Paloverde

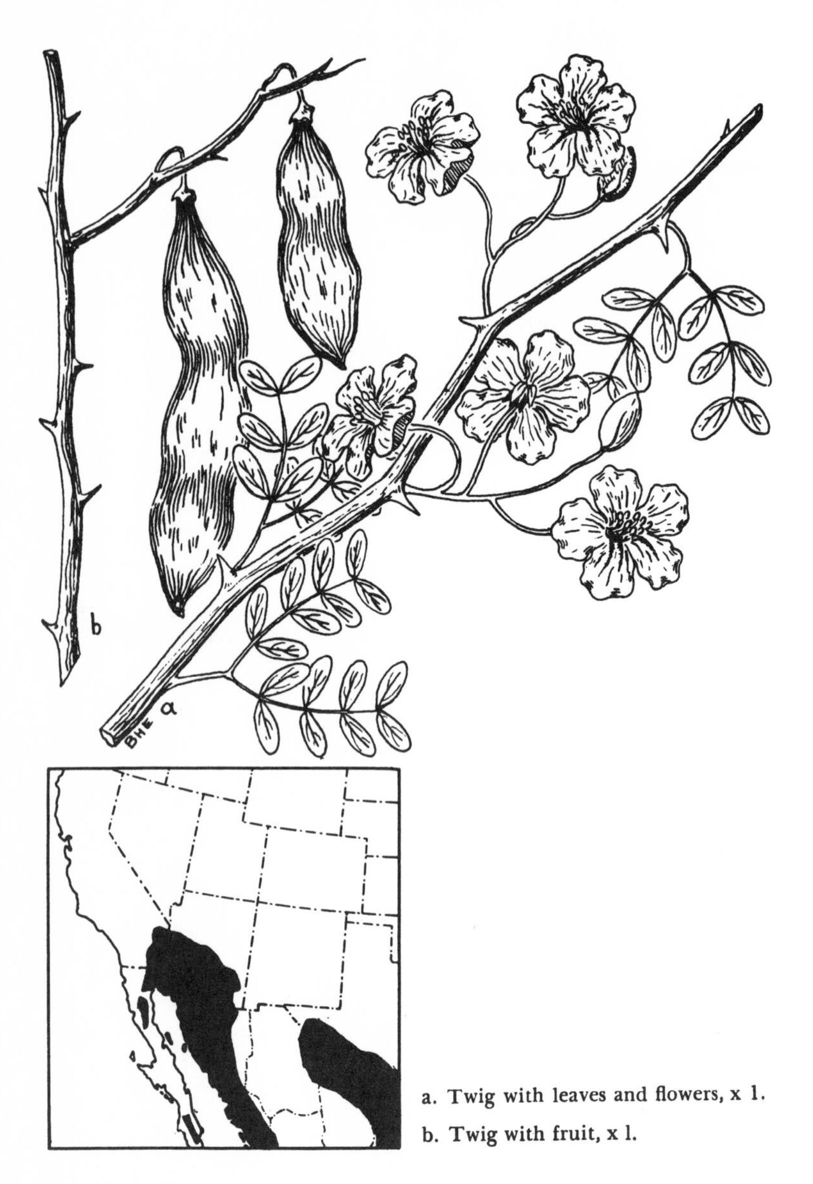

a. Twig with leaves and flowers, x 1.

b. Twig with fruit, x 1.

Blue Paloverde. Greenbarked-acacia

Cercidium floridum Benth. [*Cercidium torreyanum* (S. Wats.) Sarg.]

HABIT. A thorny shrub or small tree 15-30 feet high; wide, irregular, open crown with stout, tortuous branches.

LEAVES. Alternate; doubly and evenly pinnate with 2-3 pairs of oblong, obtuse, glaucous leaflets 1/12-1/6 inch long; few and totaling less area than green twigs; falling soon, but frequently bearing a second crop during the rainy season.

FLOWERS. Nearly regular; perfect; in conspicuous 4-5-flowered, axillary racemes; on slender pedicels 3/4-1 inch long; calyx 5-lobed, reflexed; corolla of 5, clawed, bright yellow petals, 3/4 inch in diameter; stamens 10, free, exserted.

FRUIT. Oblong, compressed or somewhat turgid legume 3-4 inches long and 1/4-1/3 inch wide; straight or somewhat contracted between the 2-8 seeds. Seed: ovoid, compressed.

TWIGS. Stout; glabrous and glaucous; light yellow or pale olive-green; armed with thin spines 1/4 inch long.

BARK. Thin (1/8 inch); smooth and pale olive-green on young trunks, becoming furrowed, scaly, and red-brown.

WOOD. Unimportant; soft; weak; close-grained; ring-porous; heartwood light brown; sapwood yellow; used as fuel.

SILVICAL CHARACTERS. Lower Sonoran zone; intolerant; abundant seeder; typical of arid, desert sinks, canyons, and depressions; in sandy or gravelly soil.

GENERAL. Two other species may reach tree size in the Southwest:

1. Yellow Paloverde, *C. microphyllum* (Torr.) R. & J., in California and Arizona has leaves with 4-6 pairs of leaflets, a 1-2-seeded legume, and twigs terminating in spines.
2. Border Paloverde, *C. macrum* Johnst., in southern Texas has slightly glandular leaflets 1/16 inch long and a 2-3-seeded legume.

The distribution map includes *floridum* and *microphyllum* in Arizona and California and *macrum* in Texas.

* * *

Jerusalem-Thorn

Parkinsonia aculeata L.

Jerusalem-thorn is a widely planted similar small tree ranging from southern Arizona and Texas to South America. It differs from paloverde in having a linear, terete legume and leaves with 25-30 pairs of small leaflets and a spinescent rachis.

MESCALBEAN

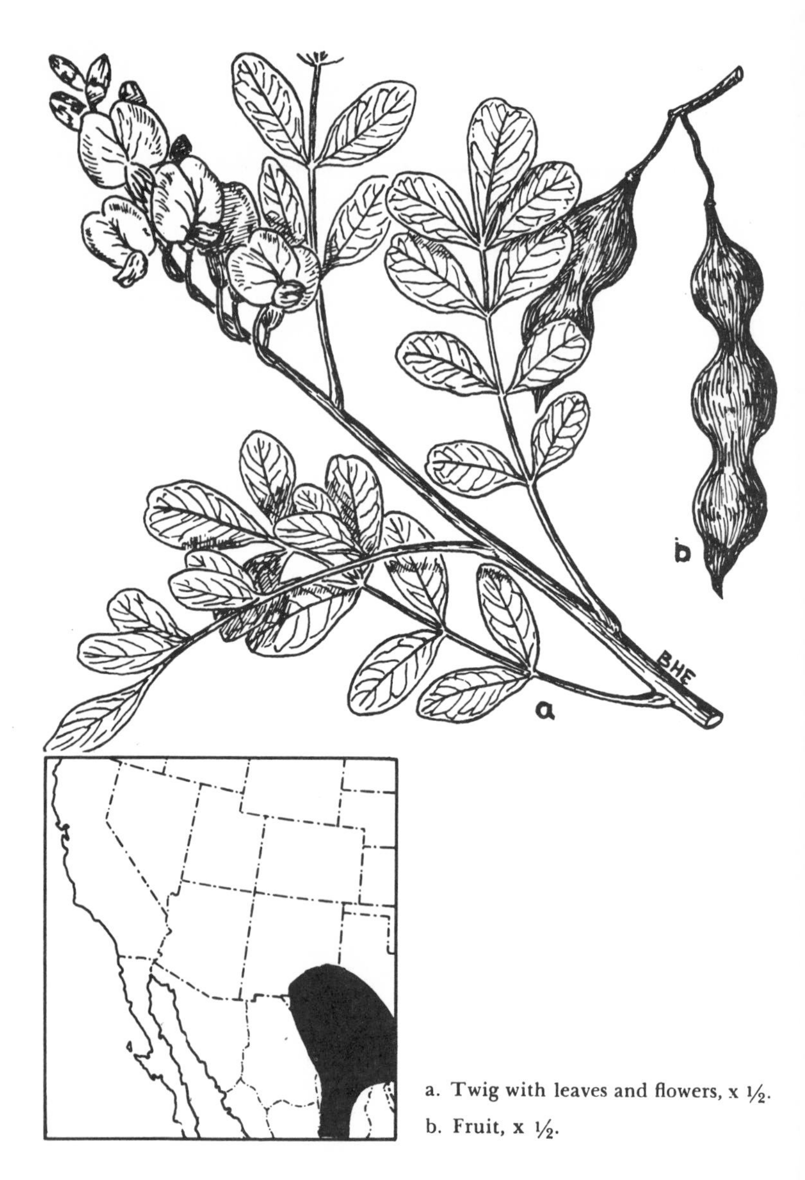

a. Twig with leaves and flowers, x ½.

b. Fruit, x ½.

Mescalbean. Frijolito

Sophora secundiflora (Ortega) Lag.

HABIT. A shrub or rarely a small tree 20 feet high and 6-8 inches in diameter; commonly thicket-forming.

LEAVES. Alternate; persistent; unequally pinnate with 7-13 elliptical, coriaceous, glabrous, entire leaflets 3/4-2 inches long; rounded; lustrous above, paler below.

FLOWERS. Irregular (papilionaceous); perfect; fragrant; showy; appearing with leaves; in terminal racemes; corolla blue or white.

FRUIT. Oblong, terete, indehiscent legume 1-7 inches long; thick woody valves; much contracted between seeds; covered with dense tomentum; tipped with remnant of style. Seed: bright scarlet; very poisonous.

TWIGS. Slender; round; unarmed; lateral buds minute.

WOOD. Heavy; hard; close-grained; orange-yellow; unimportant.

GENERAL. The Texas sophora, *S. affinis* Torr. & Gray, of Texas, Arkansas, Oklahoma, and Louisiana, differs in having thin, deciduous leaves, a fleshy legume, and white flowers in axillary racemes. The Japan Pagoda-tree, *S. japonica* L., with acute leaflets and yellow to pink flowers in terminal panicles, is a common ornamental.

* * *

Ebony Blackbead

Pithecellobium flexicaule (Benth.) Coult.

A common and beautiful shrub or small tree of the Texas Gulf Coast and banks of the lower Rio Grande into Mexico. Characterized by bipinnate, few foliolate leaves; stipular spines; fragrant, yellow flowers in cylindrical spikes; and thick legumes containing bright red-brown seeds. Huajillo, *P. pallens* (Benth.) Standl., also appears in Texas; *P. guadalupense* (Pers.) Chapm. and *P. ungis-cati* (L.) Benth. reach southern Florida.

* * *

Siberian Pea Tree

Caragana arborescens Lam.

This introduced tree has been extensively planted for shelterbelts and as an ornamental. It is characterized by even pinnate leaves with 8-12 leaflets and papilionaceous, yellow flowers, 1/2-3/4 inch long, in fascicles of 1-4.

Black Locust

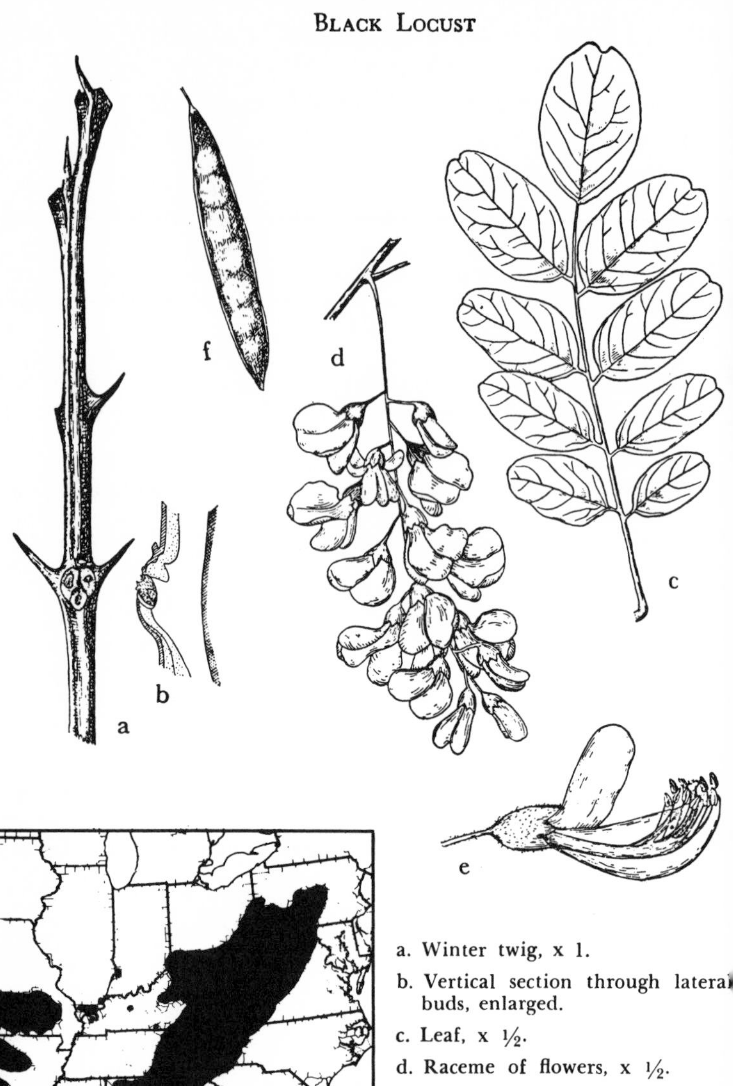

a. Winter twig, x 1.
b. Vertical section through lateral buds, enlarged.
c. Leaf, x ½.
d. Raceme of flowers, x ½.
e. Flower, with part of corolla removed, enlarged.
f. Fruit, x ½.

FABACEAE

Black Locust

Robinia pseudoacacia L.

HABIT. A medium-sized tree 40-60 feet high and 1-2 feet in diameter (max. 100 by 5 feet); open, irregular crown.

LEAVES. Alternate; pinnately compound; 8-14 inches long; 7-21 ovate-oblong, entire, glabrous leaflets, 1 1/2-2 inches long, dark blue-green above and paler below; deciduous.

FLOWERS. Papilionaceous; perfect; showy; fragrant; in racemes; calyx bell-shaped; corolla 5-petaled, white; stamens 10; ovary 1-celled; appearing after the leaves.

FRUIT. A smooth, dark brown, flat, linear-oblong legume, 2-4 inches long; 4-8 flat brown seeds about 3/16 inch long.

TWIGS. Rather stout; red-brown; usually armed with stipular spines about 1/2 inch long. Winter buds: terminal absent; lateral minute, red-hairy, submerged, 3-4 superposed.

BARK. Red-brown to nearly black; deeply fissured into rounded, scaly ridges.

WOOD. Moderately important; very heavy and hard; very durable; dark brown or golden; used for posts, ties, insulator pins.

SILVICAL CHARACTERS. Intolerant; fast-growing; seriously threatened by locust borer (*Cyllene robiniae*); prefers moist, rich soils, but hardy on poor, dry sites; a clone of this species, which propagates only by vegetative means, is shipmast locust (var. *rectissima* Raber). Widely planted and naturalized.

GENERAL. There are about 20 species of *Robinia*, all native to North America, with 4 reported to reach tree size. These tree species are keyed out below:

1. Legume, twigs, and peduncle glabrous or nearly so . . . . *R. pseudoacacia* L., Black Locust
1. Legume and peduncle hispid, glandular, or viscid.
    2. Twigs glabrous; usually a shrub; eastern Tennessee and western North Carolina . . . . *R. kelseyi* Hutchins, Kelsey Locust
    2. Twigs glandular pubescent or viscid.
        3. Glands not viscid; southern Colorado to Nevada, south to Arizona and west Texas . . . . *R. neo-mexicana* A. Gray, New Mexican Locust
        3. Glands exuding a clammy, sticky substance; mountains, Pennsylvania to Georgia and Alabama . . . . *R. viscosa* Vent, Clammy Locust

Hoptree

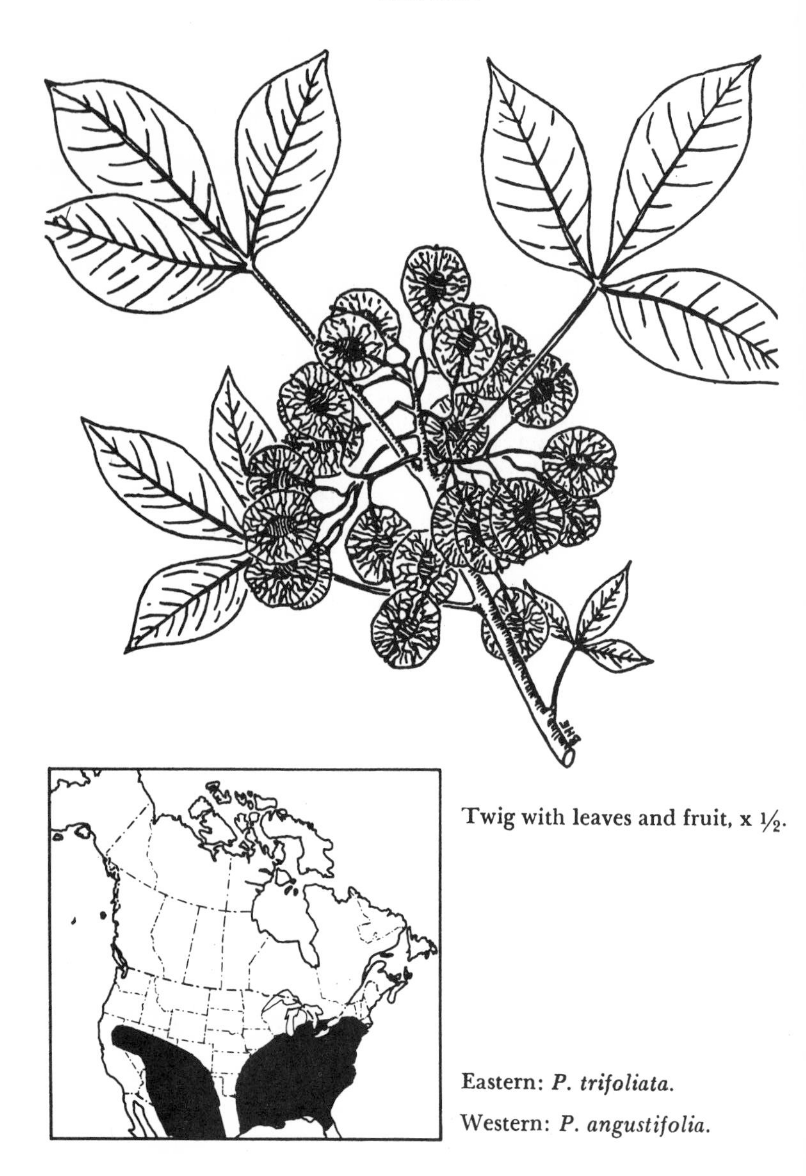

Twig with leaves and fruit, x ½.

Eastern: *P. trifoliata.*

Western: *P. angustifolia.*

Hoptree. Wafer-ash
Ptelea trifoliata L.

HABIT. An unarmed shrub or small tree 20-25 feet high and 6-8 inches in diameter; crown round-topped.

LEAVES. Alternate, rarely opposite; compound, 3 (rarely 5) subsessile, ovate to oblong leaflets; acuminate; entire or crenulate-serrate; dark green above, pale and dotted with transparent glands below; deciduous; long-petioled.

FLOWERS. Regular; polygamous; in terminal cymes or compound umbels; on pubescent pedicels; calyx 4-5-parted, pubescent; corolla green-white; 4-5-petaled; stamens 3-4.

FRUIT. Dehiscent samara; 2-3-celled; broad, thin, almost orbicular, veined wing, nearly 1 inch across; drooping clusters on slender pedicels; persisting on branches through winter. Seed: 1/3 inch long; oblong; acute; dark red-brown.

TWIGS. Slender; round; dark brown; lustrous; marked by wartlike excrescences and conspicuous leaf scars. Winter buds: terminal absent; lateral small, depressed, pale, tomentose.

BARK. Smooth or warty; thin; bitter; ill-scented; dark brown on old trunks; that of roots sometimes used as a tonic.

WOOD. Rather heavy; hard; close-grained; ring-porous; heartwood yellow-brown; sapwood thin; unimportant.

SILVICAL CHARACTERS. Rather tolerant and often an "understory" tree; dry, rocky soils.

GENERAL. The western form, also with ovate leaflets, is narrowleaf hoptree, *P. angustifolia* Benth. Shrubby *P. pallida* Greene of New Mexico and Arizona has lanceolate leaflets.

* * *

Hercules-club. Prickly-ash
Zanthoxylum clava-herculis L.

Distinctive small tree, spiny twigs, peculiar conical, corky growths 1 inch or more in diameter on smooth gray bark; leaves alternate, spiny-stemmed, late-deciduous, pinnately compound, 7-19 leathery, ovate, toothed leaflets; dioecious, clustered flowers; ovoid, brown, wrinkled capsule 1/4 inch long, single black seed hanging from it at maturity. Common prickly-ash, *Z. americanum* Mill., ranges widely through eastern North America; while commonly a shrub, reported as a small tree. *Z. fagara* (L.) Sarg. in south Florida and south Texas, ranges to South America. *Z. coriaceum* Rich. and *Z. flavum* Vahl in south Florida.

AILANTHUS

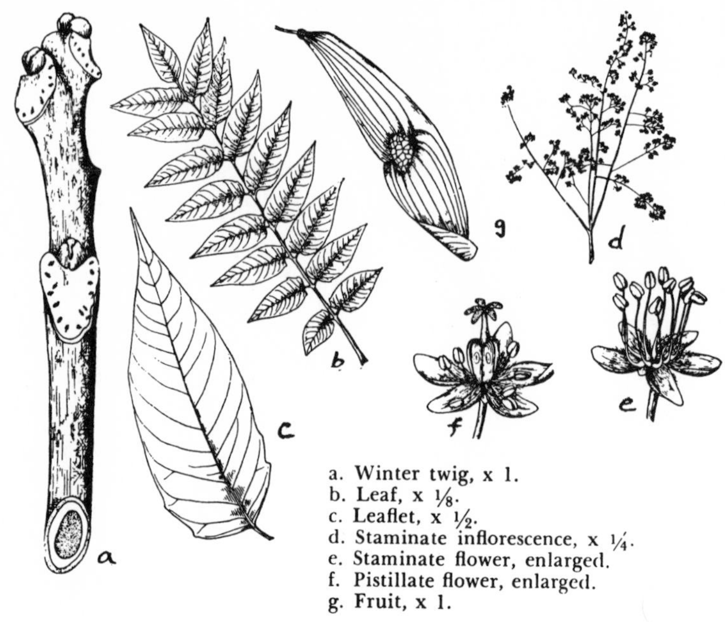

a. Winter twig, x 1.
b. Leaf, x 1/8.
c. Leaflet, x 1/2.
d. Staminate inflorescence, x 1/4.
e. Staminate flower, enlarged.
f. Pistillate flower, enlarged.
g. Fruit, x 1.

ELEPHANTTREE

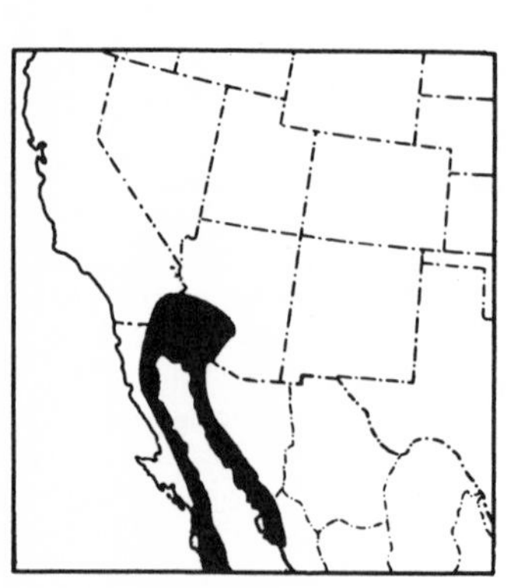

(a) Twig with fruit and leaves, x 1.

SIMAROUBACEAE

Ailanthus. Tree-of-heaven
Ailanthus altissima (Mill.) Swingle (*Ailanthus glandulosa* Desf.)

HABIT. A handsome naturalized tree 40-60 feet high and 2-4 feet in diameter; loose, open crown with very large leaves.

LEAVES. Alternate; deciduous; 1 1/2-4 feet long; pinnately compound with 13-41 leaflets; leaflets ovate-lanceolate, 3-6 inches long, entire except for 1-2 glandular teeth at the base.

FLOWERS. Regular; polygamo-dioecious; small; yellow-green; in panicles; calyx 5-lobed; petals 5; after leaves.

FRUIT. Oblong, twisted samara 1-1 1/2 inches long with seed in center; yellow-green; in crowded persistent clusters.

TWIGS. Very thick; velvety-downy; red-brown. Winter buds: terminal absent; lateral small, subglobose, brown.

BARK. Thin; gray; smooth; shallowly furrowed with age.

WOOD. Unimportant; soft; weak; pale yellow; ring-porous.

SILVICAL CHARACTERS. Intolerant; very fast-growing; short-lived; hardy on sterile sites and in smoky cities; root-suckers aggressively; a common city ornamental.

* * *

MELIACEAE

Chinaberry
Melia azedarach L.

This small Asiatic tree is commonly cultivated and naturalized from North Carolina to Oklahoma and south. It is characterized by alternate, bipinnately compound, tardily deciduous leaves 10-30 inches long with ovate, toothed or lobed, glabrous leaflets 1-2 inches long; purple, showy flowers in loose panicles; and distinctive subglobose, yellow, fleshy drupes 1/2-3/4 inch long. A low-spreading, flat-topped variety known as the umbrella Chinaberry (var. *umbraculifera* Sarg.) is popular. The West Indies mahogany, *Swietenia mahagoni* Jacq., is native to southern Florida.

* * *

BURSERACEAE

Elephanttree
Bursera microphylla A. Gray

This rarely becomes a small tree in the desert in southern California and Arizona. It is characterized by alternate, deciduous, bipinnately compound leaves with 20-40 leaflets 1/4 inch long; small, polygamous, white flowers; and a red, capsulelike drupe 1/4 inch long. The fragrant bursera, *B. fagaroides* Engelm., is a similar form in southern Arizona. Gumbo-limbo, *B. simaruba* Sarg., is native to southern Florida.

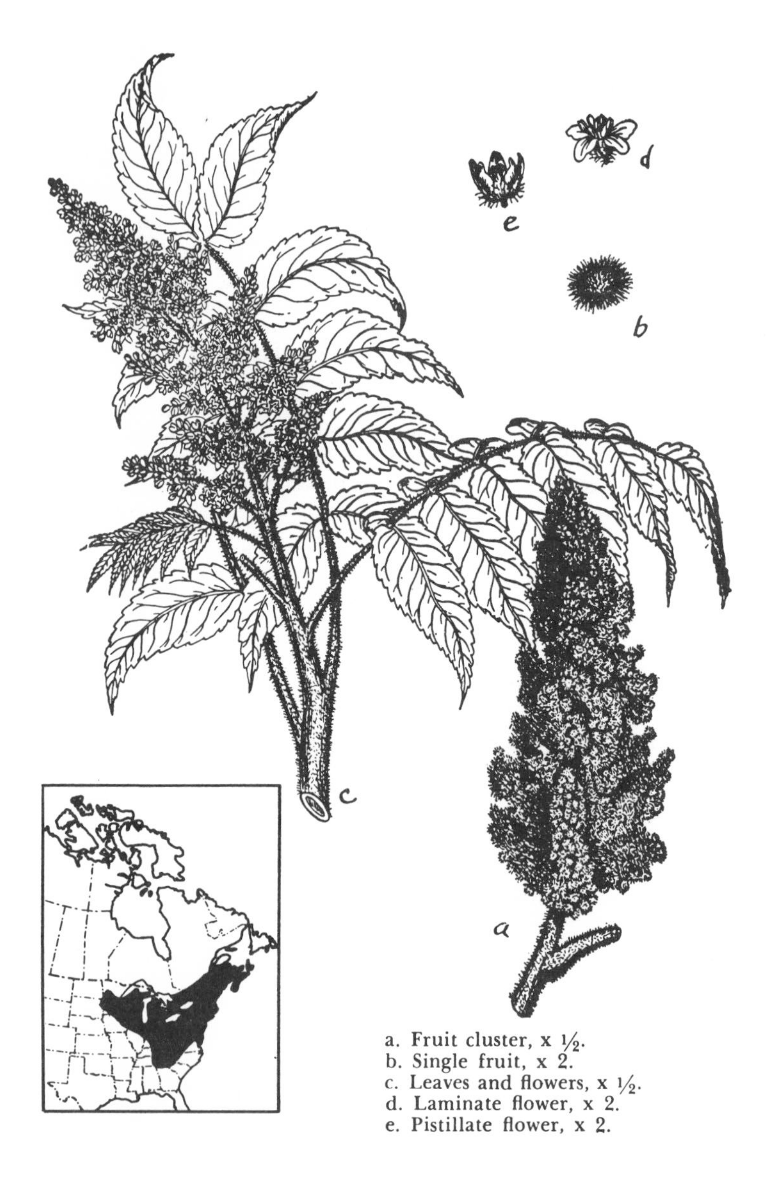

a. Fruit cluster, x ½.
b. Single fruit, x 2.
c. Leaves and flowers, x ½.
d. Laminate flower, x 2.
e. Pistillate flower, x 2.

ANACARDIACEAE

Staghorn Sumac
Rhus typhina Torn.

HABIT. A shrub or small tree to 40 feet high and 1 foot in diameter; commonly in broad thickets.

LEAVES. Alternate; deciduous; pinnately compound with 11-31 oblong, glabrous, serrate, dark green leaflets, 2-5 inches long.

FLOWERS. Regular; dioecious or polygamous; small; green-white; in dense, hairy panicles 6-12 inches long; after the leaves.

FRUIT. Compact, conelike clusters of red, hairy drupes, each 1/4 inch long.

TWIGS. Thick; velvety-hairy; milky juice; thick pith. Winter buds: terminal absent; lateral 1/4 inch long, conical, brown, silky-hairy, hidden by leaf scar.

BARK. Thin; dark brown; smooth, becoming scaly with age.

WOOD. Unimportant; soft and weak; ring-porous; orange-colored.

SILVICAL CHARACTERS. Intolerant; fast-growing; short-lived.

* * *

About 17 species are native to North America with 9 rarely reaching tree size and keyed out below. Two species, poisonous to the touch, are placed either in the genus *Rhus* or *Toxicodendron;* these are poison-sumac, *T. vernix* (L.) Ktze., and poison ivy or oak, *T. radicans* (L.) Ktze.

1. Leaves simple or rarely 3-foliolate, leathery, evergreen.
   2. Leaves broadly ovate, entire; flowers in dense spikes; fruit red; California and Arizona . . . . *R. ovata* S. Wats., Sugar Sumac
   2. Leaves oval to ovate, rounded at apex.
      3. Fruit white in branched panicles; southern California . . . . *R. laurina* Nutt., Laurel Sumac
      3. Fruit red.
         4. Leaves toothed or entire; southern coastal California . . . . *R. integrifolia* (Nutt.) B. & H., Mahogany Sumac
         4. Leaves entire; Yuma County, Arizona . . . . *R. kearneyi* Bark., Kearney Sumac
1. Leaves pinnately compound, deciduous.
   5. Flowers in dense, terminal panicles; fruit red, hairy.
      6. Twigs and petioles velvety-hairy; leaflets serrate . . . . *R. typhina* Torn., Staghorn Sumac
      6. Twigs and petioles glabrous or hairy.
         7. Rachis terete; twigs glabrous and glaucous; leaflets serrate; most of North America . . . . *R. glabra* L., Smooth Sumac
         7. Rachis winged; twigs pubescent; leaflets nearly entire.
            8. Leaflets ovate-lanceolate; New Hampshire to Michigan and Kansas, south to Georgia and Texas . . . . . . *R. copallina* L., Shining Sumac
            8. Leaflets narrow-lanceolate, often falcate; south Oklahoma to Texas and New Mexico . . . . . . *R. lanceolata* Britt., Prairie Shining Sumac
   5. Flowers not numerous in terminal or axillary panicles.
      9. Fruit red; rachis winged; leaflets less than 1 inch long; west Texas to Arizona . . . . *R. microphylla* Engelm., Littleleaf Sumac
      9. Fruit white; rachis terete; leaflets 1/2-2 1/2 inches long, entire; twigs yellow-brown, mottled; Quebec to Minnesota, south to Gulf . . . . *R.* (or *T.*) *vernix* (L.) Ktze., Poison Sumac

American Holly

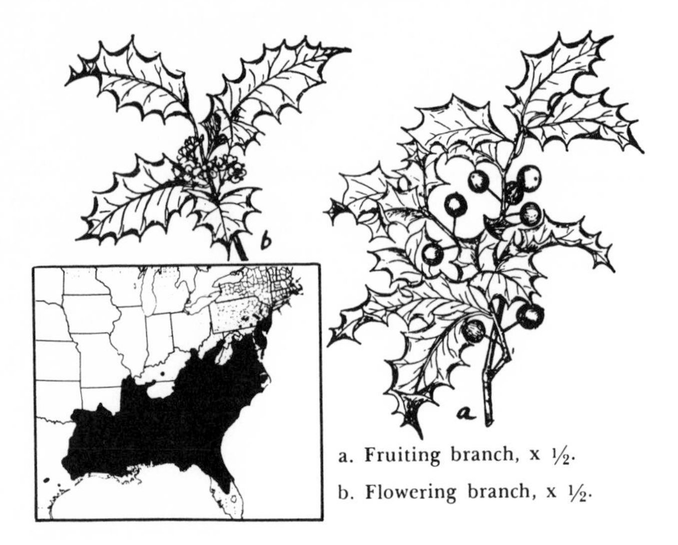

a. Fruiting branch, x ½.

b. Flowering branch, x ½.

Mountain Winterberry

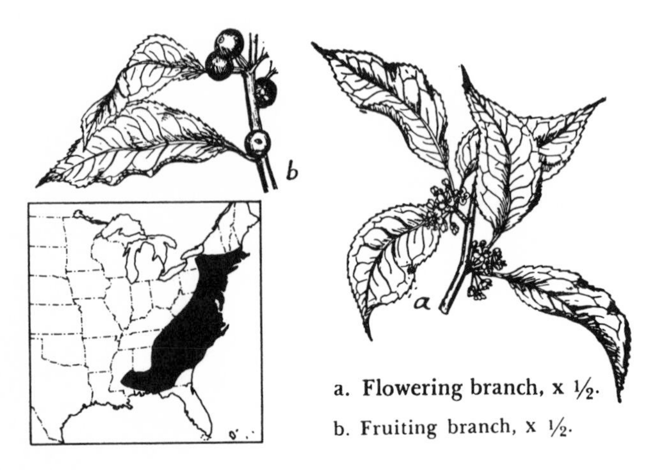

a. Flowering branch, x ½.

b. Fruiting branch, x ½.

AQUIFOLIACEAE

American Holly

Ilex opaca Ait.

HABIT. The largest native holly forming a tree 40-60 feet high and 1-2 feet in diameter (max. 100 by 3 1/2 feet); narrow, dense crown.

LEAVES. Alternate; simple; persistent 3 years; elliptical; 2-4 inches long; spiny-toothed or sometimes entire; with a spiny tip; leathery; yellow-green above; paler or yellow below.

FLOWERS. Regular; dioecious; small; solitary or in short cymes; green-white; appearing with the leaves.

FRUIT. A berrylike drupe persisting into the winter; bright red; 1/4 inch long; bony nutlet in each of 3-8 cells of ovary.

TWIGS. Slender; glabrous; green with single bundle trace. Winter buds: terminal present, scaly, 1/8-1/4 inch long, ciliate.

BARK. Thin; smooth; gray-white; with warty growths.

WOOD. Unimportant; hard and heavy; white; fine-textured; diffuse-porous; used for turning, novelties, finish.

SILVICAL CHARACTERS. Tolerant; slow-growing; reproduction sparse; commonly cultivated; rare near cities because of Christmas demand.

* * *

About 14 species of *Ilex* are native to the eastern or southeastern parts of the United States, six of these commonly forming trees, and keyed out below. Most of the others, while typically shrubs up to 6 feet high, have been reported as rarely reaching tree size. The English holly, *I. aquifolium* L., is grown in plantations for decorations.

1. Leaves persistent, thick and leathery; pedicels with bractlets.
    2. Leaves with spiny teeth or a spiny tip.
        3. Flowers and fruits usually clustered on last year's twigs; cultivated in warm areas . . . . *I. aquifolium* L., English Holly
        3. Flowers and fruits usually single on current growth.
            4. Leaves less than twice as long as broad . . . . *I. opaca* Ait., American Holly
            4. Leaves more than twice as long as broad; coastal plain, Virginia to Louisiana . . . . *I. cassine* L., Dahoon
    2. Leaves without spiny teeth or tip, entire to serrate.
        5. Fruit red; coastal plain, Virginia to Texas.
            6. Leaves oblanceolate to oblong-obovate, entire . . *I. cassine* L., Dahoon
            6. Leaves oval with few crenate teeth . . . . . . *I. vomitoria* Ait., Yaupon
        5. Fruit brown-purple; southern Florida . . . . *I. krugiana* Loes., Tawnyberry Holly
1. Leaves deciduous, not leathery; pedicels without bractlets.
    7. Leaves sharply toothed; apex pointed; Massachusetts and New York, south to Florida and Louisiana . . . . . . *I. montana* T. & G., Mountain Winterberry
    7. Leaves with few rounded teeth; apex obtuse; Maryland to Kansas and south . . . . *I. decidua* Walt., Possumhaw

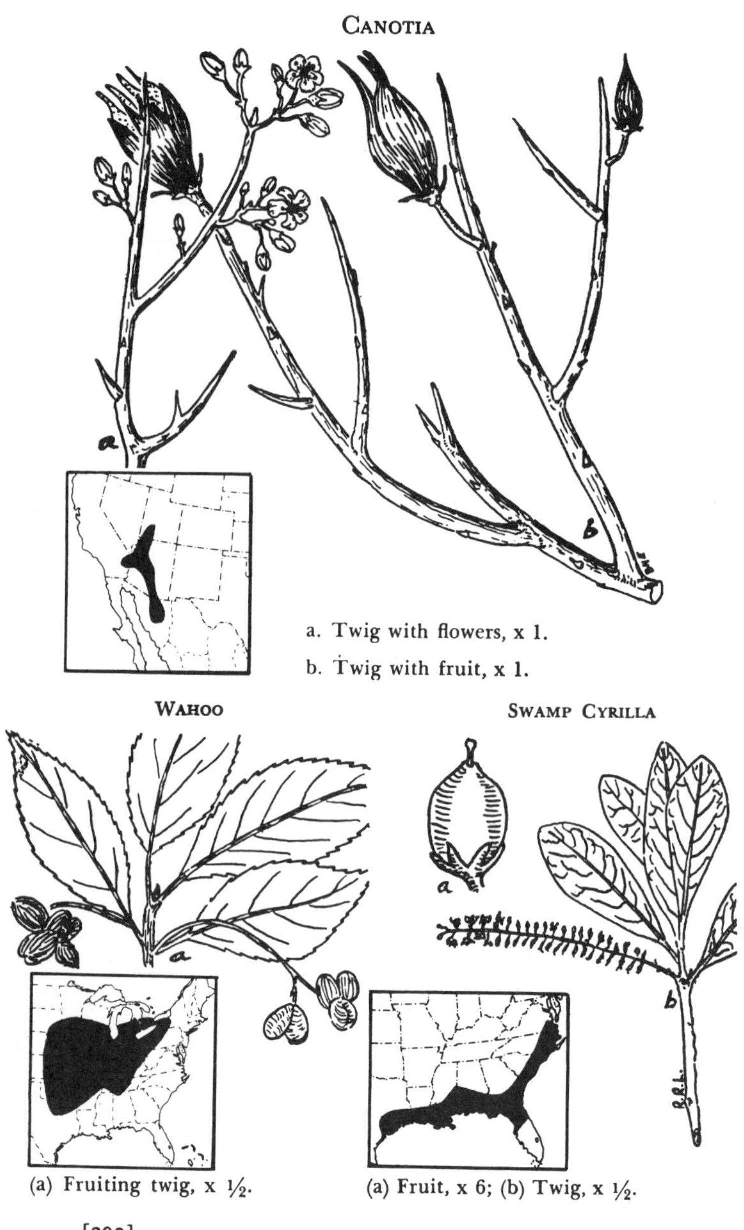

Canotia

a. Twig with flowers, x 1.

b. Twig with fruit, x 1.

Wahoo

(a) Fruiting twig, x ½.

Swamp Cyrilla

(a) Fruit, x 6; (b) Twig, x ½.

CELASTRACEAE

Canotia. Mohavethorn

Canotia holacantha Torr.

HABIT. An odd, leafless shrub or small shrublike tree 20-30 feet high; trunk short and stocky; branches rushlike.

FLOWERS. Regular; perfect; small; in 3-7-flowered fascicles; calyx 5-lobed, persistent, minute; corolla 5-petaled, white.

FRUIT. Dry, woody, ovoid, acuminate capsule 1 inch long; crowned with subulate, persistent style; 5-valved, splitting open at top. Seed: solitary or paired; about 3/4 inch long.

TWIGS. Slender; rushlike; round; alternate; glabrous; rigid; spine-tipped; pale green and carrying on photosynthetic functions; characteristic black, triangular, cushionlike processes located at base of each twig and flower cluster.

BARK. Light brown; deeply furrowed.

WOOD. Heavy; hard; close-grained; light brown.

SILVICAL CHARACTERS. Intolerant; monotypic genus; on dry mountain slopes and mesas between 2,000-4,000 feet.

* * *

Wahoo. Burning Bush

Euonymus atropurpureus Jacq.

This widespread shrubby species reaches tree size in Arkansas and eastern Texas. It can be distinguished by its winged twigs; simple, opposite, petioled leaves; 4-parted flowers; and fleshy, capsular fruit enclosed in a thin scarlet aril. The shrubby western wahoo, *E. occidentalis* Nutt., is found in the Pacific Coast states.

* * *

CYRILLACEAE

Swamp Cyrilla

Cyrilla racemiflora L.

This small tree of wet sites in the coastal plain from Virginia to Texas is characterized by simple, alternate, oblong, entire, thick leaves 2-3 inches long; small, regular, perfect, white flowers in slender axillary racemes; 2-celled, ovoid, unwinged capsule 1/8 inch long. *C. arida* Small is a small tree in south-central Florida.

The buckwheat-tree, *Cliftonia monophylla* (Lam.) Britt., is a similar species, ranging from South Carolina to Louisiana. It differs, having glandular-punctate leaves; flowers in terminal racemes; 3-4-celled capsule 1/4 inch long, 2-4 wings.

ACERACEAE

The Maples

Characteristics of the Genus *Acer* L.

HABIT. Deciduous trees or shrubs, with handsome foliage, usually assuming brilliant colors in autumn.

LEAVES. Opposite; simple or compound; deciduous; long-petioled; without stipules; simple leaves palmately 3-7-lobed; compound leaves pinnate, 3-7 leaflets.

FLOWERS. Regular; polygamous, dioecious, or rarely perfect; small; borne either in lateral fascicles from separate flower buds and appearing before the leaves, or in lateral and terminal racemes, panicles, or corymbs and appearing with or after the leaves; calyx colored, generally 5-parted; corolla usually 5-petaled or absent; stamens 4-12, usually 7-8; ovary 2-celled, 2-lobed, compressed, 2 styles; ovules 2 in each cell, ascending.

FRUIT. Double samara united at base (key); each nutlike carpel laterally compressed and produced into large, obovate wing. Seed: usually solitary by abortion; ovoid; compressed.

TWIGS. Slender to moderately stout; round; pith homogeneous, round; marked at base by bud scales with ringlike scars; leaf scars more or less U-shaped, with 3 (rarely 5-7) bundle scars. Winter buds: valvate or imbricated scales; inner scales accrescent; terminal buds larger than lateral.

BARK. Astringent and variable.

WOOD. Variable from soft to heavy and hard; diffuse-porous; pores all small and not crowded; rays distinct on cross section without lens; widely used for interior finish, etc.; sap of some species manufactured into sugar.

SILVICAL CHARACTERS. Mostly tolerant; fibrous root systems; widely used for ornamental and shade trees.

GENERAL. This genus contains over 100 species widely scattered through the Northern Hemisphere with one extending into Sumatra and Java. In the United States there are 13 native species. Several introduced species are planted extensively. Norway and sycamore maples are very common and better known than many native species (both included in the key). The distinctive Japanese maple, *A. palmatum* Thunb., has small, deeply 5-9-lobed leaves.

KEY TO THE SPECIES OF MAPLES

1. Leaves pinnately compound with 3-7 leaflets; twigs stout, smooth, green, with encircling leaf scars . . . . *A. negundo*, Boxelder, p. 307
1. Leaves simple, palmately lobed (rarely compound in Rocky Mountain Maple).
  2. Leaves with closely, often doubly, toothed margins and sharp-angled sinuses between lobes . . . . SOFT MAPLES
    3. Winter buds bright green; leaves 4-7 inches across, 5-lobed, bluntly toothed; fruit wings 1 1/2-2 inches long, diverging at right angle or less; Eurasian ornamental . . . . *A. pseudoplatanus* L., Sycamore Maple
    3. Winter buds red to brownish; native.
      4. Flowers appearing before the leaves from lateral, leafless buds; fruits maturing in late spring; buds with 4-8 visible scales; upper trunk bark smooth, light gray; eastern.
        5. Leaves very deeply 5-lobed, sides of middle lobe diverging, silvery-white beneath; young fruit woolly, wings 1 1/2-2 inches long, widely diverging, petals absent; twigs with rank odor . . . . *A. saccharinum*, Silver Maple, p. 299
        5. Leaves 3-5-lobed, sides of middle lobe usually converging toward apex, pale green beneath; fruit smooth, wings 1/2-1 inch long, slightly diverging; petals present; twigs no rank odor . . . . *A. rubrum*, Red Maple, p. 299
      4. Flowers with petals, appearing with or after the leaves from leaf buds; fruit maturing in late summer; 2 (rarely 4) visible bud scales.
        6. Leaves essentially glabrous; flowers in corymbs; western.
          7. Leaves 7-9-lobed; fruit wings widely divergent; Pacific Coast . . . . *A. circinatum*, Vine Maple, p. 301
          7. Leaves 3-lobed (rarely 3-parted); fruit wings not divergent; through West . . . . . . *A. glabrum*, Rocky Mountain Maple, p. 301
        6. Leaves hairy beneath; flowers in racemes; eastern.
          8. Leaves coarsely toothed, 3-5-lobed; fruit wings 1/2 inch long, diverging at about right angles; bark red-brown . . . . *A. spicatum*, Mountain Maple, p. 303
          8. Leaves finely doubly toothed, 3-lobed al apex; fruit wings 3/4 inch long, widely diverging; bark green, white stripes . . . . *A. pensylvanicum*, Striped Maple, p. 303
  2. Leaf lobes entire or with few remote teeth, or slightly lobed; sinuses rounded between lobes; buds, several scales; flowers, with leaves . . HARD MAPLES
    9. Petioles red with milky juice; fruit wings 1 1/2-2 inches long, widely diverging; leaves 5-7-lobed, 5-7 inches across; naturalized Eurasian species. . . . . *A. platanoides*, L., Norway Maple
    9. Petioles with watery juice; fruit wings usually diverging at right angles or less; leaves 3-5-lobed; native.
      10. Buds red, scales hairy on margins; western.
        11. Leaves 8-12 in. across. . *A. macrophyllum*, Bigleaf Maple, p. 305
        11. Leaves 2-5 in. across . *A. grandidentatum*, Bigtooth Maple, p. 305
      10. Buds brown; eastern.
        12. Leaves 1 1/2-3 inches across, hairy beneath; bark smooth, pale; Virginia to Missouri and south.
          13. Leaves pale beneath; fruit wings diverging at right angles . . . . *A. barbatum* Michx., Florida Maple
          13. Leaves green; fruit wings diverging at more than right angles . . . . *A. leucoderme* Small, Chalk Maple
        12. Leaves 3-6 inches across; bark rough and dark; eastern.
          14. Twigs red-brown, large warty lenticels; buds light brown; leaves 5-lobed, glabrous beneath . . . . *A. saccharum*, Sugar Maple, p. 297
          14. Twigs orange, small lenticels; buds dark; leaves mostly 3-lobed, usually hairy beneath . . *A. nigrum*, Black Maple, p. 297

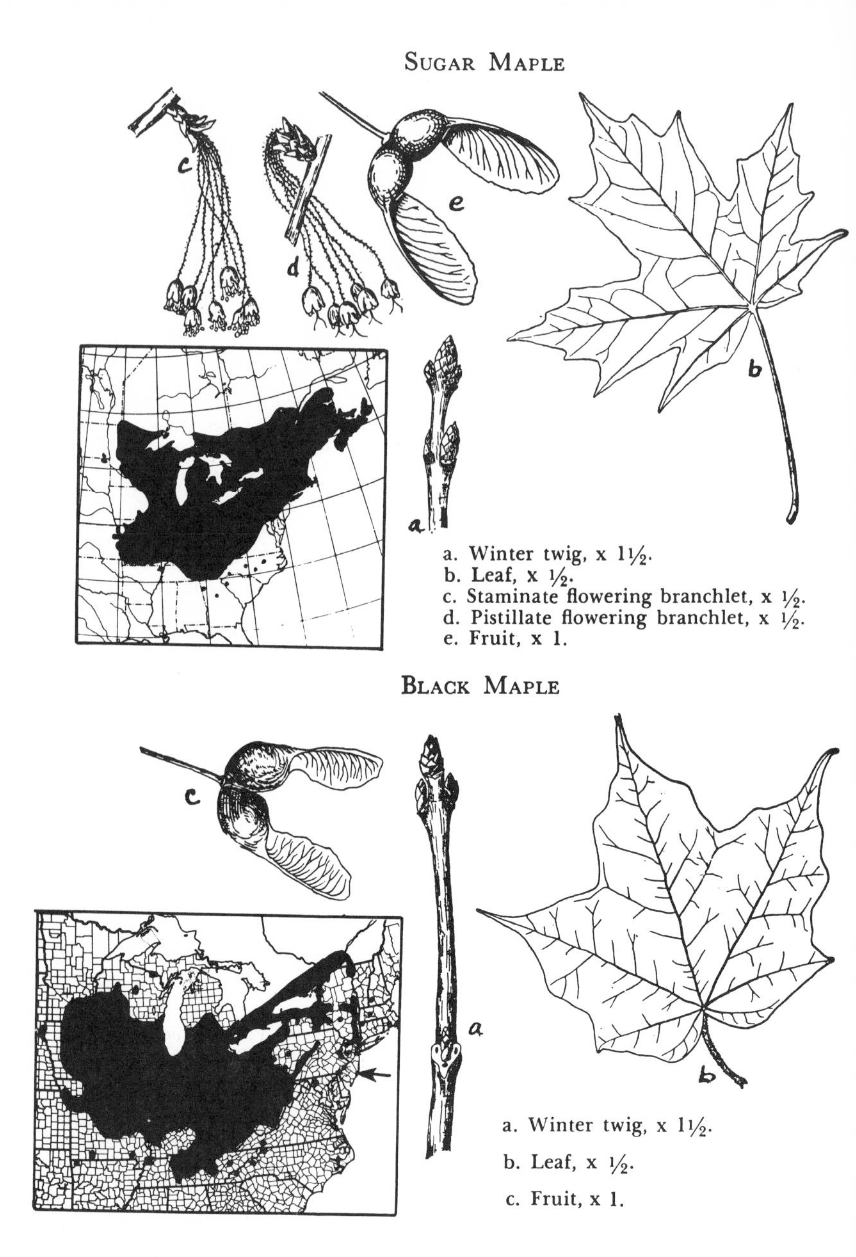

Sugar Maple

a. Winter twig, x 1½.
b. Leaf, x ½.
c. Staminate flowering branchlet, x ½.
d. Pistillate flowering branchlet, x ½.
e. Fruit, x 1.

Black Maple

a. Winter twig, x 1½.

b. Leaf, x ½.

c. Fruit, x 1.

Sugar Maple

Acer saccharum Marsh. (*Acer saccharophorum* K. Koch)

HABIT. A handsome tree 60-80 feet high and 2-3 feet in diameter (max. 135 by 5 1/2 feet); dense, broad, rounded crown.

LEAVES. Simple; orbicular; palmately 5-lobed (rarely 3-lobed); 3-5 inches long; lobe margins entire or sparingly wavy-toothed with rounded sinuses; thin and firm; glabrous; bright green above; paler below; turning bright red or yellow in autumn.

FLOWERS. Polygamous; in crowded corymbs; apetalous; yellow-green; appearing with the leaves.

FRUIT. Key U-shaped, with nearly parallel wings about 1 inch long; red-brown; glabrous; maturing in autumn.

TWIGS. Slender; smooth; red-brown; shiny. Winter buds: terminal 1/4-3/8 inch long, acute, red-brown, nearly glabrous, 4-8 pairs of visible scales.

BARK. Smooth and gray on young stems; becoming dark gray, thick, and deeply furrowed; often with long, scaly plates.

WOOD. Very important; hard; heavy; strong; flooring, furniture, etc.

SILVICAL CHARACTERS. Tolerant; rather slow-growing; long-lived; wide-spreading, lateral roots; important as an ornamental and as a source of maple syrup and sugar. The forms designated *A. barbatum* and *A. leucoderme* in the key are closely related and possibly should be considered varieties of *A. saccharum.*

* * *

Black Maple

Acer nigrum Michx.

This species very closely resembles the sugar maple and by many authors is considered to be a variety. It is distinguished from the sugar maple by the following characters: leaves mostly 3-lobed and lower leaf surface and petioles downy, leaf blades thicker and drooping at sides; twigs orange-brown; buds hoary-pubescent; bark often nearly black and more deeply and sharply furrowed.

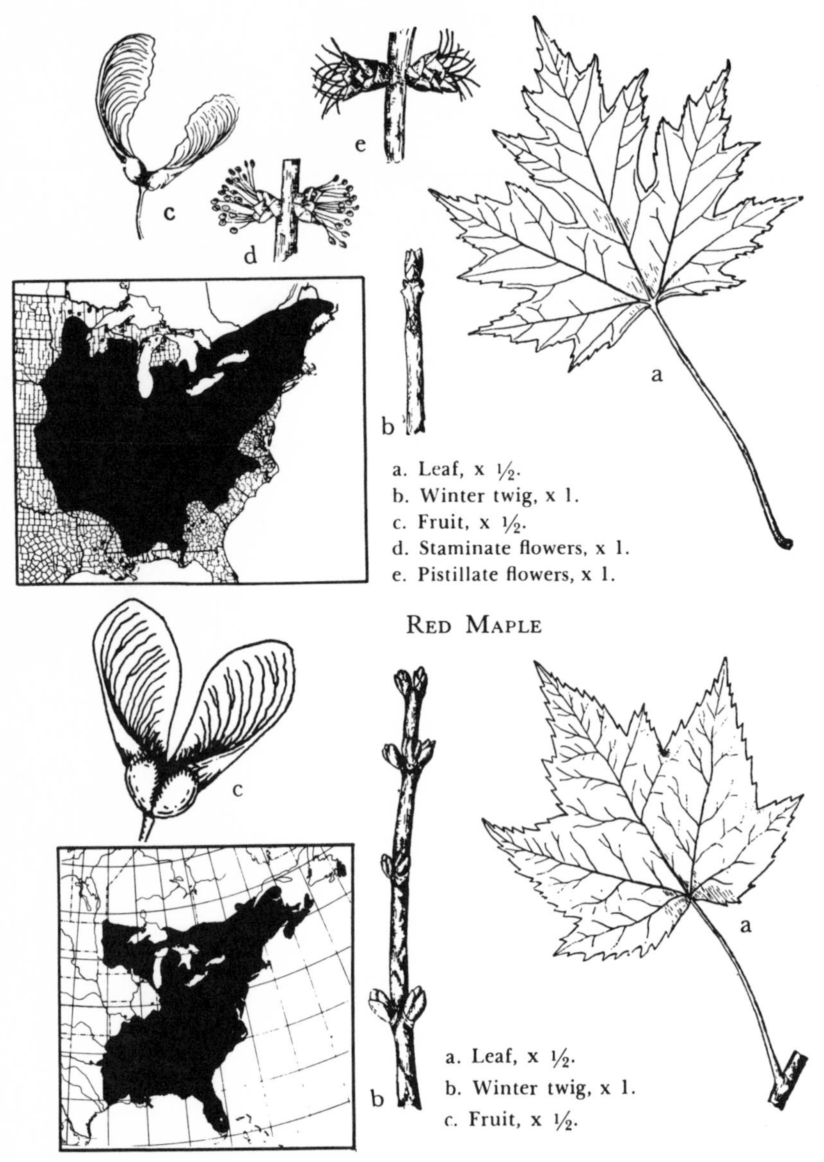

SILVER MAPLE

a. Leaf, x ½.
b. Winter twig, x 1.
c. Fruit, x ½.
d. Staminate flowers, x 1.
e. Pistillate flowers, x 1.

RED MAPLE

a. Leaf, x ½.
b. Winter twig, x 1.
c. Fruit, x ½.

Silver Maple

Acer saccharinum L.

HABIT. A handsome tree 60-80 feet high and 2-3 feet in diameter (max. 120 by 7 1/2 feet); trunk often separating near ground into several upright branches; crown broad, rounded.

LEAVES. Simple; orbicular; deeply 5-lobed, side of terminal lobe diverging toward apex; sinuses sharp-angled; 4-7 inches long; lobe margins sharply toothed; light green above, glabrous; silvery white below; turning pale yellow in autumn.

FLOWERS. Polygamous; in crowded fascicles; apetalous; red to yellow-green; appearing before the leaves.

FRUIT. Key with widely divergent wings 1 1/2-2 inches long; greenish; glabrous; one samara often aborted; maturing in late spring and germinating immediately.

TWIGS. Slender; dark red; lustrous; with fetid odor when bruised. Winter buds: terminal 1/8-1/4 inch long, blunt, dark red, 2-4 pairs of visible scales with ciliate margins.

BARK. Smooth and light gray on young stems; on old trunks becoming separated into scaly plates by narrow fissures.

WOOD. Moderately important; less heavy and hard than sugar maple and used as a substitute.

SILVICAL CHARACTERS. Tolerant; fast-growing; rather short-lived; moist sites; wide-spreading lateral roots; a common ornamental.

* * *

Red Maple

Acer rubrum L.

This species closely resembles silver maple and differs in the following characters:

LEAVES. Palmately 3-lobed (rarely 5-lobed) with short, broad lobes; sides of terminal lobe converging toward apex; sinuses sharp-angled; 2-6 inches long; lobe margins toothed; glabrous; light green above, paler and glaucous below; turning scarlet in autumn.

FLOWERS. Polygamous; petals present; reddish.

FRUIT. Key with slightly divergent wings 1/2-1 inch long.

TWIGS. Without fetid odor when bruised.

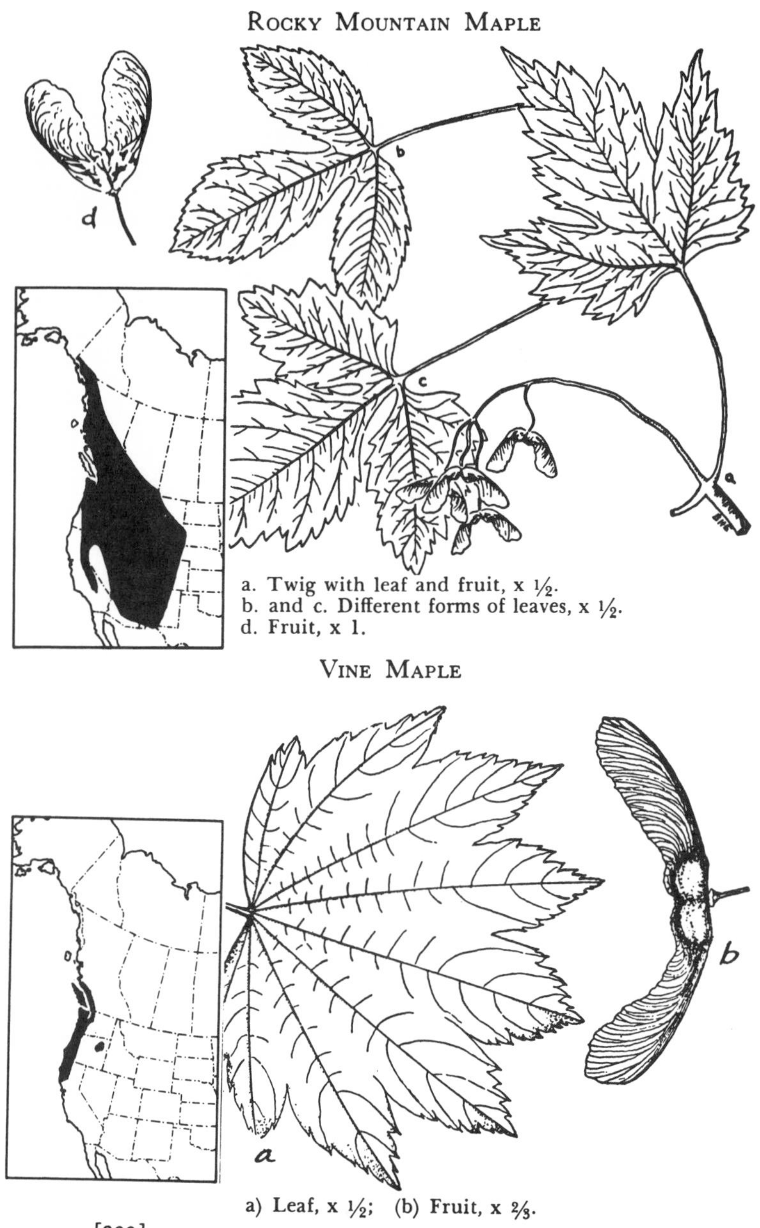

Rocky Mountain Maple

a. Twig with leaf and fruit, x ½.
b. and c. Different forms of leaves, x ½.
d. Fruit, x 1.

Vine Maple

a) Leaf, x ½; (b) Fruit, x ⅔.

ACERACEAE

Rocky Mountain Maple. Dwarf Maple

Acer glabrum Torr.

HABIT. A shrub or small tree rarely 20-30 feet high and 6-12 inches in diameter; narrow crown with small branches.

LEAVES. Mostly 3-5-lobed and simple, but sometimes divided into 3 leaflets; 3-5 inches long; sharply and doubly serrate; glabrous; thin; dark green above; paler below.

FLOWERS. Mostly dioecious; in loose, racemose corymbs; petals yellow-green; appearing after the leaves.

FRUIT. Key with slightly spreading or nearly erect wings; 3/4-7/8 inch long; glabrous; often rose-colored; maturing in summer.

TWIGS. Slender; glabrous; becoming bright red-brown. Winter buds: acute; 1/8-1/4 inch long; bright red.

BARK. Thin; smooth; dark red-brown.

WOOD. Unimportant; heavy; hard; close-grained.

SILVICAL CHARACTERS. Rather tolerant; in moist locations; the northwestern form (var. *douglasii* Dibb.) has shallower leaf sinuses.

* * *

Vine Maple

Acer circinatum Pursh

HABIT. A shrub or small tree rarely 35 feet high; often vine-like or prostrate, forming dense thickets; irregular crown.

LEAVES. Simple; orbicular; 7-11-lobed; 2-6 inches long; lobes sharply doubly toothed; glabrous; light green above; paler below; turning red or orange in autumn.

FLOWERS. Polygamo-monoecious; in loose corymbs; petals greenish-white, shorter than red calyx; appearing with the leaves.

FRUIT. Key with widely divergent wings 1 1/2 inches long; glabrous; red; maturing in late autumn.

TWIGS. Slender; glabrous; green to red-brown. Winter buds: terminal 1/8 inch long, obtuse, bright red.

BARK. Thin; smooth or with shallow fissures; bright red-brown.

WOOD. Unimportant; heavy and hard; used for handles and fuel.

SILVICAL CHARACTERS. Very tolerant; moist stream banks.

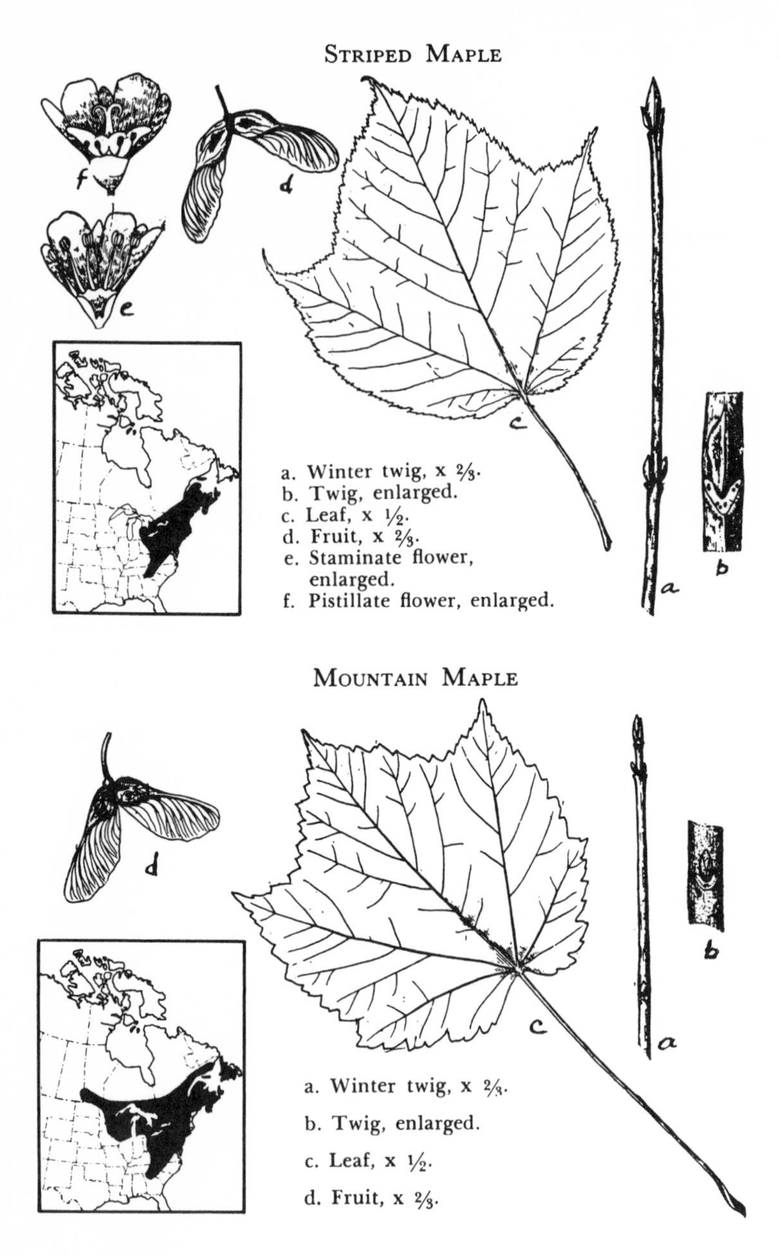
STRIPED MAPLE
f
d
e
c
a
b
a. Winter twig, x 2/3.
b. Twig, enlarged.
c. Leaf, x 1/2.
d. Fruit, x 2/3.
e. Staminate flower, enlarged.
f. Pistillate flower, enlarged.
MOUNTAIN MAPLE
d
c
a
b
a. Winter twig, x 2/3.
b. Twig, enlarged.
c. Leaf, x 1/2.
d. Fruit, x 2/3.

ACERACEAE

Striped Maple. Moosewood

Acer pensylvanicum L.

HABIT. A shrub or small tree rarely 25 feet high and 8 inches in diameter; striped, upright branches with compact crown.

LEAVES. Simple; orbicular; shallowly 3-lobed above the middle; 5-6 inches long; sharply doubly toothed; glabrous; yellow-green above; paler below; turning pale yellow in autumn.

FLOWERS. Usually monoecious; in pendulous racemes; petals bright yellow; appearing after the leaves.

FRUIT. Key with widely divergent wings 3/4 inch long; glabrous; red-brown; maturing in autumn.

TWIGS. Slender; smooth; red to green-brown; mottled. Winter buds: terminal 1/2 inch long, bright red, short-stalked.

BARK. Thin; red-brown to bright green; marked by vertical, broad, white stripes.

WOOD. Unimportant; rather light and soft; pinkish brown.

SILVICAL CHARACTERS. Tolerant; cool, moist, shaded sites.

* * *

Mountain Maple

Acer spicatum Lam.

HABIT. A shrub or small tree rarely 35 feet high and 8 inches in diameter; upright branches, forming a compact, rounded crown.

LEAVES. Simple; orbicular; shallowly 3-lobed (rarely 5-lobed) above the middle; 3-5 inches long; coarsely toothed; dark green and glabrous above; white-downy below; turning red to orange in autumn.

FLOWERS. Polygamous; in erect, terminal racemes; petals yellow-green; appearing after the leaves.

FRUIT. Key with widely divergent wings 1/2-1 inch long; glabrous; bright red; maturing in autumn.

TWIGS. Slender; slightly hairy; red to brown. Winter buds: terminal 1/8-1/4 inch long, bright red, more or less hairy.

BARK. Very thin; red-brown; smooth or slightly furrowed.

WOOD. Unimportant; rather light and soft; light brown.

SILVICAL CHARACTERS. Tolerant; cool, moist, shaded sites but less demanding of shade than striped maple; a common ornamental.

Bigleaf Maple

(a) Leaf, x 1/5; (b) Fruit, x 1.

Bigtooth Maple

(a) Leaf, x 1; (b) Fruit, x 1.

ACERACEAE

Bigleaf Maple

Acer macrophyllum Pursh

HABIT. A medium-sized tree often 80-100 feet high and 3-4 feet in diameter (max. 9 feet); narrow, compact crown.

LEAVES. Simple; orbicular; palmately 5-lobed; 8-12 inches long; lobe margins entire or sparingly wavy-toothed; glabrous; bright green above; paler below; turning bright orange in autumn.

FLOWERS. Polygamous; in pendulous racemes; petals present, bright yellow, 1/4 inch long; appearing with the leaves.

FRUIT. Key with slightly diverging wings 1 1/4-2 inches long; wings densely hairy over seed cavity, nearly glabrous elsewhere; maturing in autumn.

TWIGS. Stout; smooth; red to green-brown. Winter buds: terminal 1/4 inch long; blunt; 3-4 pairs of red-green, hairy-margined scales.

BARK. Smooth and gray-brown on young stems; becoming red-brown, deeply fissured, and broken into square scales on the surface.

WOOD. One of few important hardwoods on West Coast; medium hard and strong; veneer, furniture, flooring, etc.

SILVICAL CHARACTERS. Tolerant; moist sites; fast-growing; wide, shallow roots; vigorous reproduction; a common ornamental.

* * *

Bigtooth Maple

Acer grandidentatum Nutt.

HABIT. A small tree 30-40 feet high and 8-10 inches in diameter.

LEAVES. Simple; orbicular; 3-5-lobed; 2-5 inches in diameter; lobe margins entire or lobulate; dark green and lustrous above; paler and hairy below; turning red or yellow before falling.

FLOWERS. Polygamous; calyx yellow; petals absent.

FRUIT. Key with slightly spreading wings 1/2-1 inch long; glabrous or sparingly hairy; green at maturity in autumn.

TWIGS. Slender; glabrous; bright red. Terminal bud red, 1/16 inch long.

BARK. Thin; dark brown; with platelike scales.

SILVICAL CHARACTERS. Rather tolerant; moist sites; shallow roots.

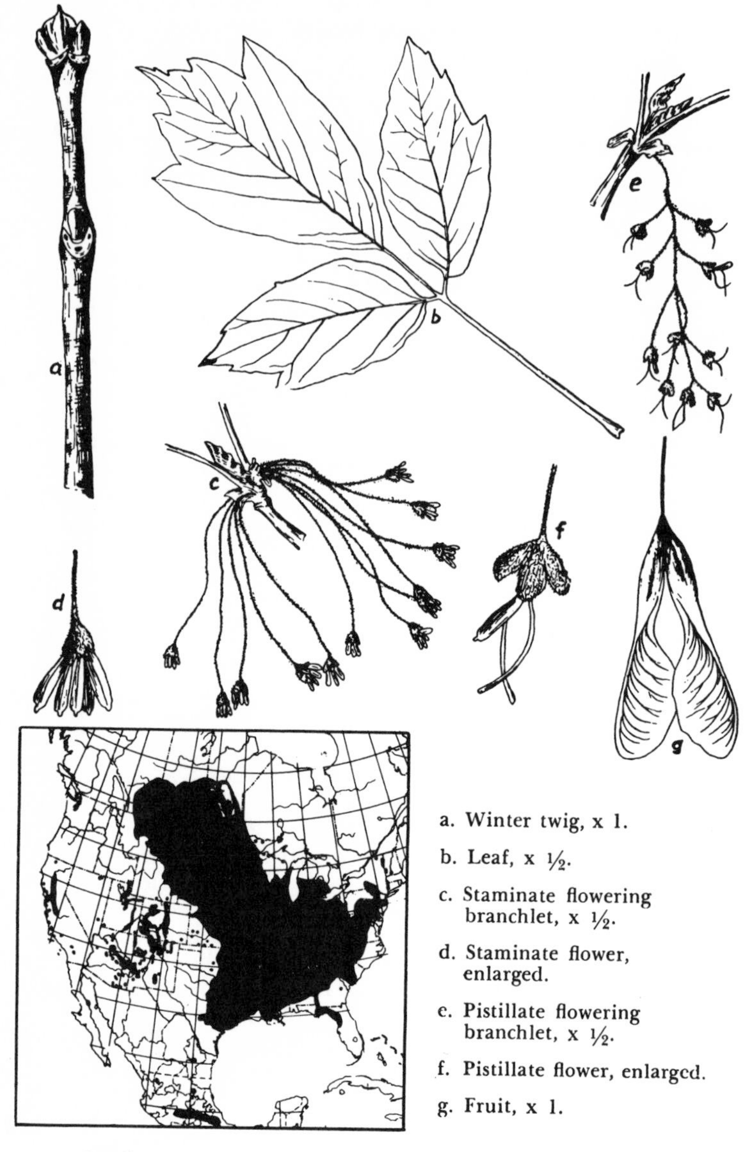

a. Winter twig, x 1.

b. Leaf, x ½.

c. Staminate flowering branchlet, x ½.

d. Staminate flower, enlarged.

e. Pistillate flowering branchlet, x ½.

f. Pistillate flower, enlarged.

g. Fruit, x 1.

ACERACEAE

Boxelder

Acer negundo L.

HABIT. A small tree, rarely 75 feet high and 4 feet in diameter; trunk usually irregular and dividing near ground into several stout, wide-spreading branches.

LEAVES. Compound; 3 (rarely 5-7) leaflets; petiole long, slender, puberulous; leaflets ovate to lanceolate; 3-4 inches long, and 1 1/2-4 inches wide; glabrous; acuminate; coarsely serrate and sometimes 3-lobed at base.

FLOWERS. Dioecious; minute; the male fascicled; the female in drooping racemes; appearing with leaves or a little before them; calyx 5-lobed, hairy, yellow-green; corolla absent; stamens 4-6 with slender, exserted, hairy filaments; ovary hairy, partly enclosed by calyx.

FRUIT. Pendent; 1-2 inches long; glabrous; ripening in autumn; in drooping racemes 6-8 inches long. Seed: narrowed at the ends; smooth; bright red-brown; 1/2 inch long.

TWIGS. Moderately stout; greenish; pubescent, or rarely nearly glabrous; marked by conspicuous bud scale scars and crescent-shaped leaf scars that surround the twig. Winter buds: acute; 1/8-1/4 inch long; reddish; tomentose.

BARK. Thin (1/4-1/2 inch); pale gray or light brown; deeply divided by furrows into broad, rounded ridges.

WOOD. Light; soft; close-grained; weak; heartwood cream-white to yellow-brown, often streaked; sapwood thick; used occasionally for cheap furniture, woodenware, etc.

SILVICAL CHARACTERS. Upper Sonoran and transition zones; moderately tolerant; shallow-rooted, except on deep soils; hardy to extremes of climate; rapid grower but short-lived and usually of poor form; reproduction by sprout and seed plentiful.

GENERAL. Five varieties are recognized: *arizonicum* Sarg. in Arizona and New Mexico, with serrate leaves and glabrous leaves, twigs, and petioles; *californicum* Sarg. in California, with coarsely serrate or nearly entire leaflets with matted hairs on the lower surface; *interius* Sarg. through the West, with hairy twigs and petioles; *texanum* Pax through the Southeast, with woolly twigs and lower leaf surface; and *violaceum* J. & B. through the northern half of the United States, with bluish glabrous twigs and entire or dentate leaves hairy on the lower surface. The distribution of all varieties is included on the map.

HIPPOCASTANACEAE

The Buckeyes and Horsechestnuts
Characteristics of the Genus *Aesculus* L.

HABIT. Handsome shrubs to medium-sized trees with dense, rounded crowns; commonly planted as ornamentals for their flowers or foliage.

LEAVES. Opposite; palmately compound; 5-9 serrate leaflets arising at end of long stem; deciduous; stipules absent.

FLOWERS. Regular; polygamo-monoecious; bell-shaped; many-flowered, showy, upright panicles; yellow, red, or white; flowers near base of panicles perfect and fertile; calyx 5-lobed (rarely 2-lobed); petals 4-5; stamens 5-8, inserted on the disk; ovules 3-valved, 2 in each cell; appearing after the leaves.

FRUIT. A large, leathery, smooth to spiny capsule, containing one or more large, lustrous, brown, leathery-coated seeds marked by a large light-colored hilum, which gives the name buckeye to this genus.

TWIGS. Stout; round; marked by conspicuous triangular leaf scars with more than 3 bundle scars; pith large, circular, light-colored. Winter buds: terminal present, large, scaly, resin-covered in some species.

BARK. Intermediate in thickness; fissured and scaly; bitter.

WOOD. Moderately important; fine-textured; diffuse-porous; creamy white; used for containers, novelties, furniture, etc.

SILVICAL CHARACTERS. Moderately tolerant; varied sites; fast-growing; short-lived; in mixed hardwood stands.

GENERAL. This genus contains about 20 species scattered over the Northern Hemisphere, with 7 species native to the United States; the European horsechestnut, *A. hippocastanum* L., is widely planted and has escaped from cultivation in the northern states; the red buckeye, *A.* X *carnea* Hayne, is a hybrid of *A. hippocastanum* and *A. pavia* and often planted for its red flowers.

* * *

TAMARICACEAE

Tamarix. Tamarisk
Tamarix pentandra Pall.

A large shrub or small tree introduced from Europe but more or less naturalized in the south and central parts of the West. Characterized by alternate, scalelike, feathery, sparse leaves (resembling those of conifers); showy, pink flowers; and 3-5-valved, capsular fruit containing many minute seeds.

KEY TO THE SPECIES OF BUCKEYE

1. Winter buds thickly coated with resin; flowers white to pink (often red in *A.* X *carnea*).
   2. Leaflets usually 7 (rarely 5), sessile or nearly so; fruit 2-2 1/2 inches long, with short spines; widely planted and naturalized . . . . *Aesculus hippocastanum* L., Horsechestnut
   2. Leaflets usually 5, leaflets on stalks 1/2-1 inch long; fruit 2-3 inches long, smooth; California . . . . *A. californica* (Spach) Nutt., California Buckeye
1. Winter buds not resinous, or slightly so; flowers yellow to scarlet; leaflets stalked; eastern.
   3. Fruit spiny, 1-2 inches long; flowers yellow-green; bud scales prominently keeled.
      4. Leaflets 5 (rarely 7); small to medium-sized tree; Pennsylvania and Michigan to Nebraska and south . . *A. glabra* Willd., Ohio Buckeye, p. 311
      4. Leaflets 7-9; shrub or rarely small tree; southern Oklahoma and east Texas . . . . *A. arguta* Buckl., Texas Buckeye
   3. Fruit nearly smooth, not spiny; bud scales not prominently keeled.
      5. Leaflets 5-7; flowers yellow, glandular-hairy calyx; fruit 2-3 inches long; large tree; Pennsylvania to Illinois and south . . . . *A. octandra* Marsh., Yellow Buckeye, p. 311
      5. Leaflets 5; flowers red to red-yellow, calyx not glandular-hairy; fruit 1-2 inches long; shrubs or small trees.
         6. Petals hairless but glandular on margin; Virginia to Kentucky and south . . . . *A. pavia* L., Red Buckeye
         6. Petals hairy but not glandular on margin; North Carolina to Florida, Alabama . . . . *A. sylvatica* Bartr., Painted Buckeye

* * *

ANACARDIACEAE

American Smoketree

Cotinus obovatus Raf.

This handsome shrub or small tree ranges from Tennessee to Missouri and Oklahoma and south. It is characterized by simple, alternate, deciduous leaves 4-6 inches long, oval to obovate, entire, and dark green; dioecious, clustered flowers, many being abortive and hairy giving the tree its name; and a dry, compressed, oblong, brown, drupe about 1/8 inch long borne in loose panicles on stems 1-3 inches long.

* * *

Texas Pistache

Pistacia texana Swingle

This shrub or small tree extends to south Texas from Mexico. Characterized by alternate, persistent, pinnately compound leaves with 9-19 spatulate leaflets; small dioecious flowers without petals or calyx; oval, red-brown drupe 1/4 inch long.

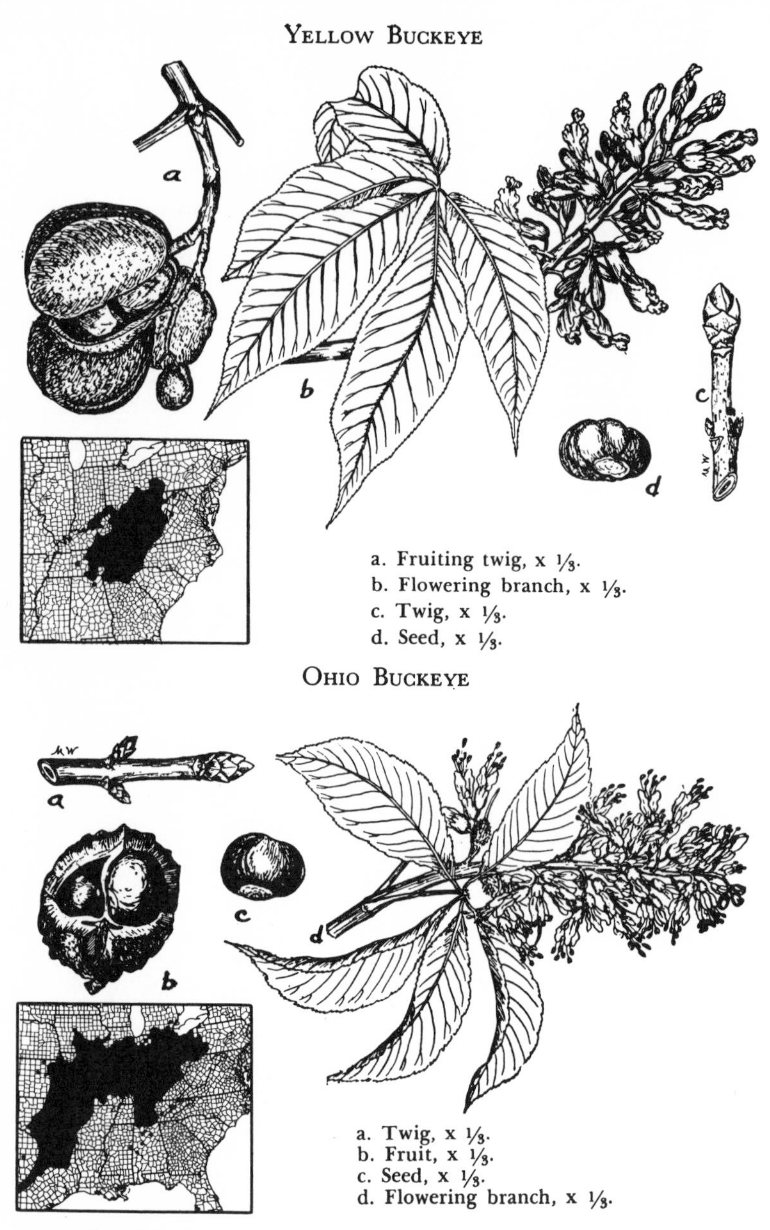

YELLOW BUCKEYE

a. Fruiting twig, x ⅓.
b. Flowering branch, x ⅓.
c. Twig, x ⅓.
d. Seed, x ⅓.

OHIO BUCKEYE

a. Twig, x ⅓.
b. Fruit, x ⅓.
c. Seed, x ⅓.
d. Flowering branch, x ⅓.

HIPPOCASTANACEAE

Yellow Buckeye

Aesculus octandra Marsh.

HABIT. A medium-sized tree 60-90 feet high and 2-3 feet in diameter (max. 110 by 5 feet); broad, rounded crown.

LEAVES. Five-foliate (rarely 7); 4-10 inches long; with elliptical, glabrous, or tomentose-serrate leaflets; dark yellow-green above, paler below.

FLOWERS. Whitish-yellow; glandular-hairy calyx and pedicel; stamens inserted within corolla.

FRUIT. Capsule pale brown; 2-3 inches long; usually smooth; containing usually 2 red-brown seeds about 1-1 1/2 inches long and poisonous.

TWIGS. Stout; hairy at first, becoming smooth and red-brown to gray. Winter buds: terminal 2/3-1 inch long, glabrous, pale brown, nonresinous, scales not prominently keeled.

BARK. Gray to dark brown; 3/4 inch thick; thin scales.

SILVICAL CHARACTERS. Tolerant; moist sites; the largest American species.

* * *

Ohio Buckeye. Fetid Buckeye

Aesculus glabra Willd.

HABIT. A small to medium-sized tree 30-50 feet high and 1-2 feet in diameter (max. 90 by 3 feet); broad, rounded crown.

LEAVES. Five-foliate (rarely 7); 3-6 inches long; with ovate or oval, glabrous, finely serrate leaflets; yellow-green above; paler below; disagreeable odor when bruised.

FLOWERS. Yellow-green; hairy; small; stamens exserted.

FRUIT. Capsule red-brown; 1-2 inches long; usually spiny; containing a single brown seed about 1 inch long.

TWIGS. Stout; hairy at first becoming smooth and red-brown to ash-gray; disagreeable odor when bruised. Winter buds: terminal 2/3 inch long, red-brown, nonresinous, prominently keeled scales.

BARK. Ash-gray; thick; deeply fissured and plated; ill-scented.

SILVICAL CHARACTERS. Tolerant; moist sites; becoming rare.

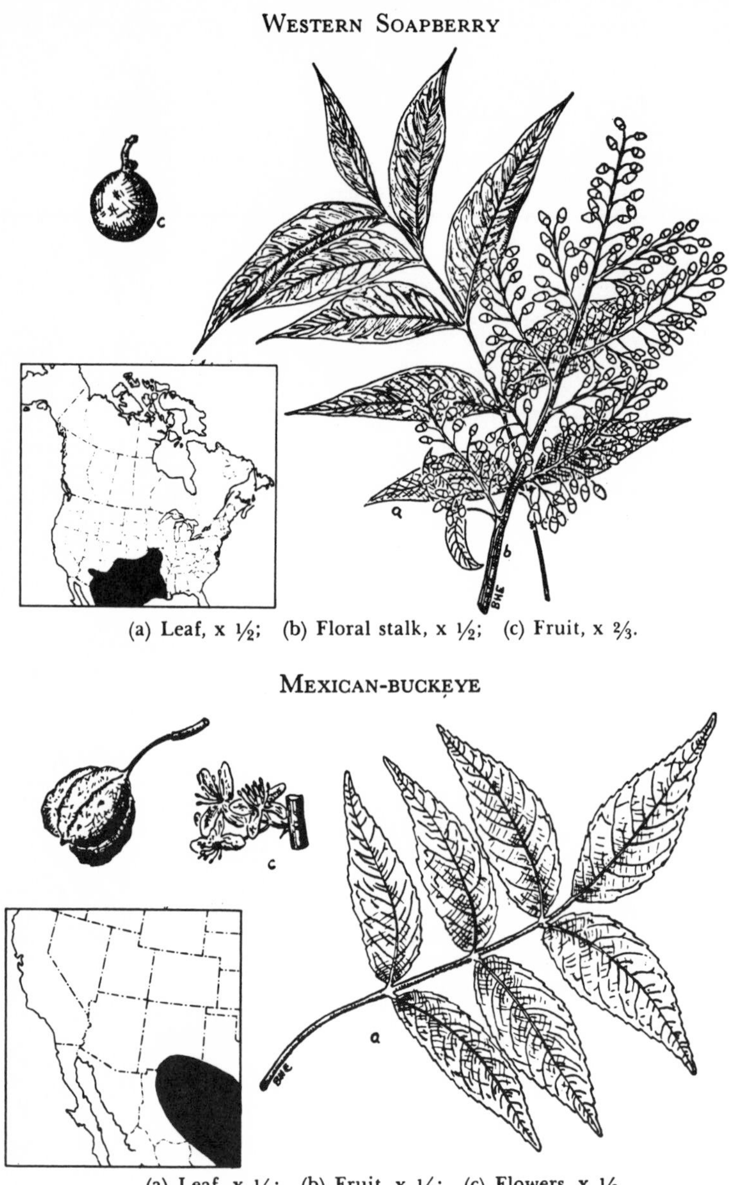

WESTERN SOAPBERRY

(a) Leaf, x ½; (b) Floral stalk, x ½; (c) Fruit, x ⅔.

MEXICAN-BUCKEYE

(a) Leaf, x ⅓; (b) Fruit, x ½; (c) Flowers, x ⅓.

Western Soapberry
Sapindus drummondii Hook. & Arn.

HABIT. A shrub or small tree rarely 40-50 feet high and 1 1/2-2 feet in diameter; branches round, usually erect.

LEAVES. Alternate; pinnately compound with 4-9 pairs of lanceolate leaflets, each 2-3 inches long; margins entire; glabrous above; pubescent beneath; yellow-green; deciduous.

FLOWERS. Regular; polygamo-dioecious; minute; in many-flowered clusters 6-9 inches long; calyx 4-5 acute sepals; corolla 4-5 white petals; stamens 8-10; ovary 2-4-celled.

FRUIT. Drupaceous; ripening into a leathery, 1-3-celled, and seeded berry; 1/2 inch in diameter; glabrous; yellow, turning black in drying; persistent on branches until spring; formerly used as soap. Seed: solitary in each carpel; obovoid; dark brown; smooth, bony coat.

TWIGS. Moderately stout; at first pubescent and pale yellow-green, becoming puberulous, gray. Winter buds: terminal absent; lateral small, globose, often superposed in pairs.

BARK. Thin (1/3-1/2 inch); red-brown; furrowed into long, superficially scaly plates; bitter and astringent.

WOOD. Unimportant; heavy; hard; strong; close-grained; ring-porous; heartwood light brown, tinged with yellow.

SILVICAL CHARACTERS. Intolerant. Florida soapberry, *S. marginatus* Willd., is a small tree from South Carolina to Florida.

* * *

Mexican-buckeye
Ungnadia speciosa Endl.

This shrub or small tree of southern Texas and New Mexico has alternate, deciduous, pinnately compound leaves with 5-7 ovate-lanceolate leaflets, 3-5 inches long, thick, and dark green; small, irregular, polygamous flowers; leathery, 3-valved, capsular, red-brown fruit 2 inches wide, containing black, shiny, leathery seeds about 1/2 inch long, reputed to be poisonous.

* * *

Goldenrain
Koelreuteria paniculata Laxm.

A small, graceful, Asiatic tree with pinnately compound leaves and crenate, serrate, or nearly lobed leaflets; yellow flowers; and a distinctive bladderlike fruit with papery walls and three black seeds.

Cascara Buckthorn

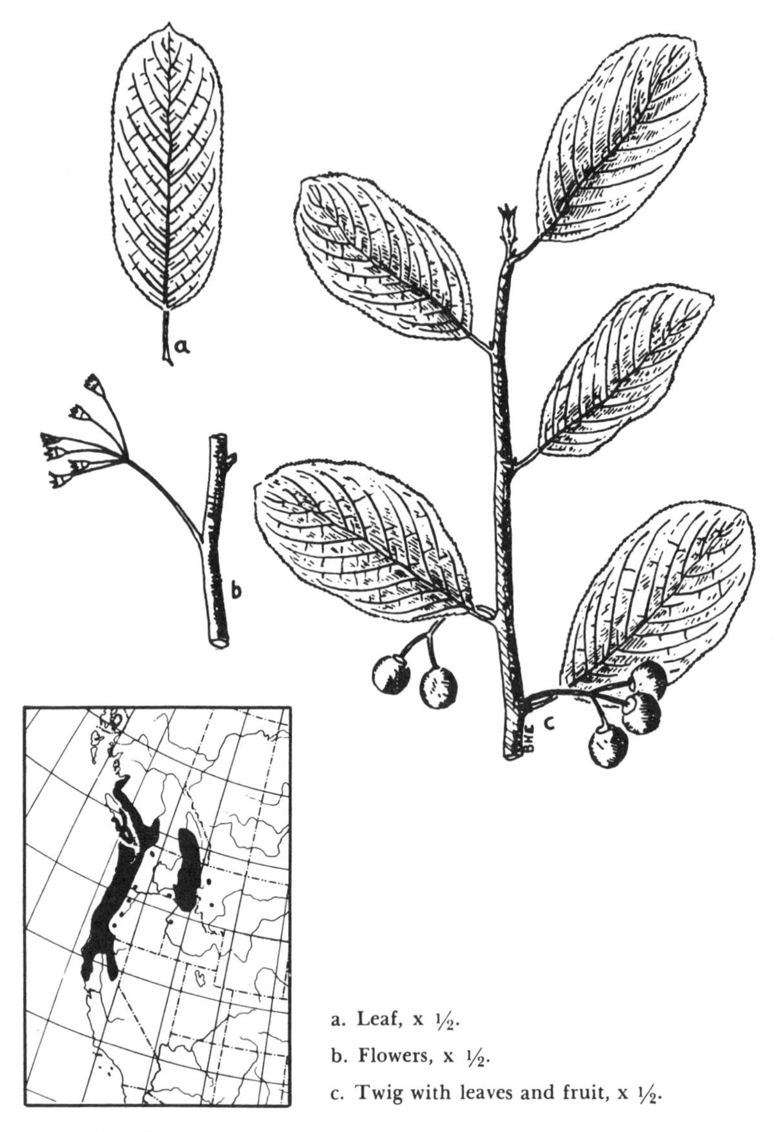

a. Leaf, x ½.

b. Flowers, x ½.

c. Twig with leaves and fruit, x ½.

RHAMNACEAE

Cascara Buckthorn
Rhamnus purshiana DC.

HABIT. A shrub or tree 20-40 feet high and 6-20 inches in diameter; wide, open crown with numerous stout branches.

LEAVES. Alternate, or rarely obliquely opposite; simple; broad-elliptic; 2-7 inches long; undulate margins finely serrate or nearly entire; thin; villous below and on veins above; deciduous; turning pale yellow before falling; petioles stout, hairy, 1/2-1 inch long; stipules minute, deciduous.

FLOWERS. Regular; perfect; small; in axillary peduncled cymes; long-pedicelled; calyx 5-lobed; corolla greenish, minute, 5-lobed; stamens 5; ovary 2-4-celled; ovules solitary, erect.

FRUIT. Drupaceous; subglobose; 1/3-1/2 inch in diameter; black; bearing remnants of style; flesh thin and juicy; 2-3 obovoid, 1-seeded nutlets with thin, gray or yellow-green shell.

TWIGS. Slender; round; pubescent, usually becoming glabrous; yellow-green or red-brown. Winter buds: terminal absent; lateral small, naked, hoary-tomentose.

BARK. Thin (1/4 inch); gray to dark brown, often tinged with red; scaly; important because of laxative properties.

WOOD. Unimportant; rather light; soft; not strong; ring-porous; heartwood brown, tinged with red.

SILVICAL CHARACTERS. Moderately tolerant; prolific seeder, coppices freely; no taproot; few enemies.

GENERAL. A related shrubby form, *R. betulaefolia* Greene, grows from southern Nevada and Utah through Arizona, New Mexico, and southeastern Texas. Three native and two naturalized species reach tree size and are keyed out below.

* * *

1. Winter buds scaly; leaves oval to broad-ovate.
   2. Leaves deciduous, thin, opposite or nearly so, finely crenate-serrate; fruit black; naturalized . . . . *R. cathartica* L., European Buckthorn
   2. Leaves persistent, thick, mostly alternate, spiny-toothed; fruit red; California and Arizona . . . . *R. crocea* var. *ilicifolia* Greene, Hollyleaf Buckthorn, p. 317
1. Winter buds naked; leaves entire or finely toothed.
   3. Leaves deciduous, thin.
      4. Flowers in umbellike cymes; leaves 2-7 inches long, 8-15 pairs of veins; native.
         5. Leaves 10-15 pairs of veins; peduncles longer than petioles; West Coast . . . . *R. purshiana* DC., Cascara Buckthorn
         5. Leaves 8-10 pairs of veins; peduncles shorter than petioles; Southeast . . . . *R. caroliniana* Walt., Carolina Buckthorn
      4. Flowers fascicled or solitary; leaves 1 1/2-2 1/2 inches long, 8-9 pairs of veins; naturalized . . . . *R. frangula* L., Glossy Buckthorn
   3. Leaves persistent thick; southeastern California to Nevada and New Mexico . . . . *R. californica* var. *ursina* (Greene) McMinn, California Buckthorn

Hollyleaf Buckthorn

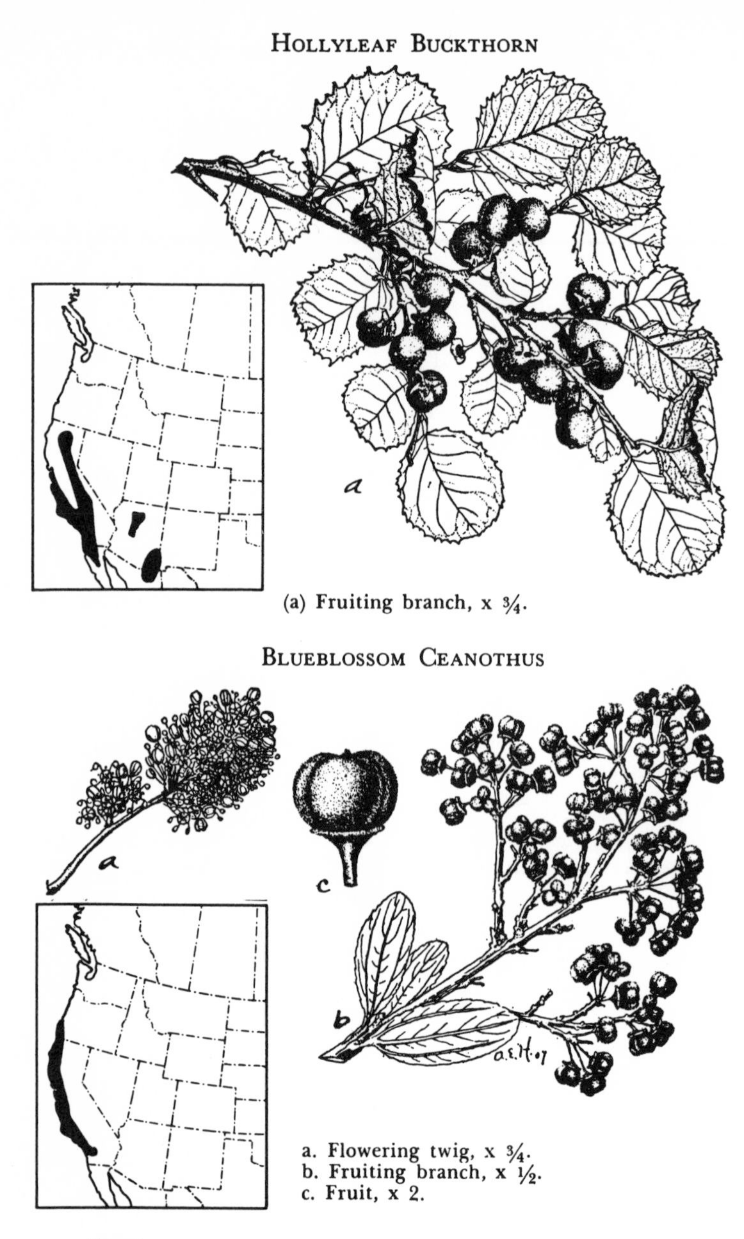

(a) Fruiting branch, x ¾.

Blueblossom Ceanothus

a. Flowering twig, x ¾.
b. Fruiting branch, x ½.
c. Fruit, x 2.

RHAMNACEAE

Hollyleaf Buckthorn

Rhamnus crocea var. *ilicifolia* (Kell.) Greene

HABIT. A shrub or small evergreen tree rarely 25 feet high and 6-8 inches in diameter; crown round; branches stout.

LEAVES. Alternate or subopposite; simple; oval or orbicular; 1-1 1/2 inches long; margin spinulose-dentate; leathery; glabrous; yellow-green above; often golden below; persistent.

FLOWERS. Regular; polygamo-dioecious; in small axillary clusters; calyx about 1/8 inch long, 4-lobed; petals absent.

FRUIT. Drupaceous; obovoid; 1/4 inch in diameter; red; flesh thin and dry; 2-3 brown nutlets 1/8 inch long, hard.

TWIGS. Slender; round; rigid, often spinescent; red-brown; glabrous. Winter buds: terminal absent; lateral 1/16 inch long, scaly, obtuse, scales hairy-fringed.

BARK. Thin (1/16-1/8 inch); dark gray; slightly roughened by minute tubercles; acrid and bitter.

WOOD. Moderately heavy and hard; brittle; fine-grained; ring-porous; heartwood light yellow-brown; unimportant.

SILVICAL CHARACTERS. Intolerant; prolific seeder; on hot, dry hillsides; in pure groups or scattered.

* * *

Bluewood. Logwood

Condalia obovata Hook.

This velvety-pubescent shrub or small tree is one of the common chaparral species of western Texas and southeastern New Mexico, commonly forming dense thickets. It is characterized by small (1/2-3/4 inch long), entire, tardily deciduous, alternate, or fascicled leaves; small axillary flowers; and a deep red, subglobose drupe 1/5 inch in diameter, to which the calyx is attached.

* * *

Blueblossom Ceanothus

Ceanothus thyrsiflorus Eschs.

This shrub or small tree is common in coastal forests from southern Oregon to central California. It is characterized by simple, alternate, minutely toothed, persistent leaves 1-1 1/2 inches long; small, perfect, blue or white flowers; a subglobose, 3-lobed, dry drupe 1/4 inch long, which separates into 3 nutlets; scaly lateral buds. *C. arboreus* Greene and *C. spinosus* Nutt. are small trees in southern California.

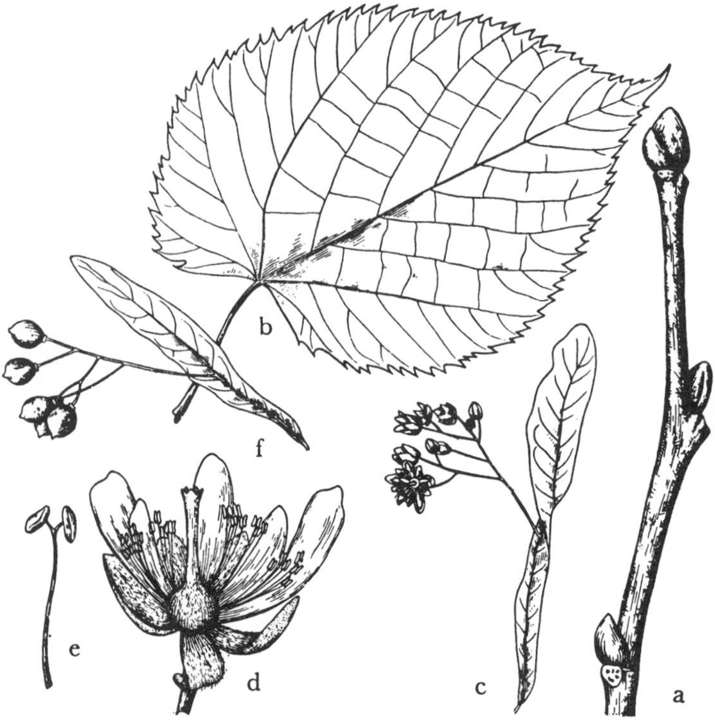

a. Winter twig, x 1.
b. Leaf, x 1/3.
c. Bract with flowers, x 1/2.
d. Flower, enlarged.
e. Stamen, enlarged.
f. Bract bearing fruit, x 1/2.

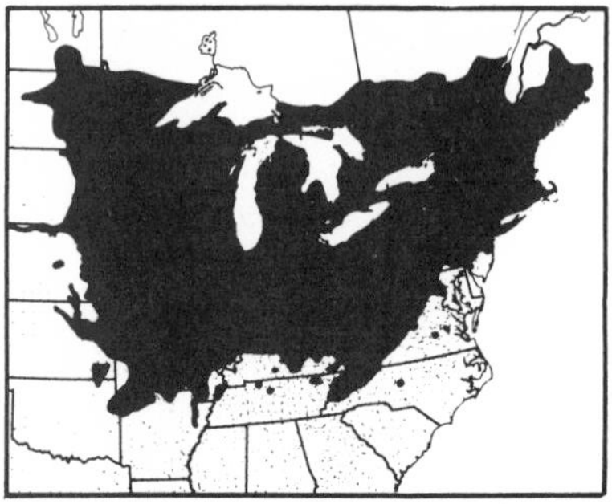

American Basswood

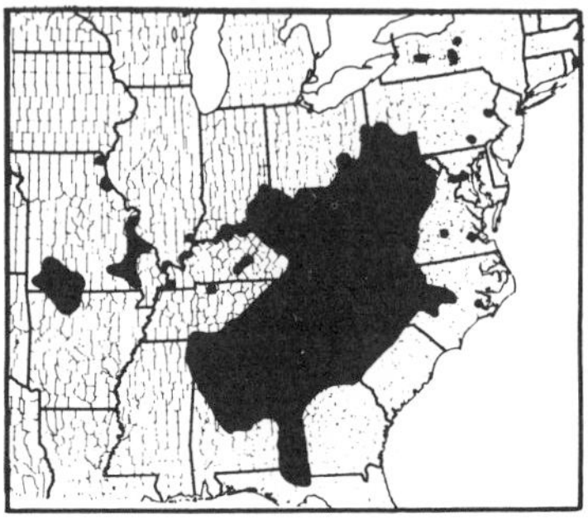

White Basswood

TILIACEAE

American Basswood. Linden

Tilia americana L. (*Tilia glabra* Vent.)

HABIT. A medium-sized tree 60-80 feet high and 2-3 feet in diameter (max. 125 by 4 1/2 feet); dense, rounded crown.

LEAVES. Alternate; simple; deciduous; 5-6 inches long; broadly ovate; coarsely serrate; unequally heart-shaped base; glabrous; dull dark green above; paler beneath; petioles slender, 1-2 inches long; stipules falling early.

FLOWERS. Regular; perfect; in loose 6-15-flowered cymes, the long stalk attached to a leafy bract for half its length; yellow-white; fragrant; 5 sepals; 5 petals; many stamens; 5-celled ovary; appearing after the leaves.

FRUIT. A gray, globose, woolly, nutlike drupe 1/3-1/2 inch long; in cymes and attached to leafy persistent bracts.

TWIGS. Generally rather stout; green to red-gray, becoming dark gray; conspicuous stipule scars; pith circular. Winter buds: terminal absent; lateral subglobose, acute, red, lopsided, 1/4 inch long, 2-3 scales usually visible.

BARK. Smooth and gray-green on young trunks; becoming thick, dark gray, furrowed, with narrow, scaly ridges.

WOOD. Important; light; soft; fine-textured; diffuse-porous; used for novelties, patterns, excelsior, containers, etc.

SILVICAL CHARACTERS. Tolerant; moist sites; abundant seed produced; sprouts vigorously; rapid growth; mixed hardwoods; deep lateral roots; prized as source of honey.

GENERAL. Some authors have recognized as many as 16 native species of *Tilia*, the differences being minor and variable. Four species are tentatively accepted in the 1953 checklist, with *T. caroliniana* Mill. and *T. floridana* Small being unimportant trees in the Southeast. White basswood, *T. heterophylla* Vent., is an important tree differing from American basswood by having leaves that are woolly on the lower surface and 10-25 flowers in a cluster.

Several European species are planted for ornamental purposes, the most common being the European linden, *T.* X *europea* L., with leaves 2 1/2-4 inches long; and the small-leaved linden, *T. cordata* Mill., with leaves 1 1/2-2 1/2 inches long.

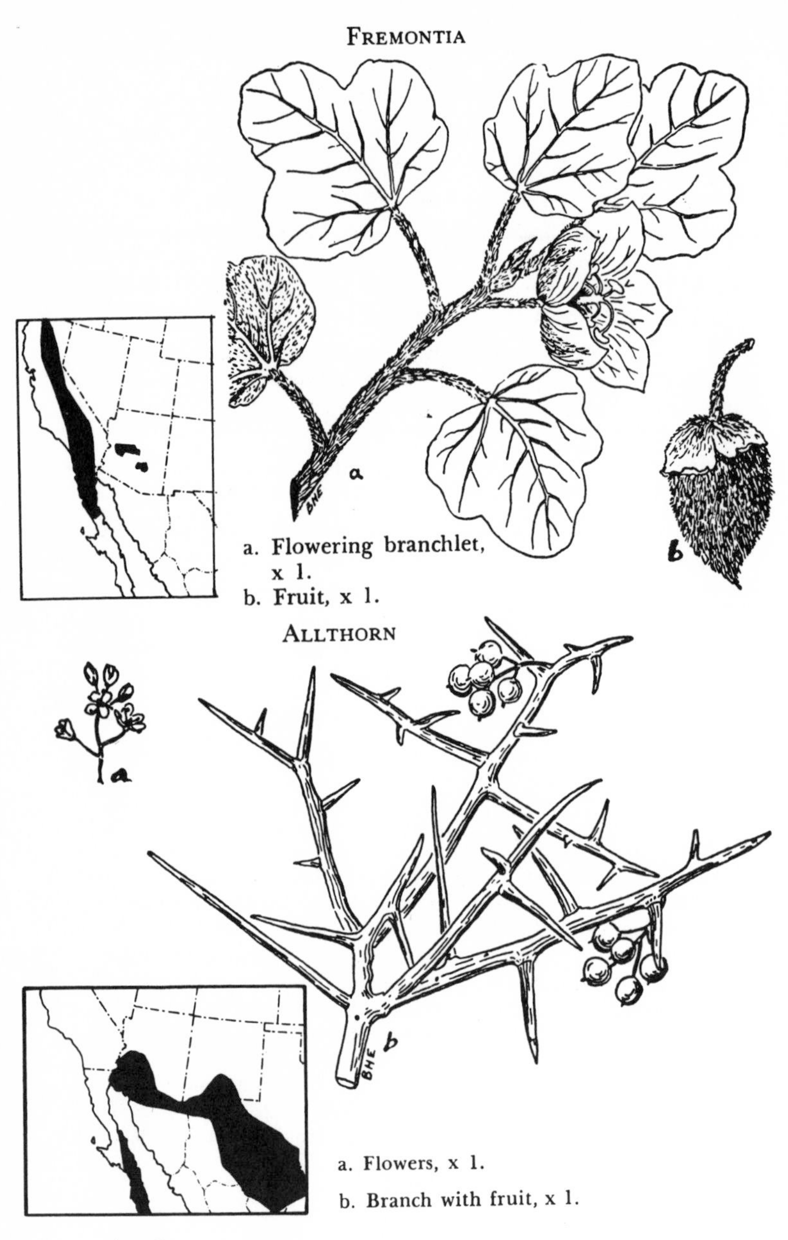

FREMONTIA

a. Flowering branchlet, x 1.
b. Fruit, x 1.

ALLTHORN

a. Flowers, x 1.
b. Branch with fruit, x 1.

STERCULIACEAE

Fremontia Flannelbush

Fremontia californica Torr. (*Fremontodendron californicum* Cov.)
[*Fremontia mexicana* (Dav.) Macb.]

HABIT. A shrub or small tree 20-30 feet high and 12-14 inches in diameter; crown open with stout branches.

LEAVES. Alternate, simple; broadly ovate; 1 1/2 inches in diameter; usually 3-lobed; thick; stellate; rusty-pubescent below; persistent 2 years; petioles stout, 1/2-2/3 inch long.

FLOWERS. Regular; perfect; solitary; calyx, deeply 5-lobed, yellow, 1 inch long; corolla absent; stamens 5; ovary 5-celled.

FRUIT. Ovoid, acuminate, 4-valved capsule; 1 inch long; densely woolly-dehiscent; inner surface villous-pubescent. Seed: oval; small (about 3/16 inch long); very dark brown.

TWIGS. Stout; round; stellate; rusty-pubescent at first, becoming glabrous and light red-brown. Winter buds: naked.

BARK. Thin; furrowed; dark red-brown.

WOOD. Unimportant; heavy; hard; ring-porous; red-brown.

SILVICAL CHARACTERS. Intolerant; on very poor, dry foothills; forming dense thickets; a monotypic genus.

* * *

KOEBERLINACEAE

Allthorn. Corono de Cristo

Koeberlinia spinosa Zucc.

HABIT. A shrub or small bushy tree rarely 20-25 feet high; often appearing to bear neither leaf, flower, nor fruit.

LEAVES. Alternate; simple; early deciduous, the tree usually leafless; scalelike and minute (not over 1/8 inch long).

FLOWERS. Regular; perfect; small; in short umbellike racemes; petals 4, green-white, much longer than sepals.

FRUIT. Small (3/16-1/4 inch), subglobose, 2-celled, black berry; flesh thin and succulent; cells 1-2 seeded. Seed: coiled and shell-shaped; seed coat brittle and wrinkled.

TWIGS. Stout; glabrous; terminating in sharp, rigid spine; pale green in color. Winter buds: minute and inconspicuous.

BARK. Thin; red-brown; scaly.

WOOD. Unimportant; very hard; heavy (sp. gr. 1.12); fine-textured; diffuse-porous; heartwood dark brown.

SILVICAL CHARACTERS. Intolerant; on dry, gravelly plains and foothills; a monotypic botanical curiosity.

Gordonia

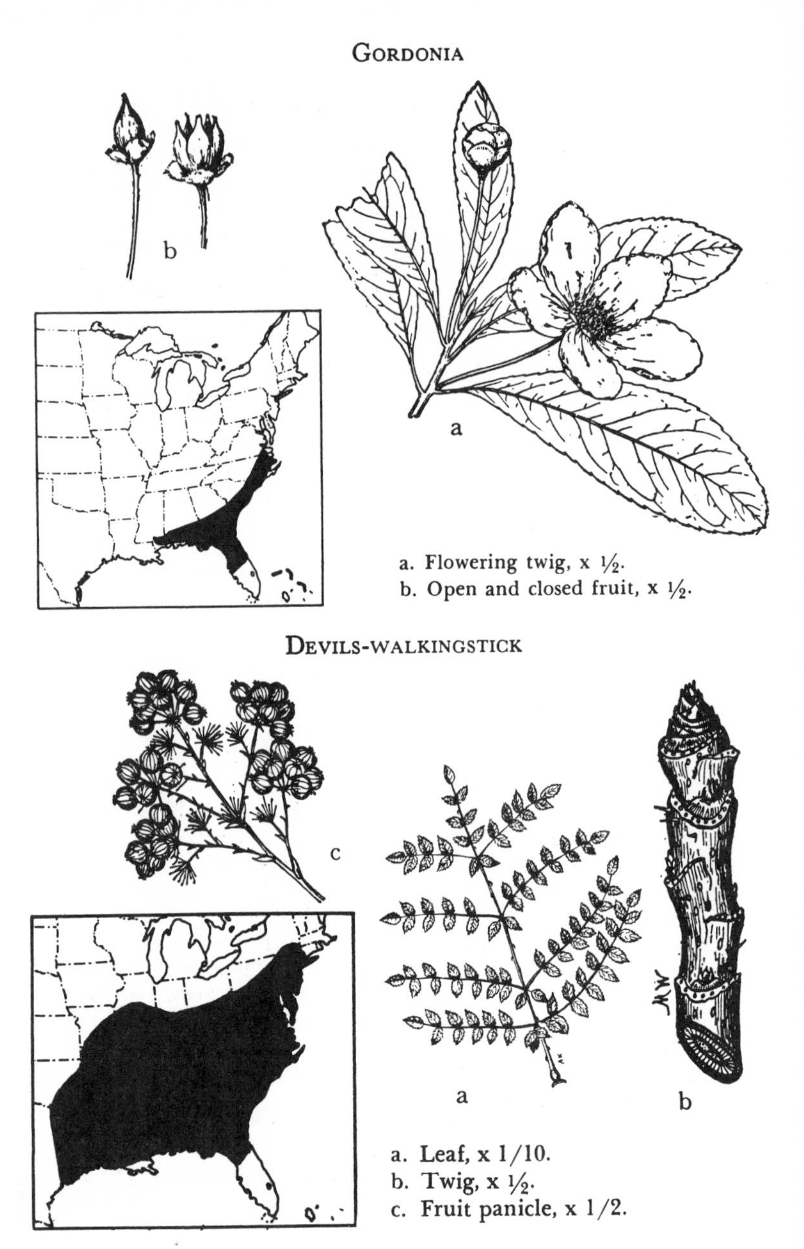

a. Flowering twig, x ½.
b. Open and closed fruit, x ½.

Devils-walkingstick

a. Leaf, x 1/10.
b. Twig, x ½.
c. Fruit panicle, x 1/2.

THEACEAE

Gordonia. Loblolly-bay
Gordonia lasianthus (L.) Ellis

HABIT. A beautiful, medium-sized tree 60-75 feet high and 1-2 feet in diameter; narrow, compact crown.

LEAVES. Alternate; simple; persistent; 3-6 inches long; lanceolate to elliptic; leathery; shallowly toothed; dark green; turning scarlet before falling.

FLOWERS. Regular; perfect; showy; white; fragrant; 2 1/2 inches across petals; sepals silky; ovary 3-5-celled.

FRUIT. A woody, oblong, 5-celled capsule 2/3 inch long; silky surface. Seed: 1/16 inch long; flat; nearly square; winged.

TWIGS. Slender; dark brown; rough. Winter buds: terminal present, 1/4 inch long, acute, silky-hairy.

BARK. Thick; red-brown; deeply fissured and ridged.

GENERAL. A rare shrub, franklinia, *Franklinia alatamaha* Bartr. [*Gordonia alatamaha* (Bartr.) Sarg.] with deciduous, lustrous, oblong leaves 6-9 inches long and creamy, cup-shaped flowers, is planted as an ornamental. Native to McIntosh County, Georgia, but not found growing wild since 1790.

* * *

The shrubby mountain Stewartia, *Stewartia ovata* (Cav.) Weath., with deciduous, hairy-margined leaves and flowers 4 inches in diameter, occurs in the mountains from Virginia to Georgia. The similar Virginia Stewartia, *S. malacodendron* L., occurs on the coastal plain from Virginia to Mississippi.

* * *

ARALIACEAE

Devils-walkingstick
Aralia spinosa L.

HABIT. A prickly grotesque shrub or small tree rarely 35 feet high with few branches and a flat crown.

LEAVES. Alternate; doubly pinnately compound; deciduous; 2-4 feet long; pinnae usually bearing 5-6 pairs of ovate, serrate, thin, dark green leaflets 2-3 inches long and prickly.

FLOWERS. Regular; perfect or staminate; small; in many-flowered panicles 3-4 feet long; green-white.

FRUIT. Small, juicy, black berry 1/4 inch long; tipped with persistent style. Seed: 2-5, oblong; compressed; red-brown.

TWIGS. Stout (1/2-1 inch thick); orange; lustrous; armed with stout prickles. Winter buds: terminal present, conical.

CACTACEAE

Characteristics of the Cactus Family

HABIT. Shrubs or seldom trees, rarely 50-60 feet high and 2 feet in diameter; stems commonly columnar, fluted, succulent, and branched; numerous spines springing from cushions of small bristles (areolae).

LEAVES. Alternate; simple; mostly reduced to spines or scales or absent; photosynthetic processes taking place in the green parts of the fleshy stems.

FLOWERS. Regular; perfect; usually single; large and showy; calyx of numerous sepals forming a tube, those of inner series petallike; corolla showy, of numerous petals; stamens many, inserted on calyx tube; ovary inferior, 1-celled, with several parietal placentae and numerous horizontal ovules, styles united into one, stigmas as many as placentae.

FRUIT. 1-celled, fleshy (rarely dry) berry, often edible. Seed: numerous, small.

WOOD. An internal, woody frame or skeleton, made up of a cylinder or a meshed network of strands.

BUDS. Modified into pulvini or cushions, which are usually depressions often consisting of a complex series of spines, wool, glands, and growing points.

SILVICAL CHARACTERS. Very intolerant; although seed generally produced abundantly, natural reproduction by seed is rather scanty because of unfavorable environment; vegetative reproduction common; typical of very dry desert areas where they are often the only woody plants.

GENERAL. This family contains about 120 genera and 1,200 species; there are 2 genera and 5 species reaching tree size in the United States in southern California, Arizona, and New Mexico. Two species, *Cereus schottii* Engelm. and *Cereus thurberi* Engelm., of southern Arizona and Mexico, have columnar branches 20 to 25 feet high but cannot be considered trees as they do not have a definite trunk.

KEY TO THE ARBORESCENT SPECIES OF CACTACEAE

1. Branches and stems columnar, ribbed, not tuberculate, continuous; areoles (growing centers) without glochids (minute bristles); leaves spinelike; tube of flower elongated; seeds dark-colored; spines not barbed; a tree often 50-60 feet high; the state flower of Arizona . . . . *Cereus giganteus* Engelm. [*Carnegiea gigantea* (Engelm.) B. & R.], Saguaro
1. Branches and stems slender, columnar, tuberculate, conspicuously jointed; areoles with both glochids and spines; leaves small, fleshy on young parts; flower tube short; seed light-colored; spines retrorsely barbed; small plants not over 15 feet high . . . . *Opuntia*
   2. Tubercles of branches broad, full, and rounded below areolae; flowers pink or purple; fruit sparingly spiny or without spines.
      3. Flowers pink; fruit green, proliferous (one growing from another), usually spineless; joints pale olive-green, readily detached, freely falling, tubercles broad and ovoid; spines yellow . . . . *Opuntia fulgida* Engelm., Cholla
      3. Flowers purple; fruit yellow, rarely proliferous, spiny; joints green or purple, not readily detached, persistent, tubercles elongated; spines white to red-brown . . . . *Opuntia spinosior* (Engelm.) Toumey, Tasajo
   2. Tubercles of branches narrow, high, flattened laterally. Flowers purple, or green-tinted with red or yellow.
      4. Fruit smooth or but slightly tuberculate, spiny, green; branch tubercles 2/3 inch long; spines 5-11, 1/8 inch long or less, dark red-brown; flowers green, tinted with red or yellow . . . . *Opuntia versicolor* Engelm., Staghorn Cholla
      4. Fruit manifestly tuberculate, naked, yellow; branch tubercles 3/4 inch long; spines 8-30, 3/4-1 1/4 inches long, brown; flowers purple . . . . . . *Opuntia imbricata* (Haw.) DC. (*Opuntia arborescens* Engelm.), Cane Cactus

* * *

LYTHRACEAE

Crapemyrtle

Lagerstroemia indica L.

A handsome shrub or small tree from China widely planted and possibly naturalized from Maryland to Texas. Characterized by profuse late summer bloom of showy pink, purple, or white flowers and dehiscent capsular fruit. Leaves are deciduous and mostly opposite, entire, elliptic to obovate or oblong, and 1-2 1/2 inches long. Twigs are 4-angled, glabrous; pointed buds with 2 outer scales. Distinctive grayish, exfoliating bark on fluted trunks.

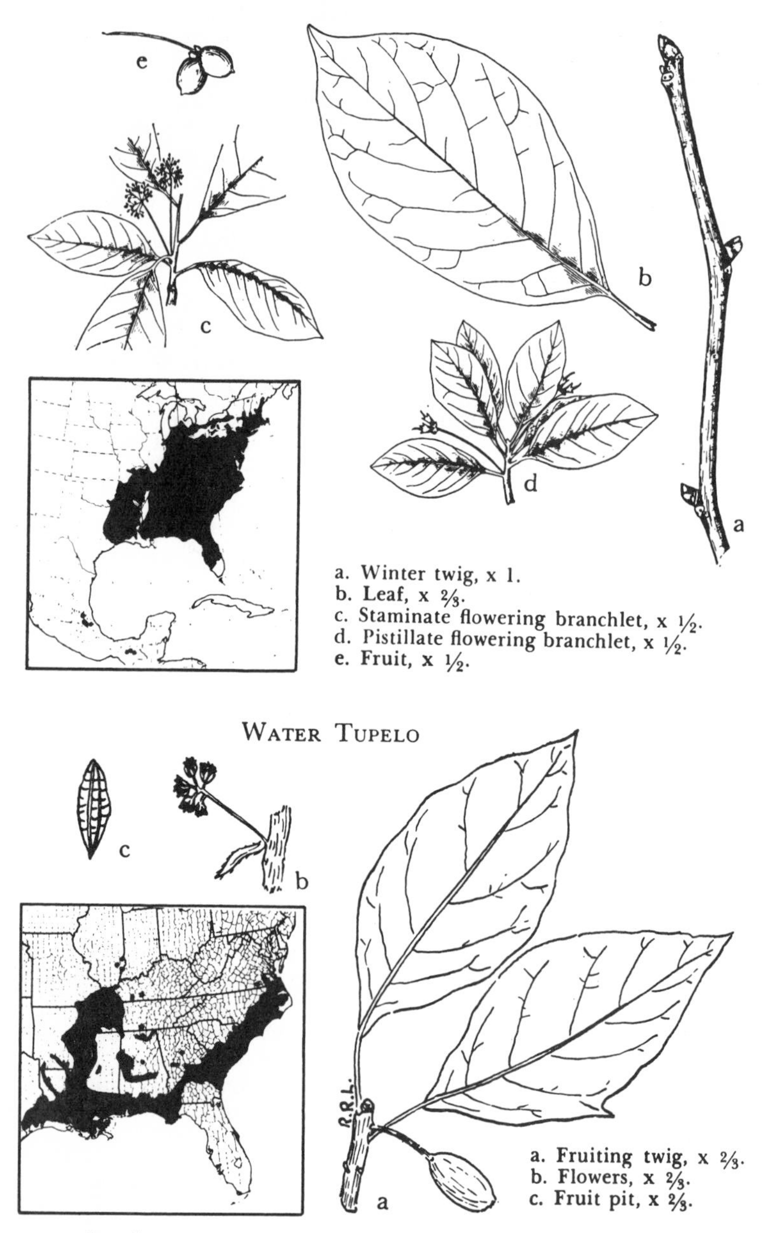
BLACK TUPELO
e
b
c
d
a
a. Winter twig, x 1.
b. Leaf, x ⅔.
c. Staminate flowering branchlet, x ½.
d. Pistillate flowering branchlet, x ½.
e. Fruit, x ½.
WATER TUPELO
c
b
R.R.L.
a
a. Fruiting twig, x ⅔.
b. Flowers, x ⅔.
c. Fruit pit, x ⅔.

NYSSACEAE

Black Tupelo. Blackgum
Nyssa sylvatica Marsh.

HABIT. A medium-sized tree 40-80 feet high and 1-3 feet in diameter (max. 125 by 6 feet); rounded crown, horizontal branches.

LEAVES. Alternate; simple; deciduous; 2-5 inches long; obovate; entire or wavy; thick, lustrous; estipulate; dark green above, paler, often hairy below; scarlet in autumn.

FLOWERS. Regular; polygamo-dioecious; small; staminate in heads; green-white; appearing with the leaves.

FRUIT. Fleshy, ovoid, blue-black drupe 1/3-2/3 inch long, indistinctly ribbed pit (distinctly ribbed in swamp tupelo).

TWIGS. Slender; green to red-brown; smooth; pith diaphragmed; leaf scars with 3 bundle scars. Winter buds: terminal present, scaly, 1/4 inch long, ovoid, yellow-brown.

BARK. Thick; red-brown; deeply fissured, broken into conspicuous square blocks.

WOOD. Important; rather heavy and hard; fine-textured; diffuse-porous; light-colored; used for furniture, containers, etc.

SILVICAL CHARACTERS. Intolerant; moist sites; rapid growth; shallow, lateral roots; scattered, never abundant. Swamp tupelo, *Nyssa sylvatica* var. *biflora* (Walt.) Sarg., with somewhat narrower leaves and more distinctly ribbed pit, is common in southern swamps. The genus *Nyssa* is placed in the family Cornaceae by some authors.

* * *

Water Tupelo
Nyssa aquatica L.

This species, important for timber, occurs in pure stands or mixed with cypress in the southern swamps, often under water. It differs from black tupelo, having leaves 5-7 inches long, oblong-obovate, entire or sometimes irregularly toothed; fruit red-purple, 1 inch long, conspicuously ribbed pit; bark thin, gray-brown, scaly ridges; buds globose, not over 1/8 inch long; butt of the tree is often conspicuously swollen.

* * *

Ogeechee Tupelo
Nyssa ogeche Bartr.

This small, rare tree of the coastal plain from South Carolina to Florida resembles water tupelo and is distinguished from it by its red fruit containing a pit with broad, thin, papery wings. Bear tupelo, *Nyssa ursina* Small, is a shrub or small tree of northwestern Florida.

Silver Buffaloberry

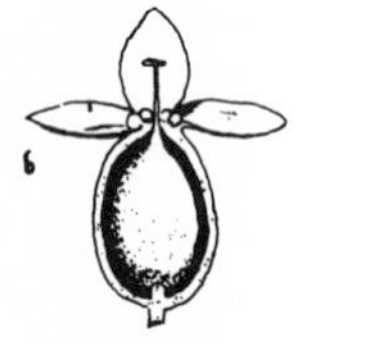

a. Twig with leaves and flower, x ½.

b. Section through flower, x 2.

c. Fruit, x 2.

d. Stone, x 2.

MYRTACEAE

Blue Gum

Eucalyptus globulus Labill.

HABIT. A tall, fast-growing tree commonly planted as an ornamental or windbreak.

LEAVES. Alternate; simple; persistent; lanceolate; 6-13 inches long; margin entire; aromatic; short-petioled; sessile and opposite on young growth.

FLOWERS. Perfect; large; white; mostly solitary.

FRUIT. Woody capsule; 3/4-1 inch long; warty; rough.

BARK. Outer deciduous and shedding; inner smooth, gray.

WOOD. Heavy; hard; cross-grained; used for timber and fuel.

GENERAL. About 500 species of eucalyptus are native to Australia and the Malayan region. Blue gum is the commonest of some 75 species planted in warmer parts of the United States.

* * *

ELAEAGNACEAE

Silver Buffaloberry. Buffaloberry

Shepherdia argentea (Pursh) Nutt.
[*Lepargyrea argentea* (Pursh) Greene]

HABIT. A silvery, often spiny shrub or small tree rarely 15-20 feet high; stoloniferous, thicket-forming.

LEAVES. Opposite; simple; oblanceolate to oblong; cuneate at base; 1-2 1/2 inches long; margin entire; densely silvery-scurfy on both sides with stellate hairs; deciduous; short, stout petioles.

FLOWERS. Dioecious; regular; small; yellowish; calyx urn-shaped; corolla lacking; ovary enveloped in receptacle.

FRUIT. Drupe; ovoid; scarlet; 1/2 inch long; juicy.

* * *

Russian-olive

Elaeagnus angustiofolia L.

An introduced shrub or small tree planted from New England to California and established in some areas. Drought and alkaline resistant, a common dryland ornamental and windbreak. Characterized by alternate, lanceolate, silvery-scurfy leaves; silvery, often spiny branches; silvery-scurfy drupe 1/2 inch long.

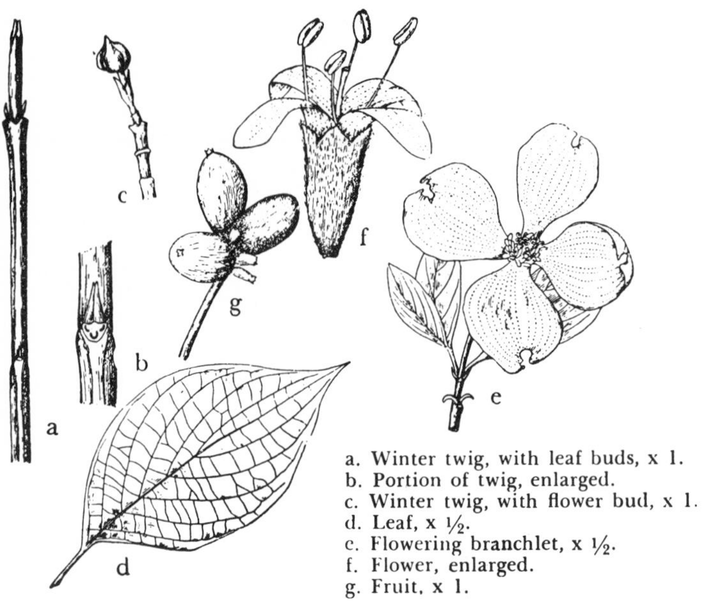

a. Winter twig, with leaf buds, x 1.
b. Portion of twig, enlarged.
c. Winter twig, with flower bud, x 1.
d. Leaf, x ½.
e. Flowering branchlet, x ½.
f. Flower, enlarged.
g. Fruit, x 1.

FLOWERING DOGWOOD

PACIFIC DOGWOOD

Flowering Dogwood

Cornus florida L.

HABIT. A shrub or small tree rarely 40 feet high and 1 foot in diameter; bushy, flat crown; valuable ornamental.

LEAVES. Opposite; simple; deciduous; 3-6 inches long; oval; entire; thick; arcuately veined; hairy; bright green above; paler beneath; turning scarlet in autumn.

FLOWERS. Perfect; small; in heads; surrounded by 4 showy, petallike, notched, white or pink bracts (red in variety *rubra* West.); appearing with leaves.

FRUIT. An ovoid, scarlet drupe 1/2 inch long; in clusters of 3-4; flesh bitter; containing a 2-celled pit.

TWIGS. Slender; glaucous; green or purplish. Winter buds: terminal present, 1/8 inch long, narrow-conical, acute, covered with 2 valvate scales; terminal flower buds subglobose, gray.

BARK. Thin; dark red-brown; broken into small, square blocks.

WOOD. Moderately important; very heavy and hard; fine-textured; diffuse-porous; in demand for shuttles.

SILVICAL CHARACTERS. Very tolerant; moist sites as "understory" species; slow-growing; lateral roots.

* * *

Pacific Dogwood

Cornus nuttallii Aud.

This small, handsome tree of the Pacific Coast is very similar to flowering dogwood in its appearance and differs in the following ways: petallike flower bracts not notched; often producing a second crop of flowers in later summer; bark remaining smooth on trunk.

* * *

Five other of the 16 species of dogwood native to North America, while typically shrubs, may form small trees. These are the alternate-leaf dogwood, *C. alternifolia* L.; the rough-leaf dogwood, *C. drummondii* Meyer; the red-osier dogwood, *C. stolonifera* Michx.; and the stiff cornel dogwood, *C. stricta* Lam., of the eastern United States; and the western dogwood, *C. occidentalis* Cov., in the Northwest.

Pacific Madrone

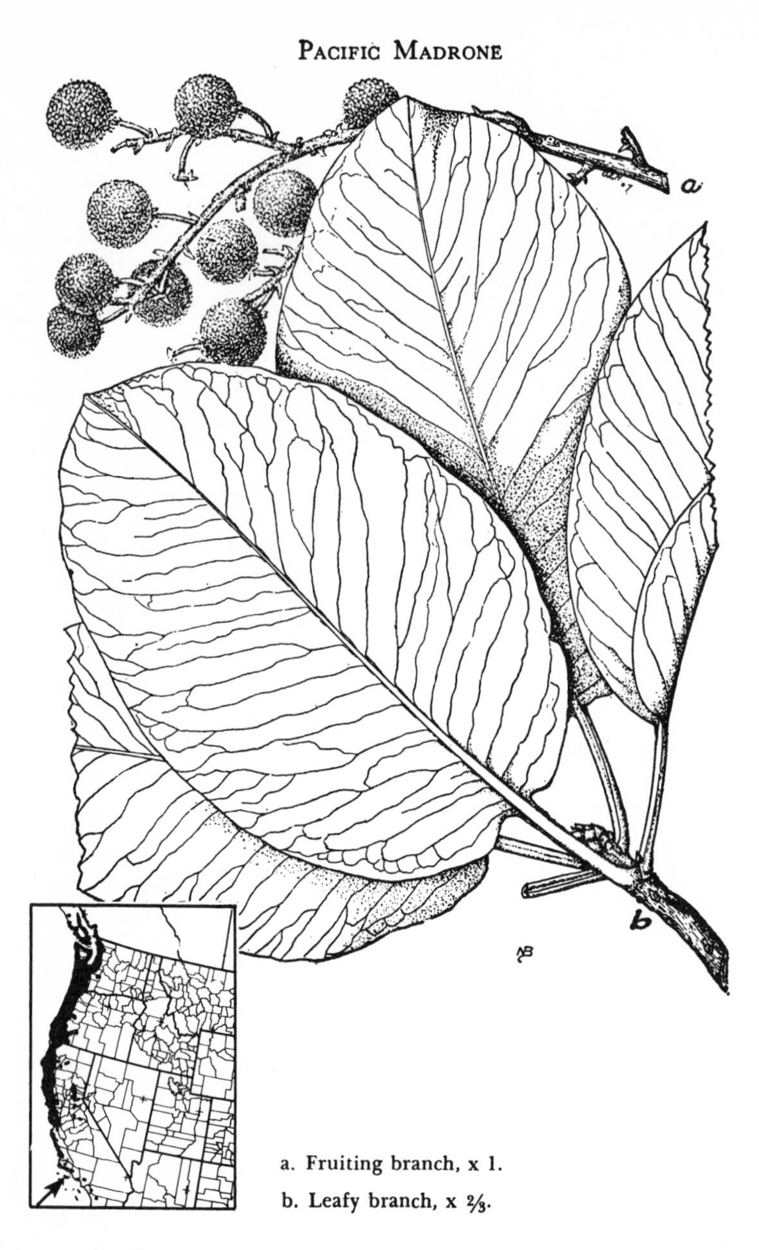

a. Fruiting branch, x 1.

b. Leafy branch, x ⅔.

ERICACEAE

Pacific Madrone

Arbutus menziesii Pursh

HABIT. A medium-sized tree 20-100 feet high and 1-4 feet in diameter (max. 125 by 9 feet); rounded crown.

LEAVES. Alternate; simple; persistent; 3-6 inches long; oval to oblong; entire or toothed on vigorous growth; leathery; glabrous; dark green and lustrous above; glaucous beneath.

FLOWERS. Regular; perfect; in terminal panicles; white; urn-shaped; 1/4 inch long; appearing with or after the leaves.

FRUIT. A globose, orange, semifleshy, glandular-coated, drupaceous berry 1/3-1/2 inch long. Seed: several in each cell; small; brown; angled.

TWIGS. Slender; green to red-brown; glabrous. Winter buds: terminal 1/3 inch long, ovoid, scaly, bright brown.

BARK. Thin; distinctive; red-brown; separating into papery scales exposing the light red inner bark.

WOOD. Unimportant; heavy; hard; diffuse-porous; red-brown.

SILVICAL CHARACTERS. Moderately tolerant; varied sites; pure or mixed stands; occasionally planted as an ornamental.

GENERAL. Two other species are native to the Southwest: Texas madrone, *A. texana* Buckl., similar to Pacific madrone except for smaller leaves (1-3 inches long) and pubescent ovary, is a small tree on dry sites in southern Texas and New Mexico; Arizona madrone, *A. arizonica* (Gray) Sarg., with gray bark and a glabrous ovary, is native to Arizona and Mexico.

* * *

Elliottia, *Elliottia racemosa* Muhl., is a rare tree now known only in eastern Georgia. It is characterized by simple, alternate, oblong, entire, deciduous leaves, 3-4 inches long; perfect flowers with 4 straplike petals; and small capsular fruit.

* * *

Tree Lyonia, *Lyonia ferruginea* Nutt., is a small tree of coastal areas from South Carolina to Florida. It is characterized by simple, alternate, persistent, entire, obovate leaves 1-3 inches long and scaly below; by fascicled flowers 1/8 inch in diameter with a white, globular corolla; and by an ovoid, many-seeded, capsular fruit 1/4 inch long.

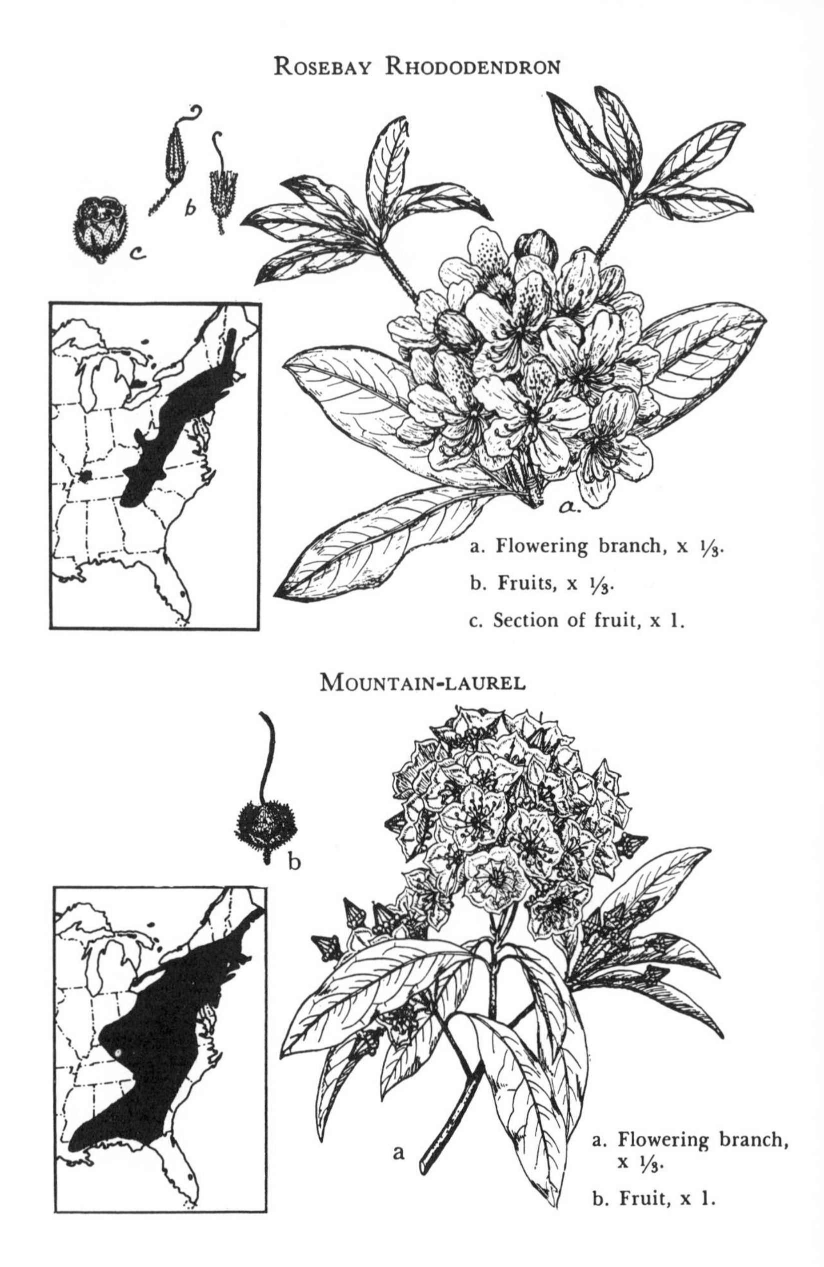

Rosebay Rhododendron

a. Flowering branch, x ⅓.

b. Fruits, x ⅓.

c. Section of fruit, x 1.

Mountain-laurel

a. Flowering branch, x ⅓.

b. Fruit, x 1.

Rosebay Rhododendron. Great Rhododendron

Rhododendron maximum L.

HABIT. A beautiful shrub or small tree rarely 35 feet high and 1 foot in diameter; short, crooked trunk; bushy, rounded crown; an important ornamental.

LEAVES. Alternate; simple; persistent 2-3 years; 4-12 inches long and 1 1/2-2 1/2 inches wide; oblong; leathery; revolute, entire margins; dark green and lustrous above; paler beneath.

FLOWERS. Regular; perfect; showy; in 16-24 terminal flower clusters; white, pink, or purple; campanulate; 1 inch across; appearing after the leaves.

FRUIT. A woody, oblong-ovoid, red-brown, sticky, glandular-hispid, 5-celled capsule 1/2 inch long. Seed: small, many.

TWIGS. Stout; glabrous; dark green to red-brown. Winter buds: terminal present, conical, green, scaly; flower buds 1-1 1/2 inches long.

BARK. Thin; red-brown; broken into thin scales.

WOOD. Unimportant; heavy; hard; diffuse-porous; fine-textured; used for pipe bowls.

SILVICAL CHARACTERS. Tolerant; moist, cool, shady sites; often in thickets; does not like limy soils.

GENERAL. Two other species reach tree size: Catawba rhododendron, *R. catawbiense* Michx., with broader leaves 3-5 inches long and ranging from Virginia to Alabama; and Pacific rhododendron, *R. macrophyllum* D. Don, ranging from British Columbia to California. The azaleas are shrubby species of this genus with deciduous leaves.

* * *

Mountain-laurel

Kalmia latifolia L.

This beautiful shrub or small tree resembles rhododendron and is a common ornamental. It can be distinguished by its flat, alternate, simple, persistent, elliptic-lanceolate leaves 3-4 inches long; its showy, white to rose-colored flowers with a saucer-shaped corolla with a short tube and 10 pouches below the 5-parted limb; and the globose, woody, glandular-hispid, 5-celled capsule 3/16 inch in diameter, which splits at maturity into 5 persistent carpels releasing the many, minute, oblong, winged seeds.

Sourwood

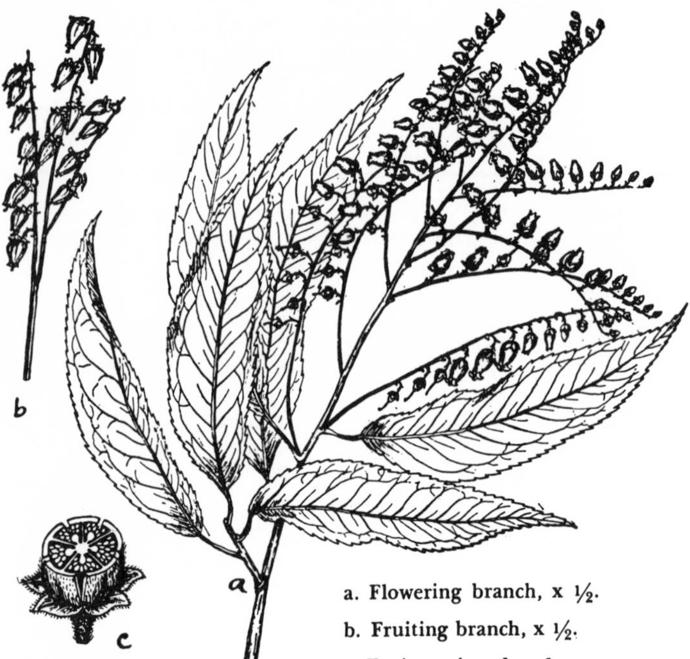

a. Flowering branch, x ½.

b. Fruiting branch, x ½.

c. Fruit sectioned, x 1.

Sourwood

Tree Sparkleberry

ERICACEAE

Sourwood

Oxydendrum arboreum (L.) DC.

HABIT. A small to medium-sized tree 20-50 feet high and 2/3-1 foot in diameter (max. 80 by 2 feet); narrow, rounded crown.

LEAVES. Alternate; simple; deciduous; 5-7 inches long; oblong to lanceolate; thin; finely toothed; nearly glabrous; dark green and lustrous above; paler beneath; sour-tasting; turning scarlet in autumn.

FLOWERS. Regular; perfect; small; in delicate panicles of racemes 6-8 inches long; corolla white, bell-shaped, 5-lobed, 1/3 inch long; appearing after the leaves.

FRUIT. A 5-valved, 5-lobed, dry, persistent capsule 1/3-1/2 inch long; terminated by a persistent style. Seed: 1/8 inch long; pointed; pale brown.

TWIGS. Slender; glabrous; yellow-green to red-brown; leaf scar with single bundle scar. Winter buds: terminal absent; lateral small, scaly, red-brown, partially embedded in bark.

BARK. Thick; gray; tinged with red; furrowed; broad, scaly ridges.

WOOD. Unimportant; heavy; hard; diffuse-porous; red-brown.

SILVICAL CHARACTERS. Rather tolerant; dry, well-drained sites; does not like limy soils; an attractive ornamental; a monotypic genus.

* * *

Tree Sparkleberry. Farkleberry

Vaccinium arboreum Marsh.

This shrub or small tree of the blueberry-cranberry genus is found on moist well-drained sites. It is characterized by alternate, simple, persistent (deciduous during winter in north) leaves obovate to oblong, entire or minutely toothed, leathery, dark green and lustrous above and paler beneath, and 1-2 1/2 inches long; small, white, bell-shaped flowers either solitary or in racemes; and a black, globose, dry, scarcely edible berry about 1/4 inch long, which persists on the branch into winter.

GUM BUMELIA

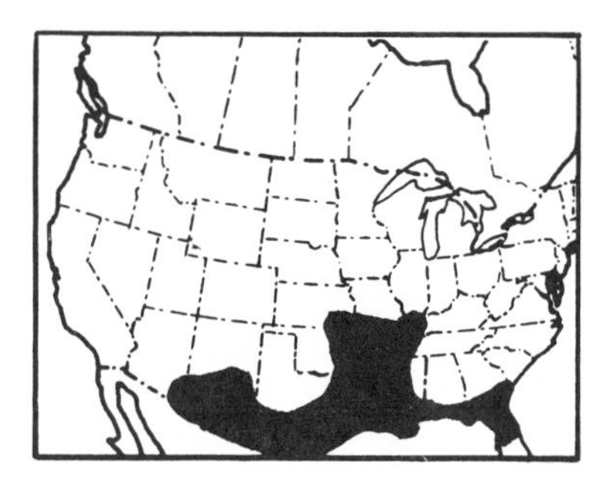

a. Flowering branchlet, x 1.

b. Fruit, x 1.

SAPOTACEAE

Gum Bumelia. Gum Elastic

Bumelia lanuginosa (Michx.) Pers.

HABIT. A shrub or small tree 40-50 feet high and 1-2 feet in diameter; narrow crown with short, spinescent, spiny branches.

LEAVES. Alternate; simple; 1-3 inches long; oblanceolate to obovate; margins entire; thin; dark green above; soft, rusty-brown hairs below; tardily deciduous in winter; stipules absent.

FLOWERS. Regular; perfect; minute; in axillary clusters; corolla white, campanulate, 5-lobed.

FRUIT. Drupe; oblong; 1/2 inch long; black; solitary or in 2-3-fruited clusters; thick flesh. Seed: 1/4 inch long, shiny.

TWIGS. Slender; rounded; spinescent or with stout spines; red-brown to ash-gray. Winter buds: scaly; small; obtuse.

BARK. Thin; dark gray-brown; divided into scaly ridges.

WOOD. Unimportant; heavy; not strong; close-grained; ring-porous; light brown; producing clear, viscid gum.

GENERAL. Four native species sometimes become trees and are keyed out below:

1. Leaves persistent, thick and leathery, dull blue-green; Florida and Texas . . . . *B. celastrina* H. B. K., Saffron-Plum
1. Leaves deciduous.
   2. Leaves glabrous or nearly so, Virginia to Illinois and south . . . . *B. lycioides* (L.) Pers., Buckthorn Bumelia
   2. Leaves hairy on lower surface.
      3. Leaves with silky, golden hairs; South Carolina to Florida . . . . *B. tenax* (L.) Willd., Tough Bumelia
      3. Leaves reddish, dull, woolly hairs . . . . *B. lanuginosa* Pers.

* * *

STYRACACEAE

Bigleaf snowbell, *Styrax grandifolia* Ait., extending from Delaware to Florida and west to Tennessee, Arkansas, and Texas, has oblong, nearly entire, alternate leaves; subglobose, tomentose drupe about 1/3 inch long; a 5-7-parted corolla; and hoary stellate pubescent twigs. Rarely reaches tree size.

Common Persimmon

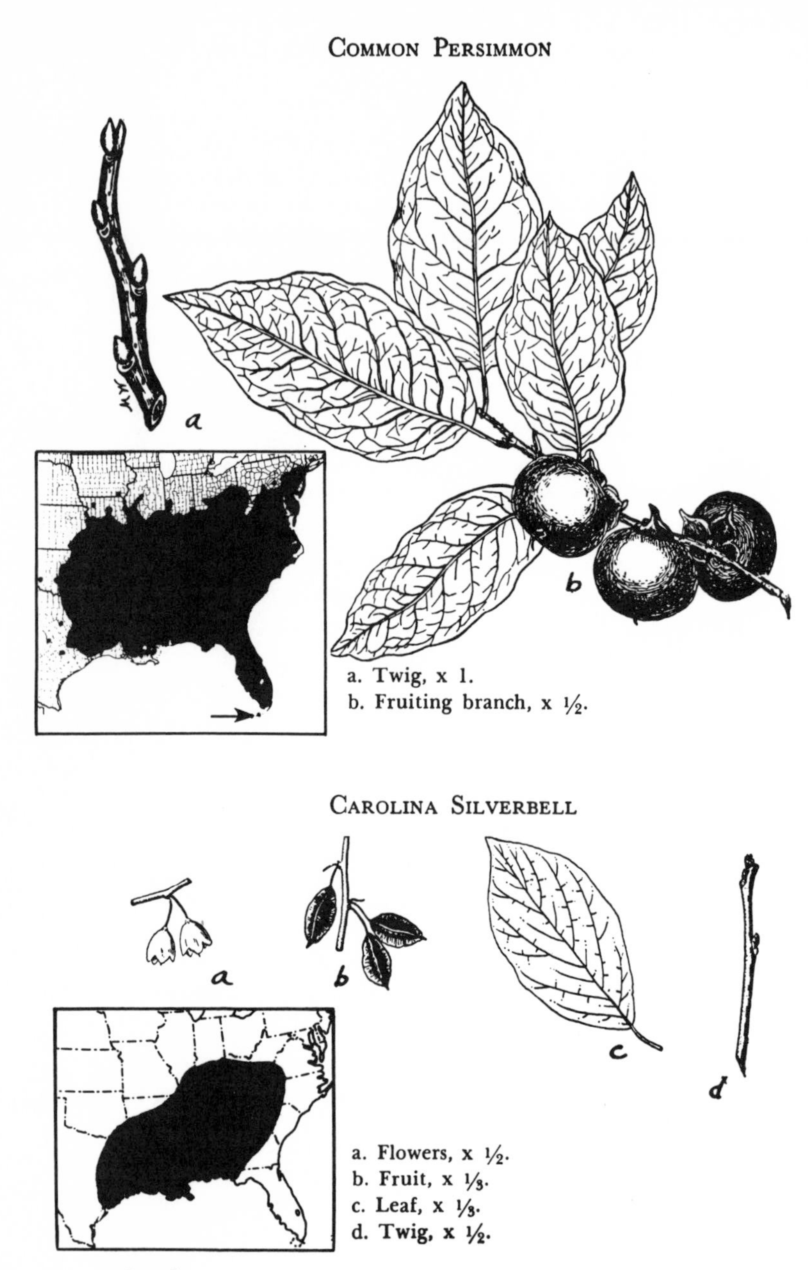

a. Twig, x 1.
b. Fruiting branch, x ½.

Carolina Silverbell

a. Flowers, x ½.
b. Fruit, x ⅓.
c. Leaf, x ⅓.
d. Twig, x ½.

EBENACEAE

Common Persimmon
Diospyros virginiana L.

HABIT. A small to medium-sized tree 25-50 feet high and 1 foot in diameter (max. 130 by 3 feet); broad, rounded crown.

LEAVES. Alternate; simple; deciduous; 3-6 inches long; oblong-ovate to oval; glabrous; entire; dark green and lustrous above; paler, sometimes hairy below.

FLOWERS. Regular; dioecious; 1/2-3/4 inch long; corolla urn-shaped or tubular; yellow-green; appearing with leaves.

FRUIT. A globose, orange to purple berry 1-1 1/2 inches long; subtended by 4 woody, persistent calyx lobes; astringent till frost, then edible. Seed: 1-8; compressed; oblong; rounded.

TWIGS. Slender; gray-brown; velvety or glabrous; pith diaphragmed or homogeneous in same twig; leaf scar with single, elongated bundle scar. Winter buds: terminal absent; lateral ovate, acute, 1/8 inch long, red, 2 greatly overlapping scales.

BARK. Thick; hard; in distinctive square, scaly blocks.

WOOD. Moderately important; heavy; hard; strong; ring-porous; used for shuttles, golf clubs, billiard cues, brushes, etc.

SILVICAL CHARACTERS. Rather intolerant; varied sites.

GENERAL. This tropical genus contains over 200 species including the ebony of commerce. *Diospyros texana* Scheele is a small Texas tree.

* * *

STYRACACEAE

Carolina Silverbell
Halesia carolina L.

HABIT. A small handsome tree; rounded crown.

LEAVES. Alternate; simple; deciduous; 3-4 inches long; oblong-ovate; finely toothed; yellow-green and glabrous above; pale and hairy below.

FLOWERS. Regular; perfect; showy; corolla white, bell-shaped, 1/2 inch long.

FRUIT. A dry, oblong, 4-winged drupe 1 1/2 inches long.

GENERAL. Twigs slender; terminal buds absent; bark thin, scaly; moist sites; stump sprouts vigorously. Three other species form trees; the related mountain species *H. monticola* Sarg. with flowers and fruit 2 inches long is included on the map; *H. parviflora* Michx. with small flowers and club-shaped fruits extends from Georgia and Mississippi into Florida; and *H. diptera* Ellis with a 2-winged fruit extends from South Carolina to Oklahoma and south.

OLEACEAE

The Ashes

Characteristics of the Genus *Fraxinus* L.

HABIT. Deciduous trees, or rarely shrubs; ornamental with handsome foliage; several species are important timber trees.

LEAVES. Opposite; odd-pinnately compound (rarely reduced to a single leaflet); without stipules; petiolate; deciduous; leaflets serrate or entire, sessile or petiolulate.

FLOWERS. Regular; perfect; dioecious or polygamous; small, but quite conspicuous in slender-branched panicles; appearing before or with the leaves; calyx 4-lobed or wanting; corolla usually 4-lobed or wanting; stamens usually 2 (rarely 3 or 4); single 2-celled ovary (rarely 3-celled); ovules suspended in pairs from inner angle of the cell.

FRUIT. Samara; 1-seeded (rarely 2 or 3); with an elongated terminal wing. Seed: oblong; compressed; filling cavity in the fruit; chestnut-brown; albuminous.

TWIGS. Slender to stout; glabrous or pubescent; pith thick, rounded, homogeneous; leaf scars suborbicular to semicircular, sometimes notched on the upper edge; bundle scars numerous. Winter buds: terminal larger than lateral; both with 1-3 pairs of scales, the inner accrescent.

BARK. Thick and furrowed or rarely thin and scaly.

WOOD. Ring-porous; late wood with rather few pores not in distinct radial lines and with tangential bands of parencyhma; tough; straight-grained; not structural timber, but important for specialty purposes; sapwood not durable.

SILVICAL CHARACTERS. Rather intolerant trees; rapid-growing; fibrous root system; reproducing well naturally and artificially; comparatively free from destructive attacks by insects and fungi.

GENERAL. This genus contains about 65 species of trees scattered through the Northern Hemisphere and extending into the tropical forests of Java and Cuba. In North America there are 18 recognized native species.

KEY TO THE SPECIES OF ASHES

1. Flowers with corolla, perfect, showy, in terminal panicles; leaves 5-7 inches long, 3-7 lanceolate, stalked leaflets; southwestern . . . . *F. cuspidata*, Fragrant Ash, p. 345
1. Flowers without corolla, dioecious or polygamous, axillary panicles.
   2. Leaflets 1/2-3/4 inch long, spatulate; leaves 1 1/2-3 inches long; southwestern . . . . *F. greggii*, Gregg Ash, p. 345
   2. Leaflets 1-6 inches long; leaves (unless simple) over 3 inches long.
      3. Twigs 4-angled; fruit compressed, oblong wing extending to base.
         4. Leaves 8-12 inches long, 5-11 leaflets; flower with minute calyx; eastern . . . . *F. quadrangulata*, Blue Ash, p. 353
         4. Leaves 1 1/2-6 inches long, 1-5 leaflets; southwestern.
            5. Leaflets 1 (rarely 2-3); fruit 1/2 inch long . . . . *F. anomala*, Singleleaf Ash, p. 345
            5. Leaflets 5 (rarely 3); fruit 1-1 1/2 inches long.
               6. California . . . . *F. dipetala*, Two-petal Ash, p. 347
               6. Arizona . . . . *F. lowellii*, Lowell Ash, p. 355
      3. Twigs terete.
         7. Body of fruit compressed, wing extending to seed base.
            8. Leaflets 7-11, sessile; wing oblong, flat; northern . . . . *F. nigra*, Black Ash, p. 351
            8. Leaflets 3-7, stalked; fruit elliptic, often 3-winged; southeastern . . . . *F. caroliniana*, Carolina Ash, p. 351
         7. Body of fruit nearly terete, wing not extending to base.
            9. Wing terminal or not decurrent to below middle of seed cavity; leaflets stalked.
               10. Leaflets crenulate-serrate to entire, glabrous; twigs nearly glabrous, leaf scars notched at top.
                   11. Leaves 8-12 inches long, 7 (rarely 5-9) leaflets; eastern . . . . *F. americana*, White Ash, p. 347
                   11. Leaves 5-8 inches long, 5 (rarely 7) leaflets; Texas and Oklahoma . . . . *F. texensis*, Texas Ash, p. 347
               10. Leaflets sharply serrate, at least above middle, glabrous to hairy below; twigs glabrous to hairy, leaf scar truncate or slightly notched . . . . *F. pennsylvanica*, Green Ash, p. 349
            9. Wing decurrent to below middle of seed cavity.
               12. Fruit 2-3 inches long, frequently 3-winged; leaflets soft-pubescent below; coastal and river swamps of eastern United States . . . . *E. profunda*, Pumpkin Ash, p. 349
               12. Fruit not over 2 inches long; western.
                   13. Lateral leaflets usually sessile; West Coast . . . . *F. latifolia*, Oregon Ash, p. 353
                   13. Lateral leaflets stalked; southwestern.
                       14. Fruit 1-1 1/2 inches long; leaflets glabrous . . . . *F. berlandieriana*, Berlandier Ash, p. 355
                       14. Fruit 3/4 inch long; leaflets tomentose below . . . . *F. velutina*, Velvet Ash, p. 355

* * *

SYMPLOCACEAE

Sweetleaf, *Symplocos tinctoria* (G.) L'Her., extending Delaware to Florida, west to Tennessee, Arkansas, Texas, has oblong, nearly entire, thick, alternate, subpersistent leaves 5-6 inches long; small, yellow, fragrant flowers in dense, axillary spikes; dry, brown drupe about 1/4 inch long. Rarely tree size.

Singleleaf Ash

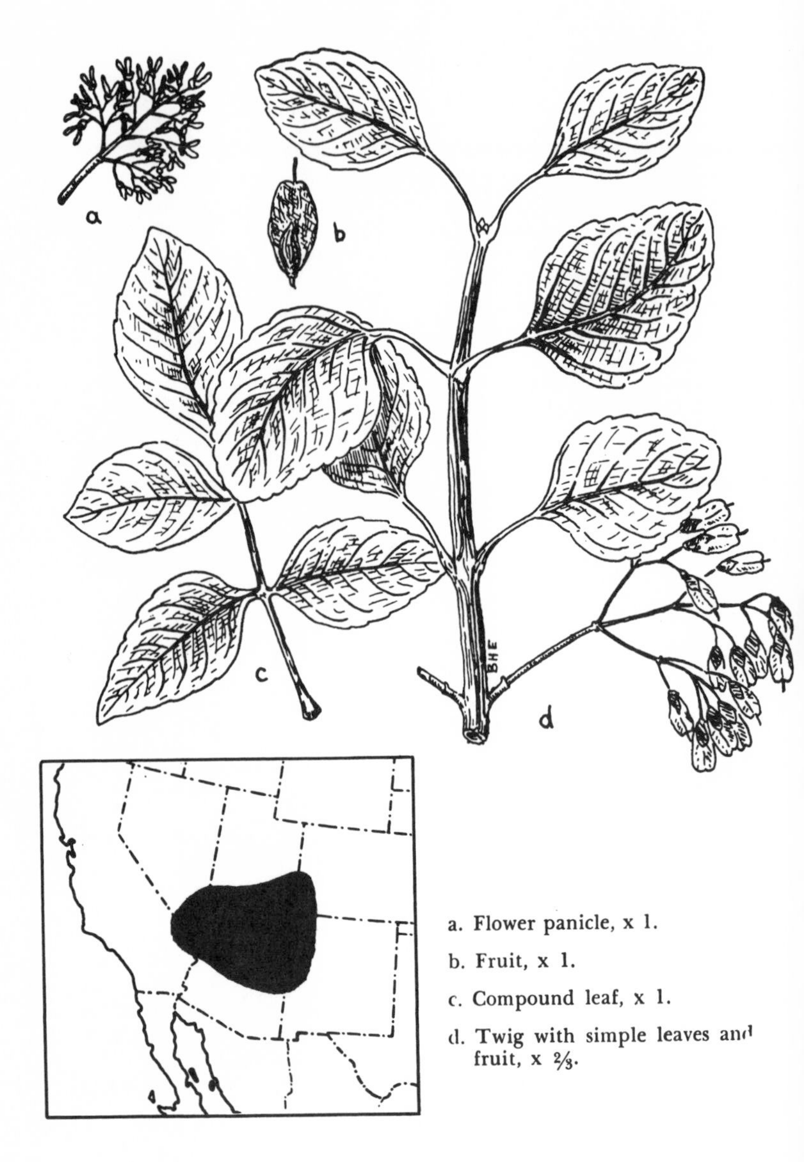

a. Flower panicle, x 1.

b. Fruit, x 1.

c. Compound leaf, x 1.

d. Twig with simple leaves and fruit, x ⅔.

OLEACEAE

Singleleaf Ash. Dwarf Ash

Fraxinus anomala Torr.

HABIT. A shrub or small tree 18-20 feet high; crown round-topped, contorted branches.

LEAVES. Usually single, but rarely 2-5 leaflets; broadly ovate or suborbicular; 1-2 inches long (smaller if compound); margins entire or sparingly crenate-serrate above middle; glabrous and dark green above, paler below.

FLOWERS. In short panicles; appearing with leaves; perfect or unisexual by abortion of stamens; calyx cup-shaped, minutely 4-toothed; corolla absent.

FRUIT. Obovate-oblong; 1/2 inch long; wing rounded or emarginate at apex, surrounding flattened seed cavity.

TWIGS. Quadrangular, slightly winged, and orange-colored at first; later round, ash-gray. Winter buds: terminal broad-ovoid, 1/8-1/4 inch long, covered by orange tomentum.

BARK. Thin (1/2 inch); dark brown, slightly tinged with red; divided by shallow furrows into narrow, scaly ridges.

WOOD. Heavy; hard; close-grained; heartwood light brown; sapwood lighter colored and thick; no importance.

SILVICAL CHARACTERS. Intolerant; in the neighborhood of streams or on dry hillsides.

* * *

Fragrant Ash. Flowering Ash

Fraxinus cuspidata Torr.

A handsome shrub or low tree with showy, white flowers and small leaves; seldom a tree in the United States; upper Sonoran zone of Texas, New Mexico, Arizona, and Mexico.

* * *

Gregg Ash. Littleleaf Ash

Fraxinus greggii Gray

This shrub or small tree, rarely 25 feet high, is characterized by 3-7 leaflets less than 1 inch long, usually entire, covered below with small, black dots, and obscurely veined; spatulate samaras, 1 inch long; twigs slender and round; bark thin, separates into papery scales; reported from Santa Cruz County, Arizona, western Texas, and Mexico.

White Ash

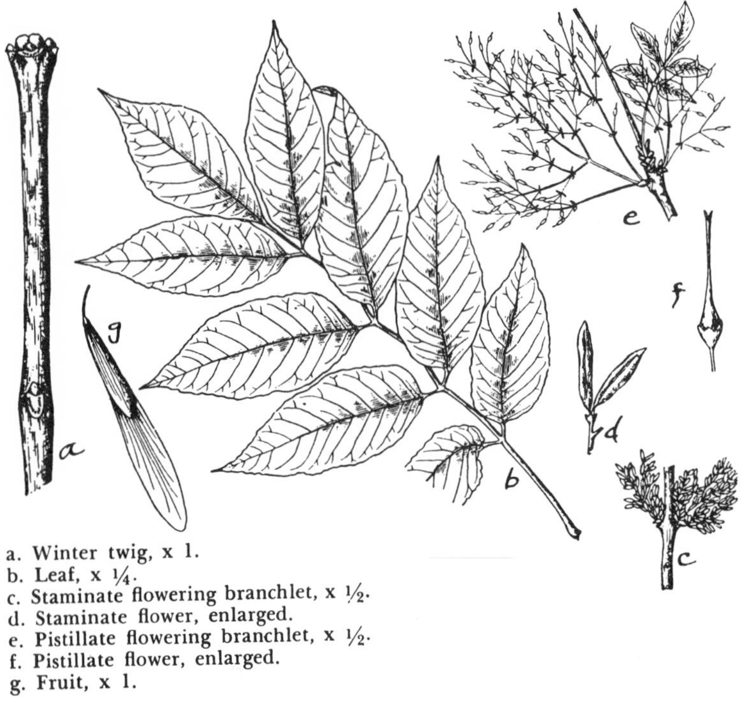

a. Winter twig, x 1.
b. Leaf, x 1/4.
c. Staminate flowering branchlet, x 1/2.
d. Staminate flower, enlarged.
e. Pistillate flowering branchlet, x 1/2.
f. Pistillate flower, enlarged.
g. Fruit, x 1.

White Ash

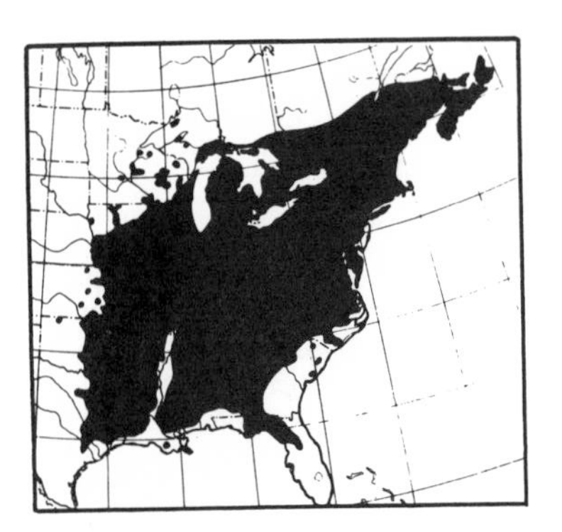

Texas Ash

White Ash

Fraxinus americana L. (*Fraxinus biltmoreana* Beadle)

HABIT. A medium-sized tree 50-80 feet high and 2-3 feet in diameter (max. 125 by 6 feet); open, pyramidal crown.

LEAVES. 8-12 inches long; 5-9 (mostly 7) leaflets, ovate to oblong-lanceolate, 3-5 inches long, entire or obscurely toothed, dark green above and paler and usually glabrous below, petiolules 1/4-1/2 inch long.

FLOWERS. In loose panicles; dioecious; corolla absent; appearing with or before the leaves.

FRUIT. Lanceolate to oblanceolate; 1-2 inches long; wing terminal or slightly decurrent along seed cavity; persistent into winter on twigs.

TWIGS. Rounded; rather stout; gray-green; lustrous; leaf scars semiorbicular to U-shaped with deep to shallow notch. Winter buds: small; rounded; dark brown; nearly glabrous; inset in leaf scar; first lateral buds at base of terminal bud.

BARK. Thick; gray; closely fissured, narrow ridges around diamond-shaped areas.

WOOD. Important; hard; heavy; strong; ring-porous; used for handles, implements, containers, etc.

SILVICAL CHARACTERS. Intermediate in tolerance; moist sites; rapid growth; the most abundant and commonly planted species.

* * *

Texas Ash

Fraxinus texensis (Gray) Sarg.

This small tree of Oklahoma and Texas is very similar to white ash, differing from it in having smaller leaves (5-8 inches long) with 5 (rarely 7) leaflets 1-3 inches long.

* * *

Two-petal Ash

Fraxinus dipetala Hook. & Arn.

This shrubby species of central to lower California forms a small tree. It is characterized by 4-angled twigs, a 2-petaled corolla, 5 leaflets, and flowers that appear before the leaves from leafless, axillary buds.

Green Ash

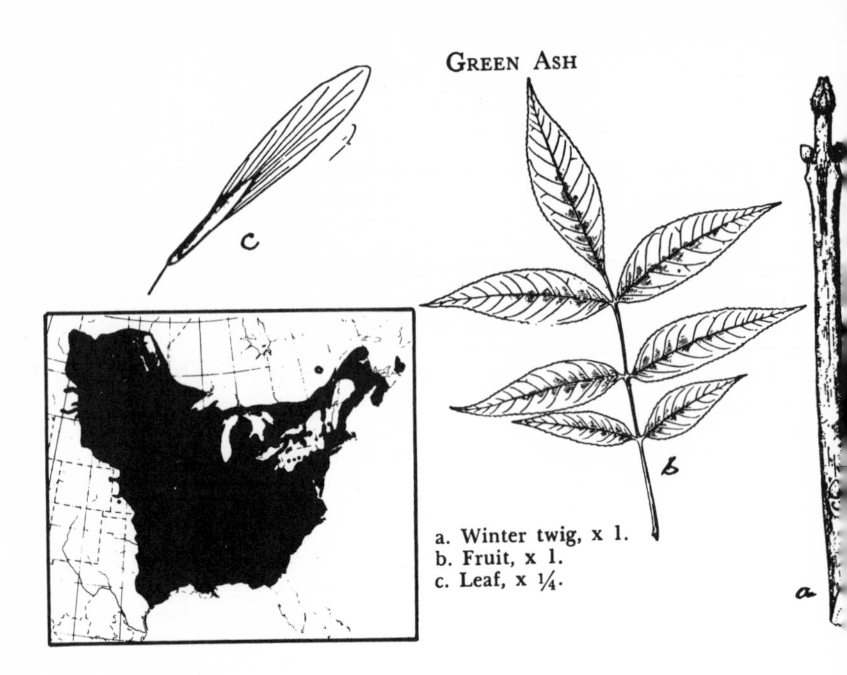

a. Winter twig, x 1.
b. Fruit, x 1.
c. Leaf, x 1/4.

Pumpkin Ash

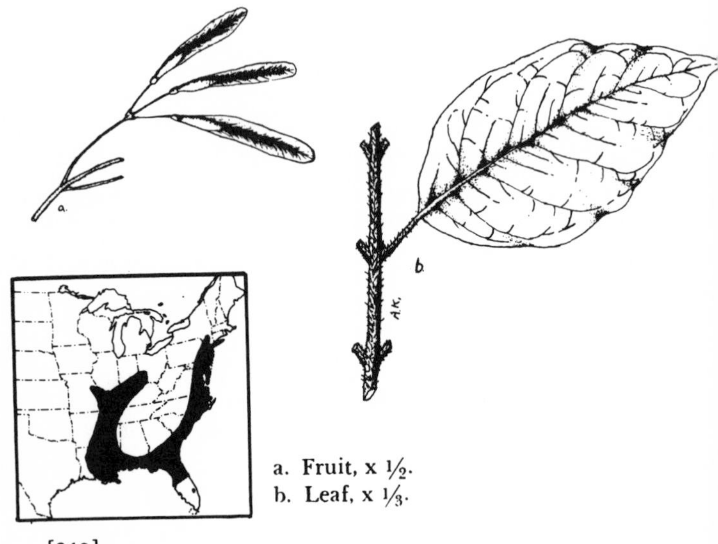

a. Fruit, x 1/2.
b. Leaf, x 1/3.

Green Ash
Fraxinus pennsylvanica Marsh.

HABIT. A small to medium-sized tree 30-60 feet high and 1-2 feet in diameter (max. 85 by 3 feet); broad, irregular crown.

LEAVES. 10-12 inches long; 7-9 leaflets, oblong-lanceolate to elliptic, 4-6 inches long, finely toothed or entire below middle, yellow-green above, paler and glabrous to silky-hairy below, petiolule 1/8-1/4 inch long.

FLOWERS. In compact panicles; dioecious; corolla absent; appearing with leaves.

FRUIT. Narrowly lanceolate; 1-2 1/2 inches long; wing terminal or extending to middle of terete, slender seed cavity; persistent through winter.

TWIGS. Rounded; rather stout; red-gray; glabrous to pale-hairy; leaf scar semiorbicular, truncate, or shallowly notched. Winter buds: small; rounded; red-brown; woolly; set above leaf scar; first laterals at base of terminal bud.

BARK. Thin; brown; shallow fissures and scaly ridges.

WOOD. Important; heavy; hard; ring-porous; like white ash.

SILVICAL CHARACTERS. Intolerant; moist sites, but hardy on dry sites; rapid growth; shallow roots.

GENERAL. This species closely resembles white ash and the glabrous form can best be distinguished by straight or slightly notched leaf scars and usually narrower leaflets and fruits. Formerly the hairy form was designated *F. pennsylvanica,* red ash, and the glabrous form *F. pennsylvanica* var. *lanceolata* (Borkh.) Sarg., green ash.

* * *

Pumpkin Ash
Fraxinus profunda (Bush) Bush

HABIT. A large, unimportant tree of coastal plain swamps and river bottoms. Characterized by prominently swollen base and a slender trunk.

LEAVES. 10-18 inches long; 7 (rarely 9) elliptic leaflets, 4-10 inches long, entire or slightly toothed, glabrous to hairy below, petiolules 1/4-1/2 inch.

FRUIT. Oblong; 2-3 inches long; wing extending from middle to base of terete seed cavity; frequently 3-winged.

TWIGS. Terete; stout; gray-brown; smooth to tomentose; leaf scars U-shaped. Winter buds: red-brown; outer bud scales with truncated tips.

BARK. Thin; light gray with shallow fissures.

Black Ash

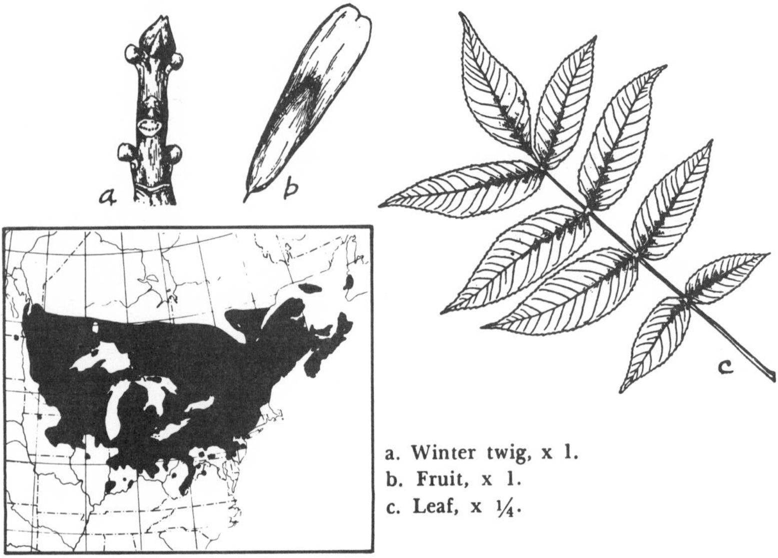

a
b
c

a. Winter twig, x 1.
b. Fruit, x 1.
c. Leaf, x ¼.

Carolina Ash

a
b

a. Fruit, x 1.
b. Leafy twig, x ⅓.

OLEACEAE

Black Ash

Fraxinus nigra Marsh.

HABIT. A medium-sized tree 40-60 feet high and 1-2 feet in diameter (max. 90 by 5 1/2 feet); narrow, small, open crown.

LEAVES. 12-16 inches long; 7-13 leaflets, oblong to oblong-lanceolate, 3-5 inches long, finely toothed, glabrous, dark green above and paler below, sessile.

FLOWERS. In loose panicles; polygamo-dioecious; corolla and calyx absent; appearing before the leaves.

FRUIT. Oblong; 1-1 3/4 inches long; wing surrounding the flat, indistinct seed cavity; falling early or late.

TWIGS. Rounded; stout; gray; glabrous. Winter buds: 1/4 inch long; ovoid-conical; nearly black; nearly glabrous; first laterals some distance below terminal bud.

BARK. Thin; gray; smooth, becoming scaly.

WOOD. Moderately important; softer and weaker than white ash.

SILVICAL CHARACTERS. Intolerant; wet sites.

* * *

Carolina Ash. Water Ash

Fraxinus caroliniana Mill. (*Fraxinus pauciflora* Nutt.)

HABIT. A small, unimportant tree of coastal or river swamps.

LEAVES. 5-12 inches long; 3-7 (usually 7) thin, mostly glabrous, long-stalked, coarsely toothed leaflets.

FLOWERS. Dioecious; appearing before the leaves.

FRUIT. Oblong-obovate to elliptic; 2-3 inches long; thin wing extending to the base of the compressed seed cavity that is more than half the length of the wing; frequently 3-winged.

SILVICAL CHARACTERS. Intolerant; wet sites.

Blue Ash

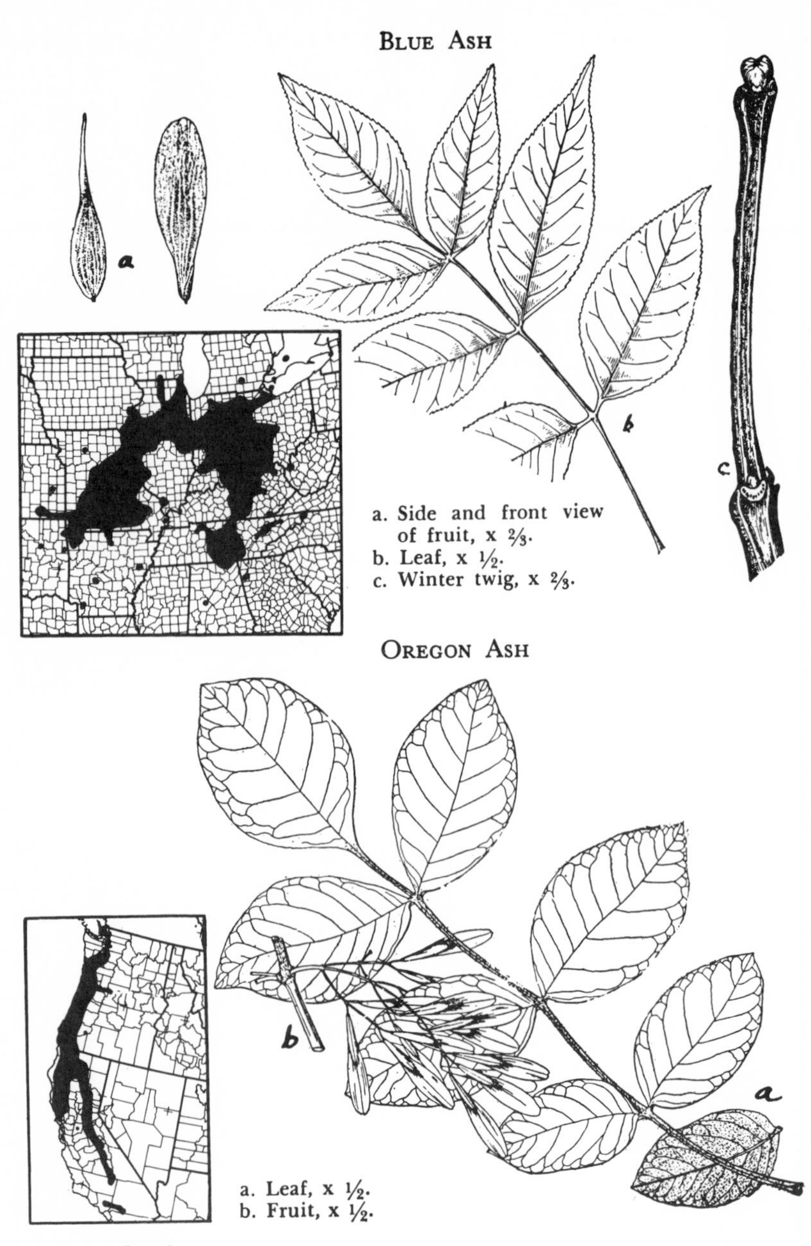

a. Side and front view of fruit, x ⅔.
b. Leaf, x ½.
c. Winter twig, x ⅔.

Oregon Ash

a. Leaf, x ½.
b. Fruit, x ½.

OLEACEAE

Blue Ash

Fraxinus quadrangulata Michx.

HABIT. A medium-sized tree 40-50 feet high and 1-2 feet in diameter (max. 120 by 3 feet); narrow, open crown.

LEAVES. 8-12 inches long; 5-11 (usually 7) leaflets, lanceolate to ovate-oblong, 3-5 inches long, coarsely toothed, thick, yellow-green above and paler below, glabrous, short petiolules.

FLOWERS. In loose panicles; perfect; corolla absent; calyx reduced to ring; appearing before the leaves.

FRUIT. Oblong-ovate; 1-2 inches long; wing surrounding the compressed seed cavity; falling soon after maturing.

TWIGS. 4-angled and corky-winged; stout; orange-brown; red-hairy. Winter buds: small; rounded; dark red-brown; somewhat hairy.

BARK. Rather thin; gray; divided into platelike scales; often shaggy; inner bark contains material that turns blue on exposure.

WOOD. Moderately important; similar to black ash.

SILVICAL CHARACTERS. Similar to green ash; a scattered tree.

* * *

Oregon Ash

Fraxinus latifolia Benth. (*Fraxinus oregona* Nutt.)

HABIT. A medium-sized tree 60-80 feet high and 2-3 feet in diameter (max. 120 by 6 feet); narrow, compact crown.

LEAVES. 5-14 inches long; 5-7 leaflets, ovate to elliptic, 3-7 inches long, entire or finely toothed, light green above and usually hairy below, sessile or with petiolules up to 1/2 inch.

FLOWERS. In compact panicles; dioecious; corolla absent; appearing with the leaves.

FRUIT. Oblong to elliptic; 1-2 inches long; wing extending to below middle of slightly compressed seed cavity.

TWIGS. Round; stout; woolly; red-brown. Winter buds: 1/8-1/4 inch long; conical; brown; hairy.

BARK. Thick; dark gray; deeply fissured with broad ridges.

WOOD. Of slight importance; similar to white ash.

SILVICAL CHARACTERS. Rather intolerant; moist sites; mixed stands; the variety *glabra* Rehd. has glabrous leaves and twigs.

Lowell Ash

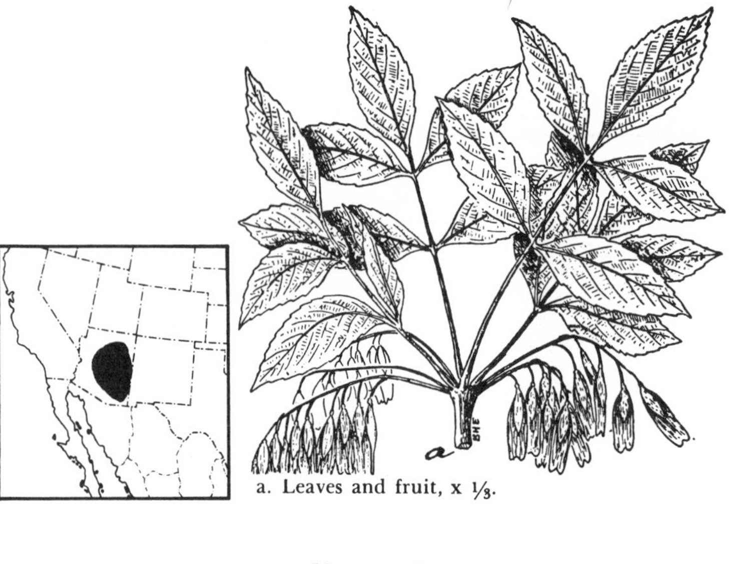

a. Leaves and fruit, x 1/3.

Velvet Ash

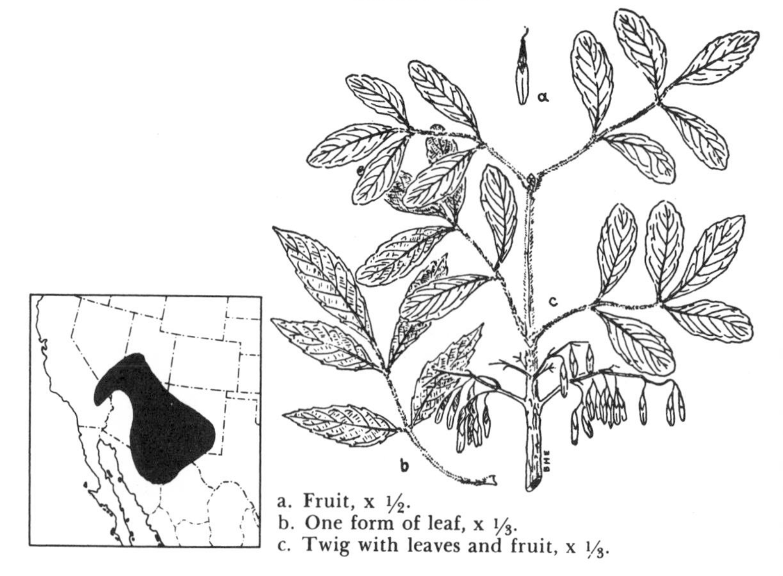

a. Fruit, x 1/2.
b. One form of leaf, x 1/3.
c. Twig with leaves and fruit, x 1/3.

Lowell Ash

Fraxinus lowellii Sarg. [*Fraxinus anomala* var. *lowellii* (Sarg.) Little]

HABIT. A small tree 20-25 feet high.

LEAVES. 3 1/2-6 inches long, stout petiole; 5 (rarely 3-7) leaflets, ovate to elliptic-ovate, 2 1/2-3 inches long, remotely and lightly serrate, yellow-green, glabrous or slightly pubescent along midrib; sometimes single.

FLOWERS. In axillary panicle; corolla absent.

FRUIT. Oblong-obovate to oblong-elliptic; 1-1 1/2 inches long; wing extending to base of compressed seed cavity.

TWIGS. Quadrangular; often winged; stout; orange-brown.

BARK. Rather thick; dark; deeply furrowed.

SILVICAL CHARACTERS. Intolerant; on dry, rocky slopes and in canyons.

* * *

Velvet Ash

Fraxinus velutina Torr. (*Fraxinus standleyi* Rehd.)

HABIT. A small, slender tree 20-50 feet high.

LEAVES. 3-6 inches long, broad, grooved, densely villous petiole; 3-9 (usually 5) leaflets, elliptic to lanceolate, 1-1 1/2 inches long, finely crenate-serrulate above middle, thick, pale green, glabrous above, tomentose below; tardily deciduous.

FLOWERS. Dioecious, in pubescent panicles; corolla absent.

FRUIT. Oblong-obovate to elliptic; 3/4 inch long; wing shorter than and extending to below middle of terete seed cavity.

TWIGS. Rounded; slender; velvety pubescent first year.

BARK. Thin; furrowed; gray, slightly tinged with red.

SILVICAL CHARACTERS. Intolerant; in dry mountain canyons and on banks of streams; hardy.

GENERAL. *F. velutina* passes into the following varieties: 1. *coriacea* Rehd., with leathery leaves. 2. *glabra* Rehd., with glabrous leaves and branches. 3. *toumeyi* Rehd., with lanceolate, acuminate leaflets, having petioles 1/8-1/2 inch long.

* * *

Berlandier Ash

Fraxinus berlandieriana A. DC.

This small tree of southern Texas and Mexico resembles *F. velutina* but has a longer fruit (1-1 1/2 inches); glabrous leaves and twigs; and thin, dark green leaflets 3-4 inches long.

FRINGETREE

a. Fruit, x ½.
b. Twig, x 1.
c. Flowering twig, x ½.

SWAMP-PRIVET

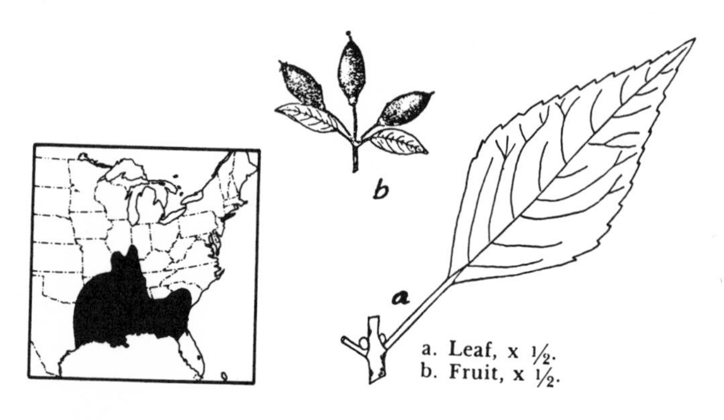

a. Leaf, x ½.
b. Fruit, x ½.

OLEACEAE

Fringetree

Chionanthus virginicus L.

HABIT. A beautiful shrub or small tree rarely 40 feet high and 1 foot in diameter; narrow, deep crown.

LEAVES. Opposite; simple; deciduous; 4-8 inches long; oval to ovate; entire; dark green above; paler and nearly glabrous below.

FLOWERS. Dioecious; showy; in drooping panicles 4-6 inches long; white, deeply divided corolla; appearing with leaves.

FRUIT. An oval, nearly black, thin-fleshed drupe 1/2-3/4 inch long; stone thick, 1/3 inch long.

TWIGS. Rather stout; green-brown; hairy. Winter buds: terminal present, ovoid, acute, 1/8 inch long, scaly.

BARK. Thin; scaly; red-brown.

WOOD. Unimportant; heavy; hard; fine-textured.

SILVICAL CHARACTERS. Tolerant; moist to wet sites; widely planted as an ornamental.

* * *

Swamp-privet. Forestiera

Forestiera acuminata (Michx.) Poir.

A rather rare shrub or small tree distributed along streams and in swamps. It is characterized by opposite, simple, deciduous, elliptic, long-pointed, glabrous, finely toothed leaves 2 1/2-4 1/2 inches long and yellow-green; by minute, dioecious or polygamous flowers without a corolla and appearing before the leaves in fascicles or panicles; and an oblong, purple, thin-fleshed drupe about 1 inch long. Desert-olive foresteria, *F. phillyreoides* Torr., is a shrubby form in southern Arizona.

* * *

Devilwood

Osmanthus americanus (L.) B.&H.

This is a shrub to medium-sized tree of the coastal plain from Virginia to Florida and Louisiana. It is characterized by opposite, simple, persistent, oblong, thick, entire, revolute leaves 3 1/2-5 inches long, lustrous bright green; small, perfect or polygamous flowers in racemes or fascicles; and oblong, purple, thin-fleshed drupe 1/2-1 inch long. The tree gets its name from the wood, which is hard and difficult to work or split. Bigfruit osmanthus, *O. megacarpus* Small, is limited to central Florida.

ANACAHUITA

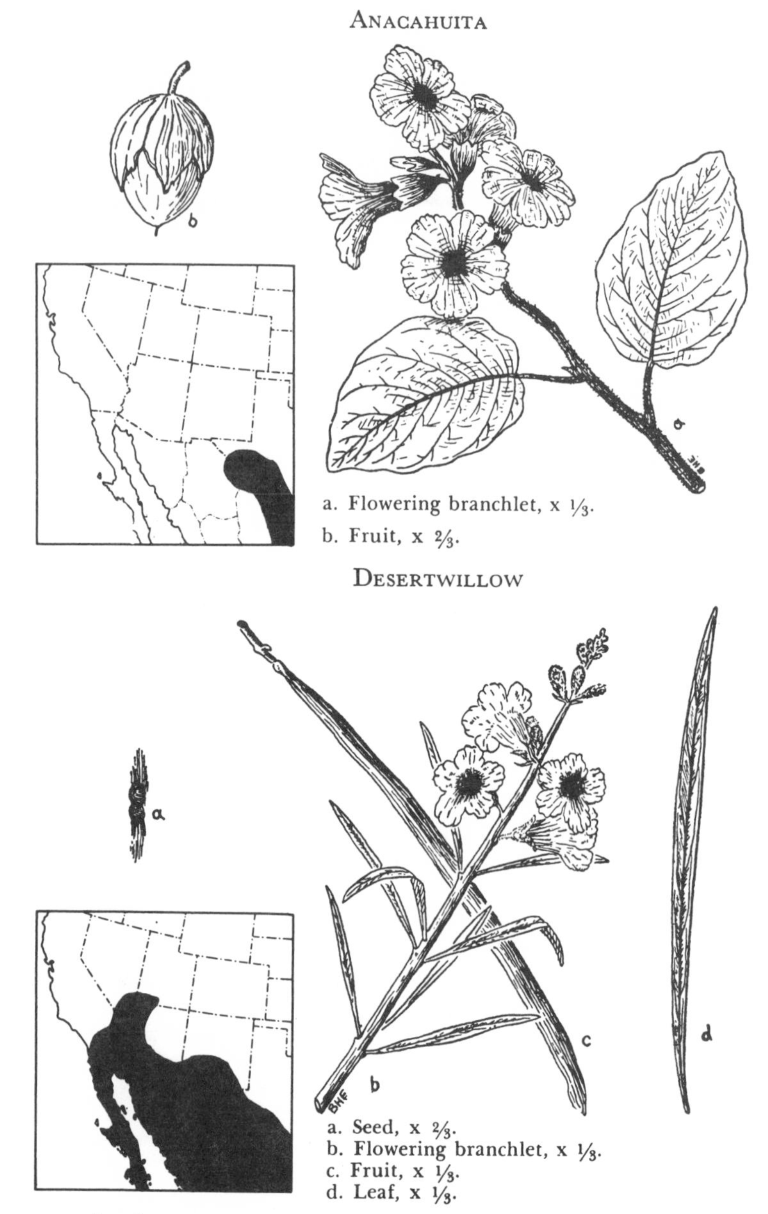

a. Flowering branchlet, x 1/3.
b. Fruit, x 2/3.

DESERTWILLOW

a. Seed, x 2/3.
b. Flowering branchlet, x 1/3.
c. Fruit, x 1/3.
d. Leaf, x 1/3.

BORRAGINACEAE

Anacahuita

Cordia boissieri A. DC.

HABIT. An aromatic shrub or small tree 20-25 feet high and 6-8 inches in diameter; crown round-topped.

LEAVES. Alternate; simple; oval to oblong-ovate; 4-5 inches long; entire or obscurely crenulate-serrate; thick and firm; dark green above; woolly below; tardily deciduous.

FLOWERS. Regular; perfect; in cymes; calyx tubular; corolla funnel-form, 2 inches across, white with yellow spot in throat.

FRUIT. Drupaceous; ovoid; 1 inch long and 3/4 inch broad; acute; lustrous, bright red-brown; enclosed by orange-brown, tomentose calyx; stone thick-walled.

TWIGS. Stout; dark gray or brown; puberulous; marked by occasional large lenticels and elevated obcordate leaf scars.

BARK. Thin; gray; divided into broad, flat ridges.

WOOD. Light; rather soft; close-grained; diffuse-porous.

SILVICAL CHARACTERS. Intolerant; dry limestone ridges and depressions; planted as ornamental.

* * *

BIGNONIACEAE

Desertwillow

Chilopsis linearis (Cav.) Sweet (*Chilopsis saligna* D. Don)

HABIT. A shrub or small tree rarely 20-30 feet high; trunk usually reclining; crown narrow with slender branches.

LEAVES. Opposite or scattered; simple; linear or linear-lanceolate; 5-12 inches long and 1/4-1/3 inch wide; acuminate; entire margins; thin; light green; deciduous during winter.

FLOWERS. Irregular; perfect; showy; in racemes; corolla white, yellow-spotted in throat, 3/4-1 1/2 inches long.

FRUIT. Slender, elongated, thin-walled capsule, 7-12 inches long and 1/4 inch thick; splitting into 2 concave valves; persistent into the winter. Seed: numerous; 1/3 inch long.

TWIGS. Slender; glabrous or densely tomentose; light brown. Winter buds: terminal absent; lateral minute, scaly.

BARK. Thin (1/8-1/4 inch); dark brown; furrowed.

WOOD. Soft; not strong; close-grained; ring-porous; brown.

SILVICAL CHARACTERS. Intolerant; short-lived; banks of water courses in desert and low mountain areas.

Northern Catalpa

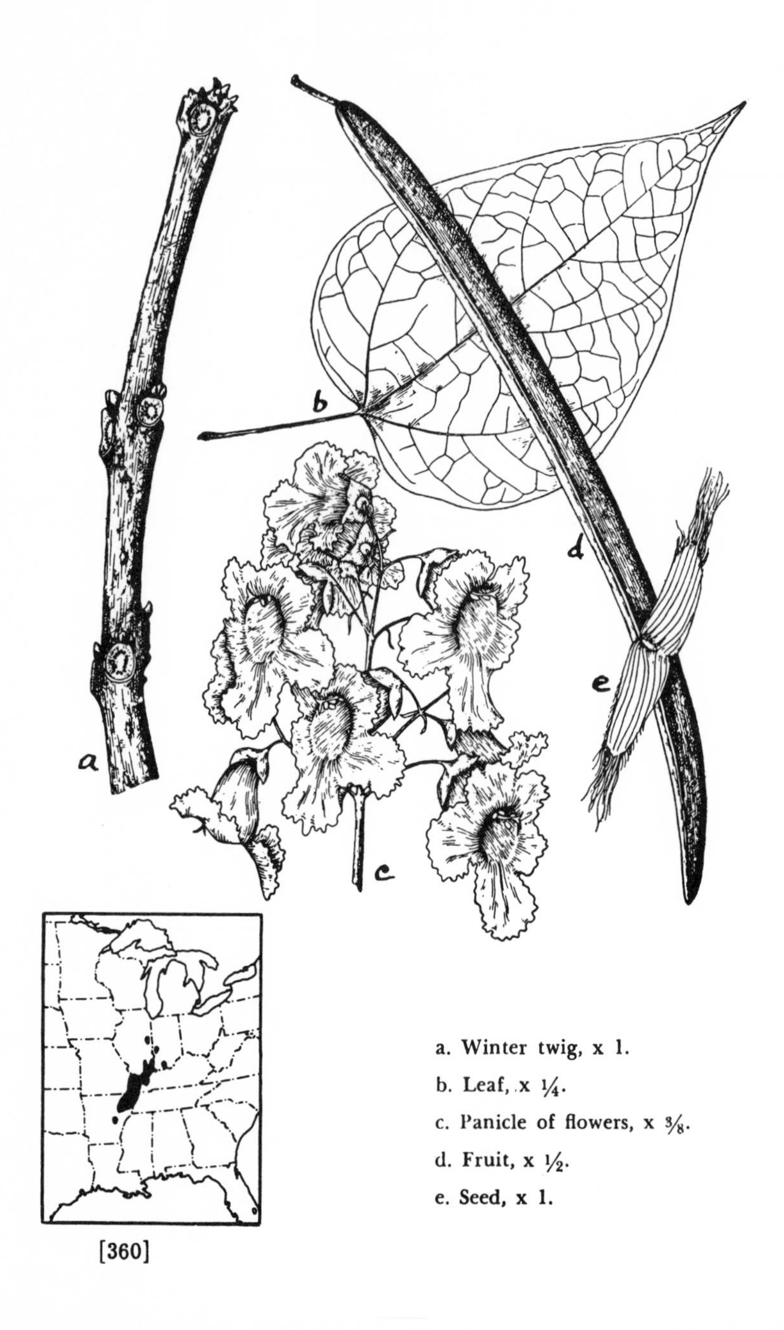

a. Winter twig, x 1.

b. Leaf, x 1/4.

c. Panicle of flowers, x 3/8.

d. Fruit, x 1/2.

e. Seed, x 1.

BIGNONIACEAE

Northern Catalpa. Hardy Catalpa
Catalpa speciosa Warder

HABIT. A medium-sized tree 30-60 feet high and 1-3 feet in diameter (max. 120 by 6 feet); broad, rounded crown.

LEAVES. Opposite or whorled; simple; deciduous; 8-12 inches long; heart-shaped; entire; thick; dark green above; paler and hairy below; petioles long, stout, round.

FLOWERS. 2-lipped; perfect; showy; in open few-flowered panicles; calyx hairy; corolla white, inconspicuous yellow spots, 2 1/2 inches across; appearing after the leaves.

FRUIT. A long, round, 2-celled, thick-walled capsule 8-20 inches long and 1/2 inch thick; persistent through winter. Seed: numerous; flattened; rounded, fringed wings 1 inch long.

TWIGS. Very stout; brown; glabrous; orbicular leaf scar. Winter buds: terminal absent; lateral small, immersed in bark.

BARK. Rather thin; brown; broken into thick scales.

WOOD. Light; soft; weak; brown; durable; ring-porous; used for posts.

SILVICAL CHARACTERS. Intolerant; varied sites; fast growth; taproot; widely planted and naturalized through all but coldest portions of United States.

* * *

Southern Catalpa
Catalpa bignonioides Walt.

This smaller and more southern tree, while native from Florida to Louisiana, has been naturalized as far north as New York. It is less hardy in cold sites than northern catalpa and is distinguished from it by having flowers 1 1/2 inches across in many-flowered, crowded panicles with glabrous calyx and corolla with many conspicuous yellow spots and a thinner-walled fruit about 1/4-1/3 inch in diameter.

* * *

Paulownia
Paulownia tomentosa (Thunb.) Sieb & Zucc.

This beautiful Chinese species has been widely cultivated in the eastern states and has become naturalized from New York to Georgia. In cold sites it winter-kills badly. It is characterized by opposite, heart-shaped, simple, deciduous leaves with entire margins, 5-8 inches long and densely hairy on the lower surface; by perfect, 2-lipped, showy, purple flowers 1 1/2-2 inches long, which appear before the leaves; and by a leathery, ovoid, beaked, persistent, brown capsule 1-2 inches long.

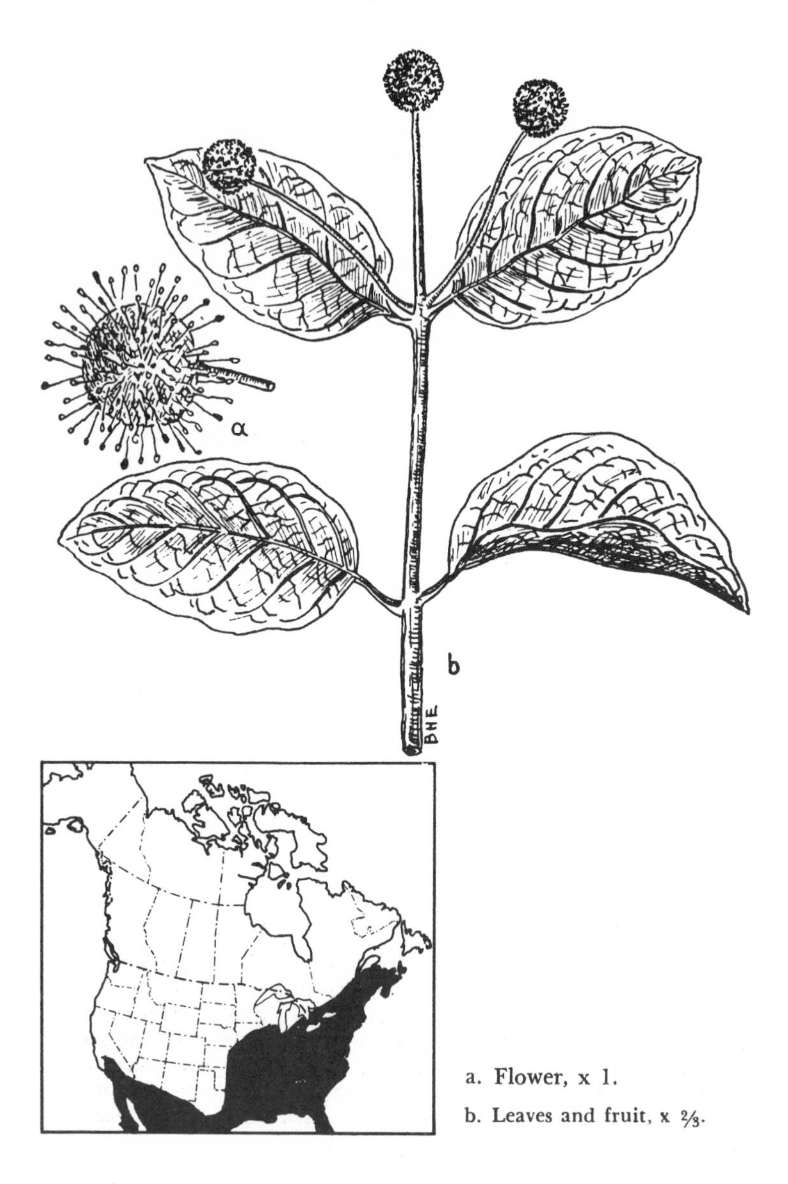

a. Flower, x 1.

b. Leaves and fruit, x ⅔.

RUBIACEAE

Common Buttonbush

Cephalanthus occidentalis L.

HABIT. A shrub or rarely a small tree up to 50 feet high and 1-2 inches in diameter; open, spreading crown.

LEAVES. Opposite or whorled in 3's; simple; ovate, lanceolate or elliptic; 2-7 inches long; acute or acuminate at apex; margins entire; thin; dark green and glabrous above; paler below, yellow midrib; tardily deciduous during winter; petioles stout, grooved, glabrous, 1/2-3/4 inch long.

FLOWERS. Regular; perfect; minute; in dense, globose heads 1-1 1/2 inches in diameter; fragrant; calyx tube 4-5 lobed; corolla cream-white, salverform, 4-5-lobed; stamens as many as and alternate with corolla lobes; ovary inferior, 2-celled, with protruding, threadlike style and capitate stigma.

FRUIT. Nutlike capsule; inversely pyramidal; splitting from base upwards into 2-4 closed, 1-seeded portions; in heads 5/8-3/4 inch in diameter; green tinged with red, becoming dark red-brown. Seed: small; oblong; pendulous.

TWIGS. Stout; glabrous; thick pith; marked by large lenticels; opposite or in whorls of 3; light green at first, becoming red-brown. Winter buds: terminal absent; lateral minute, nearly immersed in the bark.

BARK. Thin; dark brown to nearly black; broad, flat, superficially scaly ridges; contains tannin.

WOOD. Moderately heavy and hard; fine-grained; diffuse-porous; light red-brown; unimportant.

SILVICAL CHARACTERS. Rather tolerant; reproduction abundant; on moist sites or in dry stream beds; often forming dense thickets.

* * *

BIGNONIACEAE

Trumpetflower

Tecoma stans (L.) H. B. K.

A shrub or small tree extending through Mexico into the southern parts of Texas, New Mexico, and Arizona. Characterized by showy, bright yellow flowers; opposite, pinnately compound leaves with 5-13 leaflets; and a linear capsule, 4-8 inches long.

Blueberry Elder

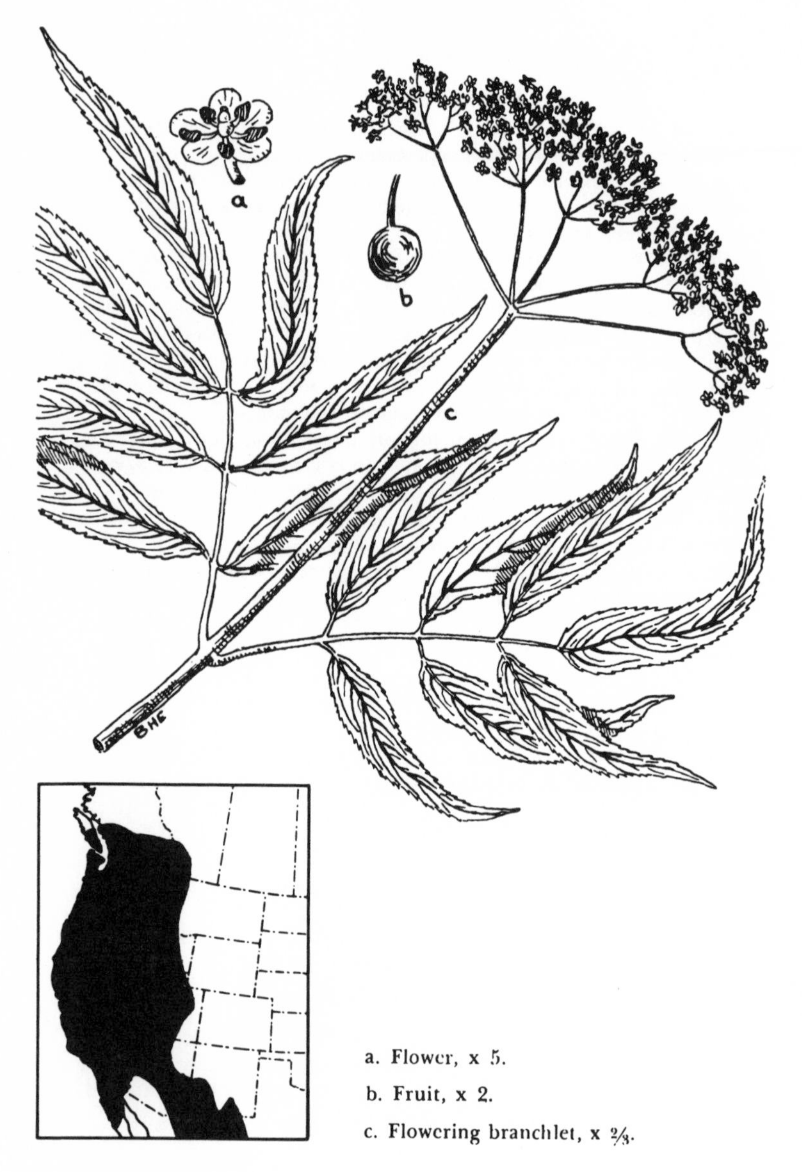

a. Flower, x 5.

b. Fruit, x 2.

c. Flowering branchlet, x ⅔.

CAPRIFOLIACEAE

Blueberry Elder

Sambucus glauca Nutt. (*Sambucus cerulea* Raf.)

HABIT. A shrub or small tree 30-50 feet high and 12-18 inches in diameter; compact, round-topped crown.

LEAVES. Opposite; unequally pinnately compound; petiolate; deciduous; 5-7 inches long; leaflets 5-9, ovate or narrow oblong, coarsely serrate margin, 1-6 inches long, green above, pale and glabrous to pubescent below.

FLOWERS. Regular; perfect; small (1/8 inch in diameter); in broad, terminal, long-branched corymbose cymes; corolla yellow-white, stamens 5; ovary inferior, 3-5-celled.

FRUIT. Dense clusters of small, blue, drupelike berries; 1/4 inch in diameter, with sweet, juicy flesh. Seed: 3-5 1-seeded nutlets in each drupe.

TWIGS. Stout; somewhat angled; pubescent first year; red-brown; nearly encircled by large, triangular leaf scars; thick, soft pith. Winter buds: terminal absent; lateral scaly, greenish.

BARK. Thin; dark brown, tinged with red.

WOOD. Light; soft; weak; coarse-grained; diffuse-porous; heartwood yellow, tinged with brown, durable; unimportant.

SILVICAL CHARACTERS. Intolerant; short-lived; reproduction abundant but scattered; coppices freely; moist porous soils; along streams, ravines, or moist hillsides.

* * *

Seven native species reach tree size:

1. Native to western North America.
    2. Cymes flat-topped; fruit blue-black, commonly glaucous.
        3. Twigs, lower surface of leaf pubescent; leaves deciduous; western Arizona and California . . . . *S. velutina* D. & H., Velvet Elder
        3. Twigs and leaves usually glabrous or nearly so.
            4. Leaves persistent; leaflets 3-5, oblong-lanceolate to ovate; western Texas to southern California . . . . *S. mexicana* Presl., Mexican Elder
            4. Leaves deciduous; leaflets 5-9, lanceolate to oblong-lanceolate . . . . *S. glauca*, Blueberry Elder
    2. Cymes pyramidal to ovoid.
        5. Fruit red; leaflets 5-9, oval, slightly pubescent below; Pacific coast from Alaska to southern California . . . *S. callicarpa* Greene, Pacific Red Elder
        5. Fruit black; leaflets commonly 5, scurfy-puberulous or villous below; Alberta and Montana south to New Mexico and California . . . . *S. melanocarpa* Gray, Black Bead Elder
1. Native to eastern North America; usually shrubs.
    6. Leaflets usually 5, obovate to oblong; Florida to Louisiana . . . . *S. simpsonii* Rehd., Florida Elder
    6. Leaflets usually 7, lanceolate to elliptic; Nova Scotia to Manitoba to Florida and Texas . . . . *S. canadensis* L., American Elder

Nannyberry

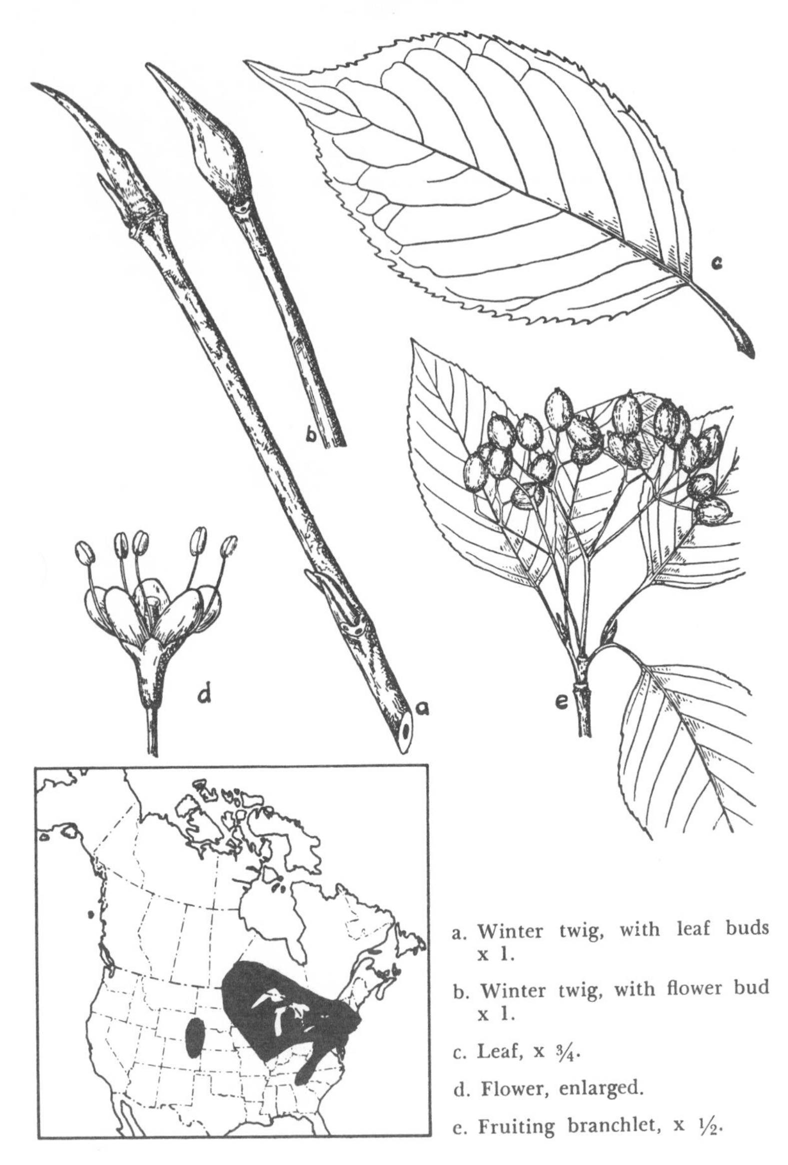

a. Winter twig, with leaf buds x 1.

b. Winter twig, with flower bud x 1.

c. Leaf, x ¾.

d. Flower, enlarged.

e. Fruiting branchlet, x ½.

Nannyberry

Viburnum lentago L.

HABIT. A shrub or small tree 15-30 feet high and 6-10 inches in diameter; bushy, compact, rounded crown.

LEAVES. Opposite; simple; ovate; 2-4 inches long and 1-2 inches wide; sharply serrate; thick and firm; lustrous and bright green above; yellow-green and marked with minute, black dots below; deciduous; turning orange and red in autumn.

FLOWERS. Regular; perfect; small (1/4 inch in diameter); fragrant; in stout-branched, scurfy, terminal cymes 3-5 inches across, corolla tubular, 5-lobed, white; stamens 5; ovary inferior, 1-celled.

FRUIT. Few-fruited, red-stemmed clusters of small, juicy, blue-black, berrylike drupes; each drupe oval or ovoid, flattened, covered with glaucous bloom. Seed: solitary within oval; rough, flattened nutlet.

TWIGS. Slender; light green and hairy, becoming dark red-brown. Winter buds: enclosed by one pair of valvate scales; flower buds 3/4 inch long, grayish, swollen; terminal leaf bud 1 inch long, light red, narrow, long-pointed.

BARK. Red-brown; irregularly broken into small, thick plates that are superficially scaly.

WOOD. Heavy; hard; close-grained; diffuse-porous; ill-scented; heartwood dark orange-brown; unimportant.

SILVICAL CHARACTERS. Intolerant; very hardy; an attractive small ornamental tree; endures city smoke.

* * *

Five species reach tree size:

1. Leaves entire or nearly so, dark green and lustrous.
    2. Leaves 4-6 inches long, broad-oval; cymes long-stalked; Connecticut to Kentucky, south to Florida and Texas. . . *V. nudum* L., Possumhaw Viburnum
    2. Leaves 1/2-2 inches long, oblong to spatulate; cymes sessile; Virginia to Florida . . . . *V. obovatum* Walt., Walter Viburnum
1. Leaves sharply toothed; cymes sessile or short-stalked.
    2. Leaves lustrous above, petioles and veins red-hairy below; buds red-hairy; New Jersey to Kansas and south . . . . . . *V. rufidulum* Raf., Rusty Blackhaw
    2. Leaves dull above, petioles and veins not hairy below; buds long and slender, scurfy or glabrous.
        3. Petioles and veins on lower leaf surface and buds red-scurfy . . . . *V. lentago* L., Nannyberry
        3. Petioles and veins glabrous; buds gray to brown, glabrous or scurfy; Connecticut to Michigan and Kansas and south . . . . *V. prunifolium* L., Blackhaw

SOAPTREE YUCCA

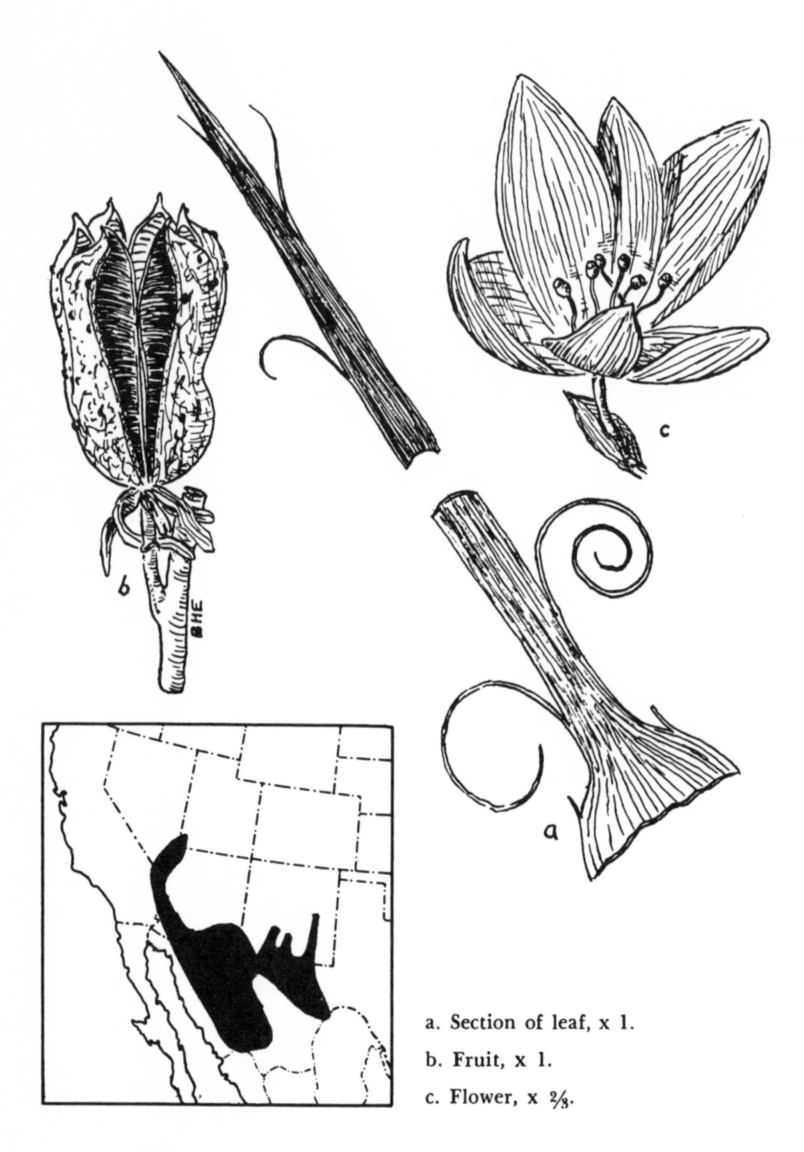

a. Section of leaf, x 1.

b. Fruit, x 1.

c. Flower, x ⅔.

LILIACEAE

Soaptree Yucca. Spanish-bayonet
Yucca elata Engelm.

HABIT. A tree often 15-20 feet high and 7-8 inches in diameter with a deep, branched, underground stem; simple or branched at top; covered with pendent, persistent dead leaves.

LEAVES. Alternate; simple; 20-30 inches long and 1/4-1/2 inch wide; thin and flat above; rounded below; glabrous; yellow-green; entire pale margins soon splitting into slender filaments; parallel-veined; persistent; stipules absent.

FLOWERS. Regular; in compound, terminal panicles, 4-6 feet high; perianth cup-shaped, white, 3 1/2-4 inches across; stamens 6; ovary 6-sided, 3-celled; ovules numerous.

FRUIT. Erect, oblong capsule, 1 1/2-2 inches long; 3-valved; light brown, thin, and woody outside; light yellow inside; edible. Seed: numerous; black; 1/3 inch wide.

BARK. Dark brown; thin plates; 1/4 inch thick.

WOOD. Light; soft; spongy; pale brown or yellow; in concentric layers.

SILVICAL CHARACTERS. Lower Sonoran zone; intolerant; on desert plateaus; large roots used as substitute for soap.

* * *

Seven species reach tree size:

1. Southwestern.
   2. Fruit erect, dehiscent; flower clusters long-stalked; leaves filamentous, thin, pale yellow-green . . . . *Y. elata*
   2. Fruit pendent, indehiscent; flower clusters sessile or short-stalked; leaves concave above the middle.
      3. Fruit with thin, dry flesh; leaves blue-green, serrate; southwest Utah to California . . . . *Y. brevifolia* Engelm., Joshua-tree
      3. Fruit with succulent flesh; leaves with no, or minute, teeth.
         4. Panicle hoary-tomentose; leaves flexible, 2 1/2-3 feet long; margins not filamentous; southern Arizona . . . . *Y. schottii* Engelm., Schotts Yucca
         4. Panicle glabrous or puberulous; margins filamentous.
            5. Leaves 2 1/2-4 feet long, dark green; Texas.
               6. Leaves concave; perianth nearly free . . . . *Y. treculeana* Carr., Trecul Yucca
               6. Leaves flat, perianth united into tube . . . . *Y. faxoniana* Sarg., Faxon Yucca
            5. Leaves 1 1/2-2 feet long, light yellow-green.
               7. Style elongated; western Texas to eastern Arizona . . . . *Y. torreyi* Shaf., Torrey Yucca
               7. Style short; southeastern Nevada, northwestern Arizona . . . . *Y. mohavensis* Sarg., Mohave Yucca
1. Southeastern coastal plain, North Carolina, and south.
   8. Leaves sharply serrate on horny margin; fruit succulent . . . . *Y. aloifolia* L., Aloe Yucca
   8. Leaves entire or nearly so; fruit with thin dry flesh . . . . *Y. gloriosa* L., Moundilly Yucca

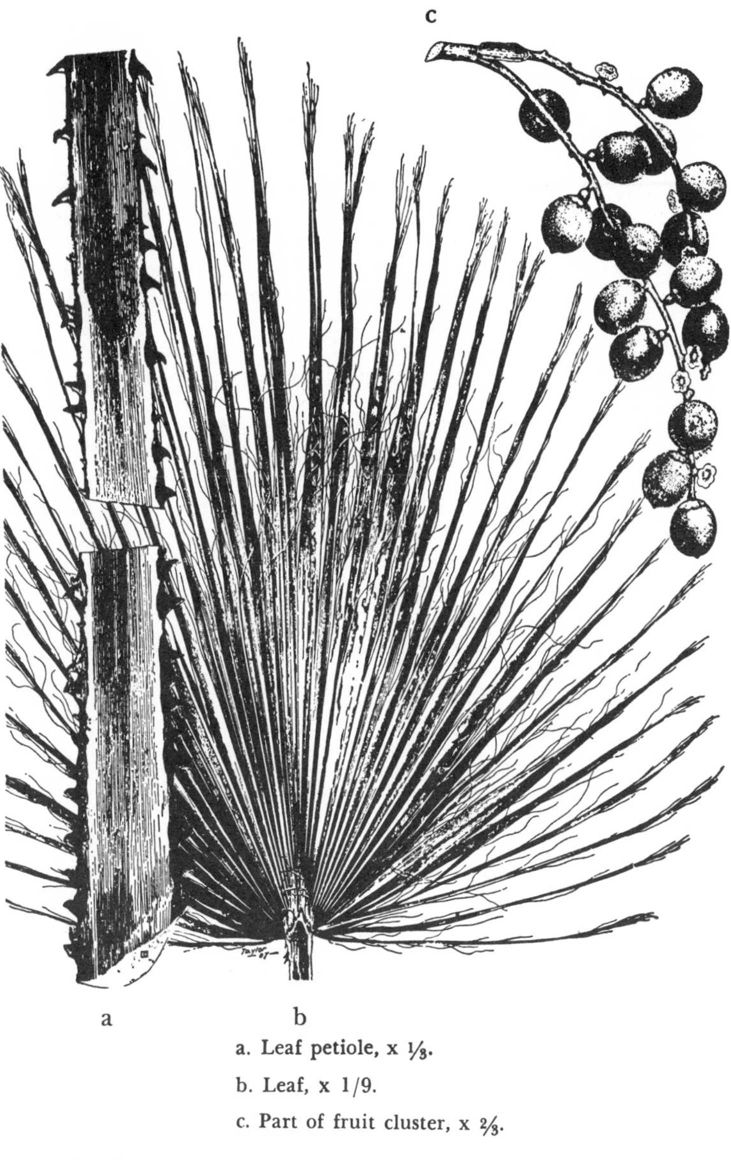

a. Leaf petiole, x 1/3.

b. Leaf, x 1/9.

c. Part of fruit cluster, x 2/3.

PALMAE

Washingtonia. California Palm

Washingtonia filifera Wendl. (*Washingtonia arizonica* Cook)
(*Washingtonia robusta* Wendl.)

HABIT. An evergreen tree 30-50 feet high and 1-2 feet in diameter; crown broad, consisting of large leaves; columnar trunk clothed with thatchlike mass of pendent dead leaves.

LEAVES. Clustered at top of stem; 3-6 feet long; fan-shaped; nearly circular; 40-70 ribbonlike folds deeply slashed 1/2-2/3 of distance to base; margins of the divisions separating into threadlike filaments; petioles 3-5 feet long, 1-3 inches wide, armed along margins with stout, hooked spines.

FLOWERS. Regular; perfect; minute; in compound clusters 8-10 feet long; appearing from axils of upper leaves; calyx tubular; corolla white, tubular; stamens 6; ovary 3-lobed, 3-celled with a single ovule in each cell.

FRUIT. Drupelike berry; 3/8 inch long; black when ripe in September; ellipsoidal; thin, dry, sweet pulp; produced in large quantities. Seed: 1/4 inch long; 1/8 inch wide; pale chestnut-brown.

BARK. Stem with thick barklike rind; narrowly furrowed; pale cinnamon to dull red-brown.

WOOD. Light; soft; spongy and fibrous; numerous, dark orange-colored, fibrovascular bundles; unimportant.

SILVICAL CHARACTERS. Moderately tolerant when young, becoming intolerant; reproduction plentiful; long, deep roots; alkaline soils; dry, warm, mountain canyons.

GENERAL. This palm is reported from southern California and southwestern Arizona, often occurring in large numbers. Considerable confusion exists as to the nomenclature and classification of the forms of *Washingtonia.*

* * *

Cabbage Palmetto

Sabal palmetto (Walt.) Lodd.

This species, of the coastal plain from North Carolina to Florida, forms a tree 40-90 feet high and 1-2 feet in diameter. It resembles *Washingtonia* in its fan-shaped leaves, flowers, and fruit but can be easily distinguished by its unarmed leaf stalks. Two other species are native to the Gulf area: Louisiana palmetto, *S. louisiana* (Darby) Bom., in Louisiana and Texas and palmetto, *S. texana* (Cook) Becc., in southern Texas.

GLOSSARY

Abortive. Imperfectly or not developed; barren.
Accrescent. Increasing in size with age.
Achene. A dry indehiscent, 1-celled, and 1-seeded fruit or carpel.
Acicular. Slenderly needle-shaped.
Acuminate. Gradually tapering to the apex; long-pointed.
Acute. Sharply pointed, but not drawn out.
Adnate. Descriptive of unlike organs or parts fused together.
Aggregate. A compound fruit developing from separate pistils of the same flower.
Alternate. Scattered singly along axis; not opposite.
Ament. A scaly, bracted spike of usually unisexual flowers, frequently deciduous in one piece.
Angiosperms. Plants with seeds borne in an ovary.
Anther. The pollen-bearing part of the stamen.
Antherozoid. Male sexual cells.
Anthesis. The time when fertilization takes place or a flower expands.
Apetalous. Without petals.
Apex. Tip.
Apiculate. Ending in a minute, short, pointed tip.
Apophysis. That part of a cone scale exposed when the cone is closed.
Appressed. Lying close and flat against.
Arborescent. Attaining the size or character of a tree.
Arcuate. Leaf veins moderately curved.
Aril. An appendage or an outer covering of a seed, growing out from the hilum or funiculus.
Attentuate. Slenderly tapering; acuminate.
Auriculate. Furnished with earlike appendage.
Awl-shaped. Tapering from the base to a slender and stiff point.
Axil. The upper angle formed by a leaf or branch with the stem.
Axillary. Situated in an axil.

Baccate. Berrylike; pulpy throughout.
Berry. A fleshy or pulpy fruit with immersed seeds.
Blade. The expanded portion of a leaf.
Bloom. A powdery or waxy substance easily rubbed off.
Bole. The stem of a tree.
Boss. A raised projection, usually pointed.
Bract. A modified leaf subtending a flower or belonging to an inflorescence.

Bractlet. The bract of a pedicel or ultimate flower stalk; a secondary bract.
Bud. The undeveloped state of a branch or flower cluster, with or without scales.
Bud scales. Modified leaves covering a bud.
Bundle (leaf). Strand of fibrovascular tissue found in cross section of leaf.

Caducous. Falling off very early.
Calyx. The flower cup or exterior part of a perianth.
Campanulate. Bell-shaped.
Canescent. Gray-pubescent and hoary.
Capitate. Shaped like a head; in dense headlike clusters.
Capsule. A dry fruit of more than one carpel that splits at maturity to release its seeds.
Carpel. A simple pistil or an element of a compound pistil.
Catkin. The same as an ament.
Caudate. Furnished with a tail or a slender tip.
Cell. The unit of structure of living things; a cavity of an ovary or anther.
Chambered. Said of pith that is interrupted by hollow spaces.
Ciliate. Fringed with hairs on the margin.
Compound. Leaves made up of several individual leaflets.
Cone. A fruit with woody, overlapping scales.
Coniferous. Pertains to cone-bearing; or the order Coniferales.
Connate. United.
Coppice. Growth arising from sprouts at the stump.
Cordate. Heart-shaped.
Coriaceous. Of the texture of leather.
Corolla. Inner part of the perianth, composed of petals.
Corymb. A flat-topped flower cluster, the flowers opening from the outside inward.
Crenate. Dentate with the teeth much rounded.
Crenulate. Diminutive of crenate, finely crenate.
Crown. The upper part of a tree, including the living branches with their foliage.
Cuneate. Wedge-shaped, or triangular with an acute angle downward.
Cuspidate. Tipped with a sharp, rigid point.
Cylindric. Shaped like a cylinder.
Cyme. A flat-topped flower cluster, the flowers opening from the center outward.

Deciduous. Not persistent; falling away as the leaves of a tree in autumn.

Decurrent. Running down, as of the blades of leaves extending down their petioles.
Decussate. In pairs alternately crossing at right angles.
Dehiscent. The opening of an anther or capsule by slits or valves.
Deliquescent. Trunk dividing into several large branches.
Deltoid. Delta-shaped, triangular.
Dentate. Toothed, with the teeth directed outward.
Denticulate. Minutely toothed.
Diadelphous. Stamens formed into two groups through the union of their filaments.
Diaphragmed. Said of pith that is solid but with more or less regularly spaced disks of horizontally elongated cells with thickened walls.
Diffuse-porous. Wood in which the pores show little difference in size throughout the seasonal growth.
Dimorphous. Occurring in two forms.
Dioecious. Unisexual, the staminate and pistillate flowers on different individuals.
Disk. A development of the receptacle at or around the base of the pistil.
Dissemination. The spreading abroad of ripe seeds from the parent plant.
Divergent. Spreading apart; pointing away.
Dorsal. Relating to the back or outer surface of an organ; the lower surface of a leaf.
Downy. Clothed with a coat of soft, fine hairs.
Drupaceous. Resembling or relating to a drupe.
Drupe. A stone fruit, such as a plum.

E. A latin prefix denoting that parts are missing.
Eglandular. Without glands.
Ellipsoidal. Of the shape of an elliptical solid.
Elliptic. Of the form of an elipse.
Emarginate. Notched at the apex.
Entire. Leaf margin without divisions, lobes, or teeth.
Erose. Descriptive of an irregularly toothed or eroded margin.
Excrescences. Warty outgrowths or protuberances.
Excurrent. Trunk extending to top of the tree.
Exfoliate. To cleave or peel off in thin layers.
Exserted. Prolonged beyond the surrounding organs, as stamens from the corolla.
Exstipulate. Without stipules.

Falcate. Scythe- or sickle-shaped.
Fascicle. Dense cluster or bundle.

Fibrovascular. Consisting of woody fibers and ducts.
Filament. The stalk of an anther.
Fluted. Regularly marked by alternating ridges and groovelike depressions.
Foliaceous. Leaflike in texture or appearance.
Follicle. A dry one-celled fruit from a simple pistil dehiscent by one suture.
Fugacious. Falling or withering away very early.
Fulvous. Tawny; dull yellow with gray.
Funiculus. The stalk of an ovule.
Furrowed. With longitudinal channels or grooves.

Gibbous. Swollen on one side.
Glabrate. Nearly glabrous or becoming glabrous.
Glabrous. Smooth, not pubescent or hairy.
Gland. Secreting surface or structure; a protuberance having the appearance of such an organ.
Glandular. Furnished with glands.
Glaucous. Covered or whitened with a bloom.
Globose. Spherical in form or nearly so.
Gymnosperms. Plants with naked seeds, i.e., not enclosed in an ovary.

Habit. The general appearance of a plant; best seen from a distance.
Habitat. The place where a plant naturally grows.
Halberdlike. Like an arrowhead, but with the basal lobes pointing outward nearly at right angles.
Hilum. The scar or place of attachment of a seed.
Hirsute. Covered with rather coarse or stiff, long hairs.
Hispid. With rigid or bristly hairs.
Hoary. Covered with a close, whitish, or gray-white pubescence.
Hybrid. A cross, usually between two related species.

Imbricate. Overlapping, like shingles on a roof.
Imperfect (flower). Containing one sex but not the other.
Indehiscent. Not splitting open; remaining closed.
Inferior ovary. Appearing to grow below the adnate calyx.
Inflorescence. Flowers appearing in clusters.
Infrastipular. Situated below the stipules.
Inserted. Attached to or growing out of.
Intolerant. Not capable of doing well under dense forest cover.
Involucre. A circle of bracts surrounding a flower cluster.
Irregular flower. Bilaterally symmetrical; similar parts of different shapes or sizes.

Keeled. With a central ridge like the keel of a boat.

Laciniate. Cut into narrow, pointed lobes.
Lanceolate. Lance-shaped.
Lateral. Situated on the side; not at apex.
Leaflet. One of the small blades of a compound leaf.
Leaf scar. Scar left on twig by the falling of a leaf.
Legume. Fruit of the pea family; podlike and splitting open by both sutures.
Lenticel. Corky growth on young bark that admits air to the interior of a twig or branch.
Linear. Long and narrow, with parallel edges.
Lobe. A somewhat rounded division of an organ.
Lobulate. Divided into small lobes.
Lustrous. Glossy, shining.

Membranaceous. Thin and somewhat translucent.
Midrib. The central vein of a leaf or leaflet.
Monoecious. The stamens and pistils in separate flowers but borne on the same individual.
Mucro. A small and abrupt tip to a leaf.
Mucronate. Furnished with a mucro (bristle-tipped).
Multiple. A compound fruit developing from ripened ovaries of separate flowers.

Naked buds. Buds without scales.
Nut. A hard and indehiscent, 1-seeded pericarp produced from a compound ovary.
Nutlet. A diminutive nut or stone.

Ob. Latin prefix signifying inversion.
Obconic. Inverted cone-shaped.
Obcordate. Inverted heart shape.
Oblanceolate. Lanceolate, with the broadest part toward the apex.
Oblique. Slanting or with unequal sides.
Oblong. About three times longer than broad with nearly parallel sides.
Obovate. Ovate with the broader end toward the apex.
Obovoid. An ovate solid with the broadest part toward the apex.
Obtuse. Blunt or rounded at the apex.
Odd-pinnate leaf. Pinnate with a terminal leaflet.
Opposite. Two leaves emerge at opposite sides from the same place on the twig.
Orbicular. A flat body circular in outline.

Oval. Broad elliptic, rounded at ends, and about 1 1/2 times as long as broad.

Ovary. The part of a pistil that contains the ovules.

Ovate. Shaped like the longitudinal section of an egg, with the broad end basal.

Ovoid. Solid ovate or solid oval.

Ovule. The part of the flower that becomes the seed after fertilization.

Palmate. Radiately lobed or divided, veins arising from one point.

Panicle. A loose, compound, or branched flower cluster.

Papilionaceous. Butterflylike; typical flower shape of legumes.

Pectinate. Comblike, with narrow closely inserted segments.

Pedicel. Stalk of a single flower in a compound inflorescence.

Pedicellate. Borne on a pedicel.

Peduncle. A general flower stalk supporting either a cluster of flowers or a solitary flower.

Peltate. Shield-shaped and attached by its lower surface to the central stalk.

Pendent. Hanging downward.

Pendulous. More or less hanging or declined.

Penniveined. Having the form of a feather; secondary veins arranged parallel to each other and arising from a main vein.

Perfect. Flower with both stamens and pistil.

Perianth. The calyx and corolla of a flower considered as a whole.

Persistent. Remaining attached, not falling off.

Petiolate. Having a petiole.

Petiole. Footstalk of a leaf.

Petiolule. Footstalk of a leaflet.

Pilose. Hairy, with soft and distinct hairs.

Pinnate. A compound leaf with leaflets arranged along each side of a common petiole.

Pistil. Female organ of a flower, consisting of ovary, style, and stigma.

Pistillate. Female flowers; descriptive of unisexual flowers.

Pith. The central, softer part of a stem.

Pollen. The fecundating grains borne in the anther.

Polygamo-dioecious. Flowers sometimes perfect, sometimes unisexual and dioecious.

Polygamo-monoecious. Flowers sometimes perfect and sometimes unisexual, the 2 forms borne on the same individual.

Polygamous. Flowers sometimes perfect and sometimes unisexual.

Pome. An inferior fruit of 2 or several carpels enclosed in thick flesh; an apple.

Prickle. A small spinelike growth from the bark or epidermis.
Prostrate. Lying flat on the ground.
Puberulous. Minutely pubescent.
Pubescent. Clothed with soft, short hairs.
Pungent. Terminating in a rigid, sharp point; acrid.
Pyramidal. Shaped like a pyramid.
Pyriform. Pear-shaped.

Raceme. A simple inflorescence of stalked flowers on a more or less elongated rachis.
Racemose. In racemes; resembling racemes.
Rachis. An axis bearing leaflets, as in a compound leaf.
Receptacle. The more or less expanded portion of an axis that bears the organs of a flower or the collected flowers of a head.
Recurved. Curving downward or backward.
Reflexed. Abruptly turned downward.
Regular flower. Radially symmetrical; similar parts of the same shape and size.
Remotely. Scattered, not close together.
Reniform. Kidney-shaped.
Repand. With a slightly sinuate margin.
Reticulate. Netted.
Retrosely. Directed backward or downward.
Revolute. Rolled backward, margin rolled toward the lower side.
Rhombic. Having the shape of a rhombus.
Ring-porous. Wood in which the pores formed in the early spring growth are much larger than those formed later.
Rufous. Red-brown.
Rugose. Wrinkled.

Salverform. Tubular corolla with a spreading limb.
Samara. An indehiscent, winged fruit.
Scabrous. Rough to the touch.
Scarious. Thin, dry, membranaceous, not green.
Scorpioid. A form of unilateral inflorescence circinately coiled in the bud.
Scurfy. Covered with small branlike scales.
Sepal. A division of the calyx, usually bractlike.
Serotinous. Late in bearing or opening.
Serrate. Toothed, the teeth pointing upward or forward.
Serrulate. Finely toothed.
Sessile. Without a stalk.
Sheath. A tubular envelope, or enrolled part or organ.
Shrub. A woody, bushy plant, branched at or near the base and usually less than 15 feet in height.

Simple. Leaves consisting of a single blade.
Sinuate. With a strong, wavy margin.
Sinus. The cleft or space between two lobes.
Spatulate. Spatula-shaped.
Spike. A simple inflorescence of sessile flowers arranged on a common, elongated axis.
Spine. A sharp, mostly woody outgrowth in the position of a leaf or stipule.
Spinescent. With short, rigid branches resembling spines.
Spinose. Furnished with spines.
Stamen. The pollen-bearing organ of the male flower.
Staminate. Male flowers; provided with stamens but without pistils.
Stellate. Star-shaped.
Sterigmata. Short, persistent leaf bases found on spruces and hemlocks.
Stigma. The part or surface of a pistil that receives pollen for the fecundation of the ovules.
Stipe. The stalklike support of a pistil or a carpel.
Stipule. An appendage at the base of the petiole, usually one on each side.
Stoloniferous. Having lower branches or runners which tend to root.
Stoma. An orifice in the epidermis of a leaf used to connect internal cavities with air.
Stomata. Plural of stoma.
Stomatiferous. Furnished with stomata.
Strobile. A cone.
Style. The attenuated portion of a pistil between the ovary and the stigma.
Sub. A Latin prefix denoting somewhat or slightly.
Suborbicular. Nearly round.
Subtend. To lie under or opposite to.
Subulate. Awl-shaped.
Succulent. Juicy; fleshy.
Superior ovary. Free from and inserted above calyx; hypogynous.
Superposed. Placed above, as one bud above another at a node.
Suture. A junction or line of dehiscence.
Syncarp. A multiple fleshy fruit.

Taproot. The primary descending root, which may be either very large or absent at the maturity of the tree.
Terete. Circular in traverse section.
Terminal. Situated at the end of a branch.
Ternate. In groups of three.

Thorn. A sharp, woody outgrowth in the position of a lateral branch.
Tolerant. Capable of enduring shade.
Tomentose. Densely pubescent with matted wool or tomentum.
Tomentulose. Slightly pubescent with matted wool.
Torulose. Cylindric, with swollen partitions at intervals.
Tree. A plant with a woody stem, unbranched at or near base, and at least 15 feet in height and 2 inches in diameter.
Truncate. Ending abruptly, as if cut off at the end.
Tubercle. A small tuber or excrescence.
Turbinate. Top-shaped.
Twig. A young woody stem; more precisely the shoot of a woody plant representing the growth of the current season.

Umbel. A simple inflorescence of flowers with pedicels all arising from the same point.
Umbo. A boss or protuberance.
Undulate. With wavy surface or margin.
Unisexual. Of one sex, either staminate or pistillate.

Valvate. Leaf buds meeting at the edges, nct overlapping.
Valve. One of the pieces into which a capsule splits.
Veins. Threads of fibrovascular tissue in a leaf or other flat organ.
Ventral. Belonging to the anterior or inner face of an organ; the upper surface of a leaf.
Vernal. Appearing in the spring.
Vesicle. A little bladder or cavity.
Villous. Hairy with long and soft hairs.
Viscid. Gluey or sticky.

Whorled. Three or more organs arranged in a circle around an axis.
Wing. A membranous or thin and dry expansion or appendage of an organ.
Woolly. Covered with long and matted or tangled hairs.

INDEX

Names in italics refer to earlier used species names that are no longer accepted as valid.